Generalized Optimal Control
of Linear Systems with Distributed Parameters

Applied Optimization

Volume 69

The titles published in this series are listed at the end of this volume.

Generalized Optimal Control of Linear Systems with Distributed Paramaters

by

Sergei I. Lyashko
Kiev National Shevchenko University,
Ukraine

SPRINGER-SCIENCE+BUSINESS MEDIA, B.V.

A C.I.P. Catalogue record for this book is available from the Library of Congress.

ISBN 978-1-4757-7567-9 ISBN 978-0-306-47571-9 (eBook)
DOI 10.1007/978-0-306-47571-9

Printed on acid-free paper

Contents

General preface

Nowadays there are many books on mathematical theory of optimal control of systems with distributed parameters, but as a rule they are devoted to the systems with regular control. The theory of optimization of systems with singular control (including important applied problems such as point, pulse, mobile optimization etc.) is much less elaborated.

The book is written by the professor of the Kiev National Taras Shevchenko University S.I.Lyashko. The author made an attempt to create the general theory of optimization of linear systems (both distributed and lumped) with a singular control. This book touches upon wide range of issues such as the solvability of boundary values problems for partial differential equations with generalized right-hand sides, the existence of optimal controls, the necessary conditions of optimality, the controllability of systems, numerical methods of approximation of generalized solutions of initial boundary value problems with generalized data, and numerical methods for approximation of optimal controls. In particular, the problems of optimization of linear systems with lumped controls (pulse, point, pointwise mobile and so on) are investigated in detail.

The book undoubtedly will awake the interest of all who is engaged in the theory of optimal control of linear systems and its application in physics, ecology, economy, medicine and other fields.

Academician
of the National Academy of Sciences
of Ukraine I.V.Sergienko

Preface

Intensive development of science and technology put the optimization of various systems in the forefront of applied mathematics and cybernetics.

The fundamental results of the optimal control theory were obtained by L.S.Pontryagin, V.P.Boltyansky, R.V.Gamkrelidze, and Y.F.Mischenko [1, 2], A.A.Feldbaum [3], R.Bellmann [4], V.M.Tikhomirov [5], N.N.Krasovsky [6], B.N.Pshenichny [7-9], B.S.Mordukhovich [10], J.Varga [11] and others mathematicians. The theory of control of systems with finite-dimensional phase space was elaborated. But in many technical applications objects have spatial length and its state is described by some classical or non-classical equations of mathematical physics (so called systems with distributed parameters) The investigation of such objects requires considerable generalization of methods of analysis of systems with distributed parameters. The solutions of such problems were obtained by A.Bensoussan [12], B.N.Bublik [13], A.G.Butkovsky [14-16], F.P.Vasilyev [17], AI.Egorov [18-20], Yu.M.Ermoliev [21-25], V.I.Ivanenko and V.S.Mel'nik [26], J.-L.Lions [27-34], Lurje K.A. [35], A.G.Nakonechny [36], Yu.S.Osipov [37], Yu.I.Samoylenko [16, 38], T.K.Sirazetdinov [39], R.P.Fedorenko [40], V.A.Dykhta [41] etc.

Many problems in physics, economics, ecology, medicine etc. are reduced to the problems with state equations, whose right-hand sides include finite order distributions (pulse, point and other controls) [14-16, 42-45].

These problems of singular optimal control yield a series of difficult problems. Although some results were obtained already, but the complete theory was not developed yet.

The problem of optimal pulse control of systems with distributed parameters was solved in [6] by introducing into consideration the Stiltjes integral and using the moment L-problem methods. This problem was investigated in [46] as a game one. The papers [47, 48] were devoted to the necessary conditions of optimality in the form of the Pontryagin maximum principle. In [12, 33, 49] the problem of synthesis of the optimal control was reduced to solving of a quasi-variational inequality. In the paper [50] the problem of pulse optimization was solved with the help of extension of variational problem and its further analysis in the class of distributions. In [51-53] the problem of point optimal control of systems with distributed parameters was investigated with the help of the semi-group theory. In the paper [20] the pulse position control was obtained for the heat transport equation as a result of solving of some linear problems. The problem of the optimal pulse control was studied in the stochastic formulation in the monograph [12] It should been noted that the introducing of noise into considered systems is equivalent to the regularization and very simplifies the investigation of the optimal controls existence. As regards the necessary conditions of optimality the situation is inverse. The introducing of noise make the formulation of the stochastic maximum principle very complicated.

In investigation of controllable systems the problem of its controllability is one of the most important. The problem of controllability of linear systems with lumped parameters, which allow generalized controls, was investigated in [6]. In this paper it was shown that the introducing of such controls does not extend the R.Kalman's conditions of complete controllability. In the case of systems with distributed parameters the state of affairs is much more complicated. It was shown in [14] that the controllability of distributed systems with point controls could be essentially dependent on the numerical nature of the point of the application of control force. Some problems of optimization of systems with generalized controls were considered in the papers [54-61].

Though many problems were solved, numerous urgent problems of generalized control of systems with distributed parameters still either unsolved (for example, problems of pulse controllability and numerical methods) or incompletely investigated (existence of optimal controls, necessary and sufficient conditions of optimality).

This monograph describes the investigations in the field of the theory of optimal control of linear systems with distributed parameters with the help of non-linear generalized impacts (including pulse). In book the theory of optimization of distributed systems elaborated on the basis of a priory inequalities in negative norms is stated. V.P.Didenko [62-64] first obtained the inequalities in negative norms and the method of its proving (integral *abc*-method) at the early 1970's. This method of investigation of generalized optimization problems is very effective. It allowed to researchers to obtain the results on existence and uniqueness of solutions of initial-boundary value problems, on existence of optimal controls, on necessary and (in some cases) sufficient conditions of controllability in the classes of various generalized impacts. Basing on this method, it is possible to construct numerical methods of generalized optimization (including the methods of solving initial boundary value problems with finite-order distributions in the right-hand sides of its state equations).

The book's content is aimed to verify the following theses:

1. In the theory of optimal control of distributed systems we often deal with non-smooth singular controls. That is why we must investigate differential equations in classes of distributions.
2. The Sobolev spaces with negative index and a priori inequalities in negative norms are very suitable for these goals.
3. These methods can be very useful not only in the field of singular control.

The first chapters of the book are devoted to the general theory of optimization of linear systems with generalized impacts, for which the a priori inequalities in negative norms hold true.

In the following chapters we consider the applications of the general theory to systems described by classical and non-classical equations of mathematical physics, proving. For these systems the validity of these inequalities in negative norms. Also, we construct and investigate numerical methods (analogues of the Galerkin method) to find approximate solutions of initial boundary value problems.

Chapter 10 is devoted to the systematic investigation of controllability of systems with pulse, point and similar control.

Last chapter contains the recent results and generalizations. We think that the section devoted to generalized solvability of operator equalities in linear topological spaces is the most interesting. These results undoubtedly will be applied to the control theory.

In the book we use the following system of numeration and references. In each section we use a separate numeration of formulae, theorems and definitions. In the frame of single chapter (section) a number of this chapter (section) in references may be omitted. For example, Theorem 1 means the first theorem of the current section in the current chapter, Lemma 1.2 is the second lemma of the first section in the current chapter, (1.2.3) is the third formula of the second section of the first chapter and so on.

The book is addressed to the broad sections of readers – students, post-graduates and scientific researchers – who deal with partial differential equations and optimal control.

The work on book teaches to be modest, as far as the author realizes, in addition, how much he depends on others people. I would to thank my colleagues from the Institute of Cybernetics of National Academy of Science of Ukraine (NANU) and from the Kiev National Taras Shevchenko University, and also others people, who stimulated (knowingly or unknowingly) my work during long time. The first place in the list of my personal thanks belong to NANU academician Yu.M.Ermoliev and professor V.P.Didenko, who are responsible for the awakening of my interest to the theory of optimization and to the issues of the solvability of differential equations with non-smooth data. Also, I am very grateful to my colleagues, who has an influence on the content of my book: NANU academicians I.V.Sergienko, B.N.Pshenichny, N.Z.Shor, correspondent-members of NANU B.N.Bublik, V.S.Mel'nik, Yu.I.Samoilenko, V.V.Skopetzky, A.A.Chikriy, professors A.G.Burkovsky, Yu.M.Danilin, A.I.Egorov, N.F.Kirichenko, A.G.Nakonechny. I would to mark with gratitude the activity and enthusiasm shown by my young disciples, especially D.A.Nomirovsky and V.V.Semenov, whose results I used in the book in them kind consent.

D.A.Nomirovsky and D.A.Klyushin translated the manuscript from Russian to English. The general editing of the book was performed by

D.A.Klyushin. Without him insistence (up to the willingness to share with author the responsibility for possible mistakes) the book would be never completed.

Finally, I am very thankful to the employees, post-graduates and students of the Department of Computational Mathematics of the Faculty of Cybernetics and the Department of Differential and Integral Equations of the Mechanics and Mathematics Faculty of the Kiev National Taras Shevchenko University, and also to all listener of my lectures and the readers of the preliminary versions of this book.

I am very grateful to the editors of the Kluwer Academic Publishers for their remarkable work. Especially, I would like to thank Ms. Angela Quilici for her kind help during the preparation of the book.

Please, inform the author about all found mistakes and misprints via e-mail addresses: sil@dialektika.com, vm214@dcp.kiev.ua.

S.I.Lyashko

Chapter 1

OPTIMIZATION OF LINEAR SYSTEMS WITH GENERALIZED CONTROL

1. FORMULATION OF OPTIMIZATION PROBLEM FOR DISTRIBUTED SYSTEMS AND AUXILIARY PROBLEMS

Consider a system which functioning is described by linear partial differential equation [11, 14-17,19, 30, 65-73]

$$Lu = F + Ah, \; u \in D(L) \tag{1}$$

in a tube domain $Q = (0,T) \times \Omega$, where $u(t,x)$ is an unknown function depended on a spatial variable $x \in \Omega$ and a time variable $t \in (0,T)$, Ω is a bounded domain in R^n with smooth boundary $\partial\Omega$, $D(L)$ is a set of the functions which are sufficiently smooth in $\overline{Q}$ and satisfy some conditions (bd) on the boundary of the domain Q.

Let us consider that operator L mapping $D(L) \subset L_2(Q)$ into $L_2(Q)$, where $L_2(Q)$ is the space of functions, which are measurable and square integrable on the set Q. Note that the set $D(L)$ is dense in $L_2(Q)$ therefore we can correctly define an adjoint by Lagrange operator $L^* : L_2(Q) \to L_2(Q)$ with the domain of definition $D(L^*)$, which is the set of functions, which are sufficiently smooth in $\overline{Q}$ and satisfy adjoint boundary conditions (bd^+).

Let the following chains of equipped Hilbert spaces are constructed with respect to $L_2(Q)$ [74,75]:

$$..\supset W_{bd}^{-n} \supset ... \supset W_{bd}^{-1} \supset L_2(Q) \supset W_{bd}^{1} \supset ... W_{bd}^{n} \supset ...$$

$$...\supset W_{bd^+}^{-n} \supset .. \supset W_{bd^+}^{-1} \supset L_2(Q) \supset W_{bd^+}^{1} \supset .. \supset W_{bd^+}^{n} \supset ...,$$

where W^n_{bd}, $n=1,2,..,$ are the completions of $D(L)$ with respect to some positive norms; $W^n_{bd^+}$, $n=1,2,..$ are the completions of $D(L^*)$ with respect to the same norms. Negative spaces W^{-1}_{bd}, $W^{-1}_{bd^+}$,... are constructed with respect to $L_2(Q)$ and to the corresponding positive spaces.

Hereinafter we shall obtain a priori estimations with respect to negative norms for various specific types of operators L:

$$\begin{gathered}\|u\|_{H^+_{bd}} \le C_1\|Lu\|_{W^{-l}_{bd^+}} \le C_2\|u\|_{W^l_{bd}} \\ \|v\|_{H^+_{bd^+}} \le c_1\|L^*v\|_{W^{-l}_{bd}} \le c_2\|v\|_{W^+_{bd^+}}\end{gathered} \tag{2}$$

where L^* is the operator adjoint to L by Lagrange, $u\in D(L), v\in D(L^*)$, c, C here and below with indices and without them denote positive constants not depending on the functions $u(t,x)$, $v(t,x)$,

$$L_2(Q)\supseteq H_1 \supseteq W^{+l}_{bd},\ L_2(Q)\supseteq H^+_{bd^+} \supseteq W^{+l}_{bd^+},$$

in addition, these embeddings are dense and operators of embedding are completely continuous.

Spaces H^+_{bd}, $H^+_{bd^+}$, $W^{-l}_{bd^+}$ are Hilbert ones endowed with the norms satisfying the inequalities

$$c_1\|u\|_{W^{+l}_{bd}} \ge c_2\|u\|_{H^+_{bd}} \ge \|u\|_{L_2(Q)},\ c_3\|v\|_{W^{+l}_{bd^+}} \ge c_4\|v\|_{H^+_{bd^+}} \ge \|v\|_{L_2(Q)}.$$

Denote by $\langle\cdot,\cdot\rangle_{W_{bd}}$, $\langle\cdot,\cdot\rangle_{W_{bd^+}}$ bilinear forms constructed by extending the inner product in $L_2(Q)$ on continuity to bilinear forms on $W^{+l}_{bd}\times W^{-l}_{bd}$ and $W^{-l}_{bd^+}\times W^{+l}_{bd^+}$, respectively. By $\langle\cdot,\cdot\rangle_{H_{bd}}$, $\langle\cdot,\cdot\rangle_{H_{bd^+}}$ we understand analogous bilinear form.

It follows from the above mentioned inequalities (2) that operator L (L^*, respectively) may be extended on continuity to an operator continuously mapping whole space W_{bd}^{+l} ($W_{bd^+}^{+l}$, respectively) into space $W_{bd^+}^{-l}$ (W_{bd}^{-l}, respectively). It should been noted that the inequalities (2) hold true for extended operators also (we shall save for them the previous denotations) for any $u(t,x) \in W_{bd}^{+l}$, $v(t,x) \in W_{bd^+}^{+l}$. For operators L and L^* the following identity holds true

$$\langle Lu, v \rangle_{W_{bd^+}} = \langle u, L^* v \rangle_{W_{bd}}, \quad \forall u \in W_{bd}^{+l}, v \in W_{bd^+}^{+l}.$$

The solutions of the original and the adjoint problems

$$Lu = F, \tag{3}$$

$$L^* v = G, \tag{4}$$

we shall mean in the sense of the following definitions.

Definition 1. *The solution of the problem (3) with a right-hand side* $F \in W_{bd^+}^{-l}$ *is a function* $u(t,x) \in W_{bd}^{+l}$, *for which there exists a sequence of such functions* $u_i(t,x) \in D(L)$, $i = 1,2,\ldots$ *that*

$$\|u_i - u\|_{W_{bd}^l} \xrightarrow[i \to \infty]{} 0, \ \|Lu_i - F\|_{W_{bd^+}^{-l}} \xrightarrow[i \to \infty]{} 0.$$

Definition 2 (strong solution). *The strong solution of the problem (3) with a right-hand side* $F \in W_{bd^+}^{-l}$ *is such a function* $u(t,x) \in W_{bd}^{+l}$, *that*

$$Lu - F = 0$$

in the space $W_{bd^+}^{-l}$.

Definition 3 (weak solution). *The weak solution of the problem (3) with a right-hand side* $F \in W_{bd^+}^{-l}$ *is such a function* $u(t,x) \in W_{bd}^{+l}$ *that the following equality*

$$\langle u, L^* v \rangle_{W_{bd}} = \langle F, v \rangle_{W_{bd^+}}$$

holds true for any functions $v \in W_{bd^+}^{+l}$.

In a similar way we introduce the definitions for solutions of the equation (4).

Lemma 1. *Let the inequalities (2) hold true for the operators L and L^*. Then Definitions 1,2,3 are equivalent.*

Proof. We shall carry out the proof following the scheme $1 \Leftrightarrow 2 \Leftrightarrow 3$.

Let $u(t,x)$ be a solution of the problem (3) in the sense of Definition 1. Then taking into consideration the inequalities (2) with the help of passing to the limit we conclude that the value of the extended operator $L : W_{bd}^{+l} \to W_{bd^+}^{-l}$ on the element $u(t,x)$ obviously equals to F, i.e. $Lu = F$ in the sense of the equality of the elements in the space $W_{bd^+}^{-l}$.

Vice versa, let $u(t,x)$ be a solution of the problem (3) in the sense of Definition 2. Choose an arbitrary sequence $u_i(t,x) \in D(L)$ such that $\|u_i - u\|_{W_{bd}^{+l}} \xrightarrow[i\to\infty]{} 0$. Further, we have

$$\|Lu_i - F\|_{W_{bd^+}^{-l}} = \|Lu_i - Lu + Lu - F\|_{W_{bd^+}^{-l}} \leq$$

$$\leq \|Lu_i - Lu\|_{W_{bd^+}^{-l}} + \|Lu - F\|_{W_{bd^+}^{-l}}.$$

As far as $Lu = F$ in $W_{bd^+}^{-l}$, the second term at the right-hand side equals to 0. Taking into account (2) and the linearity of the operator L we obtain

$$\|Lu_i - F\|_{W_{bd^+}^{-l}} \leq \|Lu_i - Lu\|_{W_{bd^+}^{-l}} \leq c\|u_i - u\|_{W_{bd}^{+l}} \xrightarrow[i\to\infty]{} 0.$$

Thus, the equivalence of the Definitions 1 and 2 is proved.

It is not difficult to prove also the equivalence of Definitions 2 and 3. Indeed, the statement $2 \Rightarrow 3$ is obvious. Let us prove that $3 \Rightarrow 2$. Let $u(t,x)$ a the solution of the problem (3) in the sense of Definition 3:

$$\langle u, L^* v\rangle_{W_{bd}} = \langle Lu, v\rangle_{W_{bd^+}} = \langle F, v\rangle_{W_{bd^+}}, \quad \forall v \in W_{bd^+}^{+l}.$$

By virtue of the arbitrariness of $v \in W^{+l}_{bd^+}$, we obtain that $Lu = F$ in $W^{-l}_{bd^+}$.

Remark. *Analogous statements hold true for solutions of the adjoint equation (4).*

Theorem 1. *Let the inequalities (2) hold true for the operators of the problems (3) and (4). Then for any element $F \in H^-_{bd^+}$ there exists a unique solution of the problem (3) in the sense of Definitions 1-3, where $H^-_{bd^+}$ is a negative space constructed with respect to $H^+_{bd^+}$ and $L_2(Q)$.*

Proof. Consider the functional $l(v) \equiv \langle F, v \rangle_{H_{bd^+}}$ on functions $v \in W^{+l}_{bd^+}$. By the Schwarz inequality and (2) we obtain

$$|l(v)| = \left| \langle F, v \rangle_{H_{bd^+}} \right| \leq \|F\|_{H^-_{bd^+}} \|v\|_{H^+_{bd^+}} \leq C \|L^* v\|_{W^{-l}_{bd}} .$$

Hence the functional $l(v)$ may be considered as a linear continuous one dependent on $L^* v \in W^{-l}_{bd}$ $\left(l(v) = \bar{l}(L^* v) \right)$. According to the Hahn-Banach theorem [76] extend the functional $\bar{l}(\cdot)$ linearly and continuously onto whole space W^{-l}_{bd}. On the basis of the theorem about common form of a linear continuous functional defined in W^{-l}_{bd} [77] there exists a function $u(t,x) \in W^{l}_{bd}$ such that $\bar{l}(w) = \langle u, w \rangle_{W_{bd}}$ for any $w \in W^{-l}_{bd}$. Consider this functional on the elements $w = L^* v, v \in W^{+l}_{bd^+}$. Then

$$\bar{l}(w) = \left\langle u, L^* v \right\rangle_{W_{bd}} = \langle Lu, v \rangle_{W_{bd^+}} = \langle F, v \rangle_{H_{bd^+}} = \langle F, v \rangle_{W_{bd^+}} .$$

Hence, $\langle Lu-F,v\rangle_{W_{bd^+}}=0$, $\forall\, v\in W^{+l}_{bd^+}$. By virtue of the arbitrariness of $v\in W^{+l}_{bd^+}$, we obtain that $Lu-F=0$ in the space $W^{-l}_{bd^+}$. The uniqueness of the solution follows from the left-hand side of the first inequality in (2) and from the embedding $W^{+l}_{bd}\subset H^{+}_{bd}$. The theorem is proved.

Theorem 2. *Let the inequalities (2) hold true. Then for any element $G\in H^{-}_{bd}$ there exists a unique solution of the problem (4) in the sense of the analogues of Definitions 1-3 for the problem (4).*

Proof of Theorem 2 is similar to the proof of Theorem 1.

It is possible to consider the generalized solution from more wide class.

Definition 4. *The solution of the problem (3) with a right-hand side $F\in W^{-l}_{bd^+}$ is a function $u(t,x)\in H^{+}_{bd}$, for which there exists a sequence of such functions $u_i(t,x)\in D(L), i=1,2,...$ that*

$$\|u_i-u\|_{H_1}\xrightarrow[i\to\infty]{}0,\ \|Lu_i-F\|_{W^{-l}_{bd^+}}\xrightarrow[i\to\infty]{}0.$$

Definition 5 (weak solution). *The weak solution of the problem (3) with a right-hand side $F\in W^{-l}_{bd^+}$ is a function $u(t,x)\in H^{+}_{bd}$, such that the equality*

$$\langle u,L^*v\rangle_{H_{bd}}=\langle F,v\rangle_{W_{bd^+}}$$

holds true for any functions $v\in W^{+l}_{bd^+}:L^*v\in H^{-}_{bd}$.

The generalized solutions of the problem (4) are defined in a similar way.

Lemma 2. *Assume that the inequalities (2) hold true for the operators L and L^*. Then Definitions 4 and 5 are equivalent.*

Proof. Let $u(t,x)$ be a solution of the problem (3) in the sense of Definition 5, i.å. $u(t,x)\in H^{+}_{bd}$ and

$$\left\langle u, L^{*}v\right\rangle_{H_{bd}} = \left\langle F, v\right\rangle_{W_{bd^{+}}}, \tag{5}$$

$\forall v\in W^{+l}_{bd^{+}} : L^{*}v\in H^{-}_{bd}$.

Choose a sequence $F_p \in H^{-}_{bd^{+}}$ such that $F_p \xrightarrow[p\to\infty]{} F$ in $W^{-l}_{bd^{+}}$. Then, if $u_p(t,x)\in W^{+l}_{bd}$ is a solution of the problem $Lu = F_p$ in the sense of Definition 1 (which exists by virtue of Theorem 1) then according to Lemma 1 we have

$$\left\langle Lu_p, v\right\rangle_{W_{bd^{+}}} = \left\langle F_p, v\right\rangle_{W_{bd^{+}}}, \quad \forall v\in W^{+l}_{bd^{+}}.$$

Hence,

$$\frac{\left|\left\langle Lu_p - F, v\right\rangle_{W_{bd^{+}}}\right|}{\|v\|_{W^{+l}_{bd^{+}}}} = \frac{\left|\left\langle F_p - F, v\right\rangle_{W_{bd^{+}}}\right|}{\|v\|_{W^{+l}_{bd^{+}}}},$$

and therefore,

$$\left\|Lu_p - F\right\|_{W^{-l}_{bd^{+}}} = \left\|F_p - F\right\|_{W^{-l}_{bd^{+}}} \xrightarrow[p\to\infty]{} 0, \tag{6}$$

i.e. the sequence $\left\{Lu_p\right\}_{p=1}^{\infty}$ is fundamental in $W^{-l}_{bd^{+}}$.

Moreover, taking into consideration (2) we obtain

$$\left\|u_{p_1} - u_{p_2}\right\|_{H^{+}_{bd}} \le c\left\|Lu_{p_1} - Lu_{p_2}\right\|_{W^{-l}_{bd^{+}}} \xrightarrow[p_1,p_2\to\infty]{} 0.$$

Thus, $\{u_p\}_{p=1}^{\infty}$ is a fundamental sequence in H^{+}_{bd}, hence, there exists $u^{*}(t,x)\in H^{+}_{bd}$ such that $\left\|u_p - u^{*}\right\|_{H^{+}_{bd}} \xrightarrow[p\to\infty]{} 0$.

Further, we have

$$\left\langle u_p, L^{*}v\right\rangle_{H_{bd}} = \left\langle u_p, L^{*}v\right\rangle_{W_{bd}} = \left\langle F_p, v\right\rangle_{W_{bd^{+}}},$$

$\forall v \in W^{+l}_{bd^+} : L^* v \in H^-_{bd}$.

Passing in the last equality to the limit as $p \to \infty$, we obtain

$$\left\langle u^*, L^* v \right\rangle_{H_{bd}} = \left\langle F, v \right\rangle_{W_{bd^+}}.$$

Taking into consideration the fact that the last equality holds true for arbitrary $v \in W^{+l}_{bd^+} : L^* v \in H^-_{bd}$ and the relations (5), we state that $u(t,x) = u^*(t,x)$ in H^+_{bd}, and as far as $\left\| u_p - u^* \right\|_{H^+_{bd}} \xrightarrow[p\to\infty]{} 0$, then also $\left\| u_p - u \right\|_{H^+_{bd}} \xrightarrow[p\to\infty]{} 0$. Granting (6), we convince oneself that $u(t,x)$ is the solution of the problem (3) in the sense of the Definition 4.

Let us prove the inverse statement. Let $u(t,x)$ is a solution of the problem (3) in the sense of Definition 4. Then

$$\begin{aligned} \left\langle u, L^* v \right\rangle_{H_{bd}} &= \left\langle u_i, L^* v \right\rangle_{H_{bd}} + \left\langle u - u_i, L^* v \right\rangle_{H_{bd}} = \\ &= \left\langle L u_i, v \right\rangle_{W_{bd^+}} + \left\langle u - u_i, L^* v \right\rangle_{H_{bd}} = \\ &= \left\langle L u_i - F, v \right\rangle_{W_{bd^+}} + \left\langle F, v \right\rangle_{W_{bd^+}} + \left\langle u - u_i, L^* v \right\rangle_{H_{bd}}. \end{aligned} \tag{7}$$

Taking into account (2), let us estimate the first and the third terms in the right-hand side of (7).

$$\left| \left\langle L u_i - F, v \right\rangle_{W_{bd^+}} \right| \le \|v\|_{W^{+l}_{bd^+}} \left\| L u_i - F \right\|_{W^{-l}_{bd^+}} \xrightarrow[i\to\infty]{} 0,$$

$$\left| \left\langle u - u_i, L^* v \right\rangle_{H_{bd}} \right| \le \left\| L^* v \right\|_{H^-_{bd}} \left\| u - u_i \right\|_{H^+_{bd}} \xrightarrow[i\to\infty]{} 0.$$

Passing to the limit in (7) as $i \to \infty$, we obtain the required equality (5).

T h e o r e m 3. *Let the inequalities (2) hold true for the operators of the problems (3), (4). Then for any element $F \in W^-_{bd^+}$ there exists a unique solution of the problem (3) in the sense of Definitions 4-5.*

P r o o f. The set $H^{-}_{bd^+}$ is dense in $W^{-l}_{bd^+}$. Hence, for any element $F \in W^{-l}_{bd^+}$ there exists a sequence $F_i \in H^{-}_{bd^+}$ such that

$$\|F_i - F\|_{W^{-l}_{bd^+}} \xrightarrow[i\to\infty]{} 0.$$

It follows from Theorem 1 that for any function $F_i \in H^{-}_{bd^+}$ there exists a unique solution $u_i \in W^{+l}_{bd}$ of the problem (3) in the sense of Definition 1. Using the inequality (2), we obtain

$$c\left\|u_i - u_j\right\|_{H^{+}_{bd}} \le \left\|Lu_i - Lu_j\right\|_{W^{-l}_{bd^+}} = \left\|F_i - F_j\right\|_{W^{-l}_{bd^+}} \xrightarrow[i,j\to\infty]{} 0.$$

Thus, by virtue of the completeness of the space H^{+}_{bd} there exists a function $u(t,x) \in H^{+}_{bd}$: $\|u_i - u\|_{H^{+}_{bd}} \xrightarrow[i\to\infty]{} 0$. On the other hand, by Lemma 1 for any function $v \in W^{+l}_{bd^+}$ the following equalities hold true

$$\langle Lu_i, v\rangle_{W_{bd^+}} = \langle F_i, v\rangle_{W_{bd^+}}$$

or

$$\left\langle u_i, L^* v\right\rangle_{H_{bd}} = \langle F_i, v\rangle_{W_{bd^+}},$$

$\forall v \in W^{+l}_{bd^+} : L^* v \in H^{-}_{bd}$.

Passing to the limit as $i \to \infty$ in the last equality, we obtain

$$\left\langle u, L^* v\right\rangle_{H_{bd}} = \langle F, v\rangle_{W_{bd^+}}, \forall v \in W^{+l}_{bd^+} : L^* v \in H^{-}_{bd}.$$

Let us prove the uniqueness. Let $\overline{u}(t,x) \in H^{+}_{bd}$ be one more generalized solution of the problem (3) in the sense of Definitions 4-5. Then $\left\langle u - \overline{u}, L^* v\right\rangle_{H_{bd}} = 0$ $\forall v \in W^{+l}_{bd^+} : L^* v \in H^{-}_{bd}$. Theorem 1 implies that $L^*\left(W^{+l}_{bd^+}\right) \supseteq H^{-}_{bd}$. So, $u - \overline{u} = 0$ in H^{+}_{bd}. The proof is complete.

Theorem 4. *Let the inequalities (2) hold true for the operators of the problems (3), (4). Then for any element* $G \in W^{-}_{bd}$ *there exists a unique solution of the problem (4) in the sense of the analogies of Definitions 4-5 for the problem (4).*

Proof of Theorem 4 is similar to the proof of Theorem 3.

Lemma 3. *Let* $u(t,x)$ *be a solution of the equation (3) with a right-hand side* $F \in W^{-l}_{bd^+}$ *in the sense of Definitions 4-5, then the following estimation holds true*

$$\|u\|_{H^+_{bd}} \leq C\|F\|_{W^{-l}_{bd^+}}. \tag{8}$$

Proof. At first, we shall show that the following positive estimations follow from the à priori estimations with respect to the negative norms (2):

$$\|u\|_{W^{+l}_{bd}} \leq C\|Lu\|_{H^-_{bd^+}}, \tag{9}$$

$$\|v\|_{W^{+l}_{bd^+}} \leq C\left\|L^*v\right\|_{H^-_{bd}}. \tag{10}$$

Indeed, by virtue of Theorem 3 there exists a unique generalized solution $u(t,x)$ of the problem (3) with the right-hand side $F \in W^{-l}_{bd^+}$ and this solution belongs to the space H^+_{bd}. It follows from Lemma 2 that for any $F \in W^{-l}_{bd^+}$ and for any $v \in W^{+l}_{bd^+}$ such that $L^*v \in H^-_{bd}$ the following equality holds true

$$\left\langle u, L^*v \right\rangle_{H_{bd}} = \langle F, v \rangle_{W_{bd^+}}. \tag{11}$$

Using the Schwarz inequality we obtain

$$\left\langle F, \frac{v}{\left\|L^*v\right\|_{H^-_{bd}}} \right\rangle_{W_{bd^+}} \leq \|u\|_{H^+_{bd}}.$$

The Banach-Steinhaus theorem [76, 78] implies that the set of functions $\left\{ \frac{v}{\|L^* v\|_{H^-_{bd}}} \right\}$ is bounded with respect to the norm of the space $W^{+l}_{bd^+}$ and this proves that the inequality (10) holds true. The inequality (9) is proved in a similar way.

Let us prove that (8) holds true.

Applying to the right-hand side of the equality (11) the Schwarz inequality and (10), we obtain

$$\left| \left\langle u, \frac{L^* v}{\|L^* v\|} \right\rangle_{H^-_{bd}} \right| \le C \|F\|_{W^{-l}_{bd^+}},$$

$v \in W^{+l}_{bd^+} : L^* v \in H^-_{bd}$, that implies (8).

Remark. *In a similar way we can prove the inequality*

$$\|v\|_{H^+_{bd^+}} \le c \|G\|_{W^{-l}_{bd}},$$

where $v(t,x)$ is a solution of the equation (4) with a right-hand side $G \in W^{-l}_{bd}$ in the sense of Definition 4.

Lemma 4. *Let $u(t,x)$ be a generalized solution of the problem (3) with a right-hand side F in the sense of Definitions 4-5 and $u(t,x) \in W^{+l}_{bd}$. Then $u(t,x)$ is a generalized solution of this problem in the sense of Definitions 1-3.*

Proof. Let $v(t,x)$ be an arbitrary smooth in $\overline{Q}$ function, which satisfies the boundary conditions (bd^+).

Consider $\langle Lu - F, v \rangle_{W_{bd^+}}$. We have

$$\left| \langle Lu - F, v \rangle_{W_{bd^+}} \right| \le \left| \langle Lu - Lu_i, v \rangle_{W_{bd^+}} \right| + \left| \langle Lu_i - F, v \rangle_{W_{bd^+}} \right|,$$

where $u_i(t,x)$ is a sequence, which determines the solution $u(t,x)$ by Definition 4. Let us estimate each terms in the right-hand side.

As far as $v(t,x)$ is a smooth function, which satisfies the conditions (bd^+) then

$$\left|\langle Lu - Lu_i, v\rangle_{W_{bd^+}}\right| \le \left|\langle u - u_i, L^* v\rangle_{W_{bd}}\right| = \left|\langle u - u_i, L^* v\rangle_{H_{bd}}\right| \le$$
$$\le \|u - u_i\|_{H^+_{bd}} \|L^* v\|_{H^-_{bd}} \xrightarrow[i\to\infty]{} 0.$$

Moreover,

$$\left|\langle Lu_i - F, v\rangle_{W_{bd^+}}\right| \le \|Lu_i - F\|_{W^{-l}_{bd^+}} \|v\|_{W^{+l}_{bd^+}} \xrightarrow[i\to\infty]{} 0,$$

i.e.

$$\left|\langle Lu - F, v\rangle_{W_{bd^+}}\right| = 0,$$

where $v(t,x)$ is an arbitrary function from a set which is dense in $W^{+l}_{bd^+}$. Hence, $Lu - F = 0$ in $W^{-l}_{bd^+}$.

Thus, $u(t,x)$ is a generalized solution of the problem (3) by Definition 2.

Remark. *The similar statement holds true for solutions of the problem (4).*

Taking into account the above-mentioned facts, we shall further consider the following optimization problem.

Let the functioning of a system is described by a partial linear differential equation:

$$Lu = f + A(h), \tag{12}$$

where $f \in W^{-l}_{bd^+}$ is a given element. The control of the system (12) is carried out by choosing the controls h which are defined on a set of admissible controls U_{ad} from a reflexive Banach space of controls H; $A(\cdot)$ is an operator, possibly non-linear, $A : U_{ad} \subseteq H \to W^{-l}_{bd^+}$. On the solutions of the equation (1) we define a functional

$J(h) = \Phi(u(h))$, which is weakly lower semicontinuous with respect to a state of the system and which must be minimized on the set $U_{ad} \subseteq H$.

2. EXISTENCE OF OPTIMAL CONTROL

Consider the optimal control problem described in Section 1. Let a state of the system is determined by the equation (1.12):

$$Lu = f + A(h).$$

The state function $u(t,x)$ depends on (via the right-hand side of the equation) from a control h defined on the set of admissible controls U_{ad}, which belongs to the space of controls H. On the solutions of the equation we define some weakly semicontinuous with respect to the system state functional $J(h) = \Phi(u(h))$, which must be minimized on the set U_{ad}. Of course, it is necessary to require that the functional $\Phi(u(h))$ should been defined correctly of the solutions of the equation. For example, if

$$\Phi(u(h)) = \int_Q u^2(t,x,h)dQ,$$

then the solution $u(t,x,h)$ must belong to the space $L_2(Q)$. If

$$\Phi(u(h)) = \int_Q u^2 + u_t^2 + \sum_{i=1}^n u_{x_i}^2 dQ,$$

then $u(t,x,h)$ must belong to the Sobolev space $W_2^1(Q)$ (such situation we denote as $\Phi(\cdot): L_2(Q) \to R^1$ and $\Phi(\cdot): W_2^1(Q) \to R^1$, respectively). It is clear that the smoothness of the solution $u(t,x,h)$ depends on the smoothness of the right-hand side of the equation (see Section 1). Therefore, the more wide space of mappings $F(t,x,h) = f + A(h)$ we consider, the more narrow class of functionals $\Phi(u(h))$ we may investigate.

Theorem 1. *Let a system state be determined as a solution of the problem (1.12) under the following assumptions:*

1) the performance criterion $\Phi(\cdot): H_{bd}^{+} \to R^{1}$ *is a functional which is weakly lower semicontinuous with respect to the system state* $u(t,x,\varphi)$ (*i.e.* $u_i \xrightarrow{W} u^*$ *in* $H_{bd}^{+} \Rightarrow \Phi(u^*) \le \underline{\lim}_{i \to \infty} \Phi(u_i)$) *and below bounded;*

2) the set of admissible controls $U_{ad} \subset H$ *is bounded, closed and convex in* H*;*

3) H *is a reflexive Banach space;*

4) $A(\cdot)$ *is a weakly continuous operator mapping* H *into* $W_{bd^+}^{-l}$*;*

5) the estimations (1.2) for the operators L *and* L^* *are valid.*

Then there exists the optimal control of system (1.12), i.e. such control $h^* \in U_{ad}$ *that* $J(h^*) = \inf_{h \in U_{ad}} J(h)$.

Proof. Choose a minimizing sequence of controls $\{h_k\}_{k=1}^{\infty}$, $h_k \in U_{ad}$, i.e.

$$J(h_k) \xrightarrow[k \to \infty]{} \inf_{h \in U_{ad}} J(h) \tag{1}$$

Since the set U_{ad} is weakly compact ($U_{ad} \subset H$ is bounded, closed and convex in the reflexive space H), we may extract from this sequence a subsequence $\{h_{k_n}\}_{n=1}^{\infty}$, which weakly convergences in H to some $h^* \in U_{ad}$. As far as operator $A(\cdot)$ is weakly continuous, the corresponding sequence $\{A(h_{k_n})\}_{n=1}^{\infty}$ is weakly convergent, and by the Eberlein-Shmulyan theorem [78] this sequence is bounded in the space $W_{bd^+}^{-l}$.

Lemma 1.3 and above-mentioned reasoning imply that the sequence $\{u_{k_n}\}_{n=1}^{\infty}$, which corresponds to the subsequence $\{h_{k_n}\}_{n=1}^{\infty}$, is bounded in the norm of the space H_{bd}^{+}, and hence, we may extract from it some weakly convergent subsequence $u_{k_{n_j}}$ ($u_{k_{n_j}} \xrightarrow{W} u^*$ in H_{bd}^{+}).

Granting both the facts that $u_{k_{n_j}} = u(t, x, h_{k_{n_j}})$ is a solution of the equation (1.12) when $h = h_{k_{n_j}}$ and the relations (1.5), we obtain

$$\left\langle u_{k_{n_j}}, L^* v \right\rangle_{H_{bd}} = \left\langle F\left(h_{k_{n_j}}\right), v \right\rangle_{W_{bd^+}} \tag{2}$$

for any function $v(t,x) \in W_{bd^+}^{+l}$ such that $L^* v \in H_{bd}^{-}$, where $F(h) = f + A(h)$.

As far as $h_{k_{n_j}} \xrightarrow[j \to \infty]{} h^*$ weakly in H, and $F(\cdot)$ is weakly continuous mapping of H into $W_{bd^+}^{-l}$, then

$$\left\langle F\left(h_{k_{n_j}}\right), v \right\rangle_{W_{bd^+}} \xrightarrow[j \to \infty]{} \left\langle F(h^*), v \right\rangle_{W_{bd^+}}.$$

In addition, due to the fact that $u_{k_{n_j}} \xrightarrow{W} u^*$, we obtain

$$\left\langle u_{k_{n_j}}, L^* v \right\rangle_{H_{bd}} \xrightarrow[j \to \infty]{} \left\langle u^*, L^* v \right\rangle_{H_{bd}}.$$

Thus, passing to the limit in (2) as $j \to \infty$, we conclude that $u^*(t,x)$ is a solution of the equation $Lu = F(h^*)$ in the sense of Definition 1.5, and hence, in the sense of Definition 1.4 also.

Since the performance criterion $J(h)$ is weakly lower semicontinuous with respect to $u(t,x)$,

$$J(h^*) \le \underline{\lim}_{i \to \infty} J(h_{k_{n_i}}) = \inf_{h \in U_{ad}} J(h).$$

Hence, h^* is the optimal control and Theorem 1 is proved.

Theorem 2. *Let a system state be determined as a solution of the problem (1.12) under the following assumptions:*

1) the performance criterion $\Phi(\cdot): W_{bd}^{+l} \to R^1$ *is a functional which is weakly lower semicontinuous with respect to the system state* $u(t,x,h)$ *and below bounded;*

2) the set of admissible controls $U_{ad} \subset H$ *is bounded, closed and convex in* H*;*

3) H *is a reflexive Banach space;*

4) $A(\cdot)$ *is a weakly continuous operator mapping* H *into* $H_{bd^+}^-$*;*

5) the estimations (1.2) for the operators L *and* L^* *are valid.*

Then there exists the optimal control of system (1.12).

Proof is similar to the proof of the previous theorem.

Remark. *As far as the operator* $A(\cdot)$ *is non-linear then the functional* $J(\cdot)$ *may be not convex and the optimal control may be non-unique.*

Consider the application of these theorems in the case of the optimization of distributed systems with point controls. The urgency of such investigations is stipulated both by the development of new technologies and by the simplicity of the control realization.

In view of specific character of control in these problems it is possible to obtain more interesting results [80-86]

Let the studied system be described by the linear partial differential equation

$$Lu = f + A(h). \qquad (3)$$

Consider the optimal control problems for the systems, which are described by the equation (3) with right-hand sides in the following forms:

1. $A_1(\cdot)=\sum_{i=1}^{s}\ddot{a}(t-t_i)\otimes\varphi_i(x)$,

$$h_1=\{(t_i,\varphi_i(x))\}_{i=1}^{s}; \tag{4}$$

2. $A_2(\cdot)=\sum_{i=1}^{s}\delta^{(k)}(t-t_i)\otimes\varphi_i(x)$,

$$h_2=\{(t_i,\varphi_i(x))\}_{i=1}^{s}; \tag{5}$$

3. $A_3(\cdot)=\sum_{i=1}^{s}\delta(x_1-x_{1,i})\otimes\varphi_i(t,x_2,\ldots,x_n)$,

$$h_3=\{(x_{1,i},\varphi_i(\cdot))\}_{i=1}^{s}; \tag{6}$$

4. $A_4(\cdot)=\sum_{i=1}^{s}\sum_{j=1}^{p}\delta(t-t_i)\otimes\delta(x_1-x_{1,j})\otimes\varphi_{ij}(x_2,\ldots,x_n)$,

$$h_4=\{(t_i,x_{1,j},\varphi_{ij}(x_2,\ldots,x_n))\}_{i,j=1}^{s,p}; \tag{7}$$

5. $A_5(\cdot)=\sum_{i=1}^{s}\delta(x_1-a_i(t))\otimes\varphi_i(t,x_2,\ldots,x_n)$,

$$h_5=\{(a_i(t),\varphi_i(\cdot))\}_{i=1}^{s}; \tag{8}$$

where $\delta(\cdot)$ is the Dirac function.

Consider every variant of the mapping $A_i(\cdot)$ separately.

1. $A_1(\cdot)=\sum_{i=1}^{s}\delta(t-t_i)\otimes\varphi_i(x)$,

where $t,t_i\in[0,T]$, and $\varphi_i(x)\in L_2(\Omega)$.

The control is

$$h_1=\{(t_i,\varphi_i(x))\}_{i=1}^{s}\in U_{ad}\subset H_1=[0,T]^s\times(L_2(\Omega))^s,$$

where U_{ad} is bounded, closed and convex set in H_1.

By $A_1(\cdot)=\sum_{i=1}^{s}\delta(t-t_i)\otimes\varphi_i(x)$ we shall mean a functional defined on smooth functions in $\overline{Q}$ in the following way:

$$l_{A_1(\cdot)}(v)=\int_{\Omega}\sum_{i=1}^{s}v(t_i,x)\varphi_i(x)d\Omega .$$

Suppose that

$$W^{+l}_{bd^+}\subset W_2^{1,0}(Q)\ \left(W^{+l}_{bd}\subset W_2^{1,0}(Q)\right)\ \text{è}\ \|\cdot\|_{W_2^{1,0}(Q)}\le c\|\cdot\|_{W^{+l}_{bd^+}}$$
$$\left(\|\cdot\|_{W_2^{1,0}(Q)}\le c\|\cdot\|_{W^{+l}_{bd}}\right), \tag{9}$$

where $W_2^{1,0}(Q)$ is a completion of the space of smooth in $\overline{Q}$ functions with respect to the norm

$$\|v\|_{W_2^{1,0}(Q)}=\left(\int_Q v^2+v_t^2\,dQ\right)^{\frac{1}{2}},$$

and also let $f\in W^{-l}_{bd^+}$.

We shall prove that in this case $A_1(h_1)\in W^{-l}_{bd^+}$, i.e. it is possible to extend $l_{A_1(h_1)}(v)$ up to a linear continuous functional on $W^{+l}_{bd^+}$. Indeed, the linearity of the functional

$$l_{A_1(h_1)}(v)=\int_{\Omega}\sum_{i=1}^{s}v(t_i,x)\varphi_i(x)d\Omega$$

is clear.

Let us prove the boundedness of this functional. By the integral Cauchy inequality we obtain

$$\left|\int_\Omega \sum_{i=1}^{s} v(t_i,x)\varphi_i(x)d\Omega\right| \le \sum_{i=1}^{s}\left|\int_\Omega v(t_i,x)\varphi_i(x)d\Omega\right| \le$$
$$\le \sum_{i=1}^{s}\left(\int_\Omega v^2(t_i,x)d\Omega\right)^{\frac{1}{2}} \cdot \|\varphi_i\|_{L_2(\Omega)}. \tag{10}$$

Since the set U_{ad} is bounded in H_1,

$$\left|\int_\Omega \sum_{i=1}^{s} v(t_i,x)\varphi_i(x)d\Omega\right| \le c\sum_{i=1}^{s}\left(\int_\Omega v^2(t_i,x)d\Omega\right)^{\frac{1}{2}}. \tag{11}$$

It is easily seen [74] that the following inequality holds true:

$$\left(\int_\Omega v^2(t_i,x)d\Omega\right)^{\frac{1}{2}} \le c\left(\int_Q v^2(t,x)+v_t^2(t,x)dQ\right)^{\frac{1}{2}}.$$

Granting this and (9), rewrite (11) in the following form:

$$\left|\int_\Omega \sum_{i=1}^{s} v(t_i,x)\varphi_i(x)d\Omega\right| \le c\sum_{i=1}^{s}\left(\int_\Omega v^2(t_i,x)d\Omega\right)^{\frac{1}{2}} \le$$
$$\le C\left(\int_Q v^2(t,x)+v_t^2(t,x)dQ\right)^{\frac{1}{2}} = C\|v\|_{W_2^{1,0}(Q)} \le C\|v\|_{W_{bd^+}^{+l}(Q)}.$$

Returning to the inequality (10), we finally obtain

$$\left|l_{A_1(h_1)}(v)\right| = \left|\int_\Omega \sum_{i=1}^{s} v(t_i,x)\varphi_i(x)d\Omega\right| \le C\|v\|_{W_{bd^+}^{+l}},$$

that proves the boundedness of the functional $l_{A_1(h_1)}(v)$ on the set of smooth in $\overline{Q}$ functions $v(t,x)$. Applying the Hahn-Banach theorem we expand on continuity the functional $l_{A_1(h_1)}(v)$ up to the linear continuous functional which is defined on the whole space $W_{bd^+}^{+l}$.

Thus, it is proved that $F(\cdot)=f+A_1(\cdot)\in W^{-l}_{bd^+}$, and hence, by Theorem 1.3 there exists the unique generalized solution of the problem $Lu=f+A_1(h)$ in the sense of Definition 1.4.

Let us prove that the mapping $A_1(\cdot)=\sum_{i=1}^{s}\delta(t-t_i)\otimes\varphi_i(x)$ is weakly continuous. Let $h^{(m)}$ be some weakly convergent in H_1 sequence $h^{(m)}\xrightarrow{W}h^*$. As far as the weak and strong convergences are equivalent in the finite-dimensional space, $t_i^{(m)}\xrightarrow[m\to\infty]{}t_i^*$ strongly in R^1 under arbitrary $i=\overline{1,s}$. Put $A_1^{(m)}=\sum_{i=1}^{s}\delta(t-t_i^{(m)})\otimes\varphi_i^{(m)}(x)$. Let us prove that for every $i=\overline{1,s}$

$$\left\|\delta(t-t_i^{(m)})\otimes\varphi_i^{(m)}(x)-\delta(t-t_i^*)\otimes\varphi_i^{(m)}(x)\right\|_{W^{-l}_{bd^+}}\xrightarrow[m\to\infty]{}0.$$

Indeed,

$$\left\|\delta(t-t_i^{(m)})\otimes\varphi_i^{(m)}(x)-\delta(t-t_i^*)\otimes\varphi_i^{(m)}(x)\right\|_{W^{-l}_{bd^+}}=$$

$$=\sup_{\substack{v\in W^{+l}_{bd^+},\\ v\neq 0}}\frac{\left|\left\langle\delta(t-t_i^{(m)})\otimes\varphi_i^{(m)}(x)-\delta(t-t_i^*)\otimes\varphi_i^{(m)}(x),v\right\rangle_{W^{-l}_{bd^+}}\right|}{\|v\|_{W^{+l}_{bd^+}}}.$$

Since the set of smooth in $\overline{Q}$ functions $v(t,x)$ from $W^{+l}_{bd^+}$ is dense in $W^{+l}_{bd^+}$, we may take supremum only on these functions. Therefore, the last expression may be rewritten as

$$\left\|\delta\left(t-t_i^{(m)}\right)\otimes\varphi_i^{(m)}(x)-\delta\left(t-t_i^*\right)\otimes\varphi_i^{(m)}(x)\right\|_{W^{-l}_{bd^+}}=$$

$$=\sup_{v\in C^\infty_{bd^+},v\neq 0}\frac{\left|\int_\Omega\left(v(t_i^{(m)},x)-v(t_i^*,x)\right)\varphi_i^{(m)}(x)d\Omega\right|}{\|v\|_{W^{+l}_{bd^+}}}. \tag{12}$$

Let us estimate the numerator

$$v(t_i^{(m)},x)-v(t_i^*,x)=\int_{t_i^*}^{t_i^{(m)}} v_t(\eta,x)d\eta .$$

Applying the integral Cauchy inequality to the right-hand side of this identity, we obtain

$$\left|v(t_i^{(m)},x)-v(t_i^*,x)\right|\le\left|t_i^{(m)}-t_i^*\right|^{\frac{1}{2}}\left|\int_{t_i^*}^{t_i^{(m)}} v_t^2(\eta,x)d\eta\right|^{\frac{1}{2}}\le$$

$$\le\left|t_i^{(m)}-t_i^*\right|^{\frac{1}{2}}\left(\int_0^T v_t^2(\eta,x)d\eta\right)^{\frac{1}{2}}. \qquad (13)$$

Taking into account (13), let us apply the Cauchy inequality once more:

$$\left|\int_\Omega\left(v(t_i^{(m)},x)-v(t_i^*,x)\right)\varphi_i^{(m)}(x)d\Omega\right|\le$$

$$\le\left(\int_\Omega\left(v(t_i^{(m)},x)-v(t_i^*,x)\right)^2 d\Omega\right)^{\frac{1}{2}}\left(\int_\Omega\left(\varphi_i^{(m)}(x)\right)^2 d\Omega\right)^{\frac{1}{2}}\le$$

$$\le\left(\int_\Omega\left|t_i^{(m)}-t_i^*\right|\left(\int_0^T v_t^2(\eta,x)d\eta\right)d\Omega\right)^{\frac{1}{2}}\left\|\varphi_i^{(m)}(x)\right\|_{L_2(\Omega)}\le \qquad (14)$$

$$\le\left|t_i^{(m)}-t_i^*\right|^{\frac{1}{2}}\left\|v_t\right\|_{L_2(Q)}\left\|\varphi_i^{(m)}(x)\right\|_{L_2(\Omega)}.$$

By virtue of the inequalities $\|v_t\|_{L_2(Q)}\le\|v\|_{W_2^{1,0}(Q)}\le c\|v\|_{W^{+l}_{bd^+}}$ and the relations (12) and (14), we obtain:

$$\left\|\delta(t-t_i^{(m)})\otimes\varphi_i^{(m)}(x)-\delta(t-t_i^*)\otimes\varphi_i^{(m)}(x)\right\|_{W^{-l}_{bd^+}}\le$$

$$\le c\left|t_i^{(m)}-t_i^*\right|^{\frac{1}{2}}\left\|\varphi_i^{(m)}\right\|_{L_2(\Omega)}.$$

As far as $\left\{\varphi_i^{(m)}\right\}$ weakly converges to φ_i^* in $L_2(\Omega)$, by Eberlein-Shmulyan theorem [78] the sequence of the norms $\left\|\varphi_i^m\right\|_{L_2(\Omega)}$ is bounded, and hence

$$\left\|\delta(t-t_i^{(m)})\otimes\varphi_i^{(m)}(x)-\delta(t-t_i^*)\otimes\varphi_i^{(m)}(x)\right\|_{W_{bd^+}^{-l}}\leq$$

$$\leq C\left|t_i^{(m)}-t_i^*\right|^{\frac{1}{2}}\xrightarrow[m\to\infty]{}0.$$

So,

$$\left\|A_1^{(m)}-\sum_{i=1}^{s}\delta(t-t_i^*)\otimes\varphi_i^{(m)}(x)\right\|_{W_{bd^+}^{-l}}\xrightarrow[m\to\infty]{}0.$$

Now, let us prove that for an arbitrary smooth in $\overline{Q}$ function $v(t,x)$ from $W_{bd^+}^{+l}$ we have

$$\left\langle\sum_{i=1}^{s}\delta(t-t_i^*)\otimes\varphi_i^{(m)}(x),v\right\rangle_{W_{bd^+}}\xrightarrow[m\to\infty]{}\left\langle A_1(t,x,h^*),v\right\rangle_{W_{bd^+}}.$$

Indeed,

$$\left\langle\sum_{i=1}^{s}\delta(t-t_i^*)\otimes\varphi_i^{(m)}(x),v\right\rangle_{W_{bd^+}}=\sum_{i=1}^{s}\int_{\Omega}v(t_i^*,x)\varphi_i^{(m)}(x)d\Omega=$$

$$=\sum_{i=1}^{s}\left(v(t_i^*,\cdot),\varphi_i^{(m)}(\cdot)\right)_{L_2(\Omega)}\xrightarrow[m\to\infty]{}\sum_{i=1}^{s}\left(v(t_i^*,\cdot),\varphi_i^*(\cdot)\right)_{L_2(\Omega)}=$$

$$=\sum_{i=1}^{s}\int_{\Omega}v(t_i^*,x)\varphi_i^*(x)d\Omega=\left\langle A_1(t,x,h^*),v\right\rangle_{W_{bd^+}}.$$

Passing to the limit in the last equality is correct as far as

$$\varphi_i^m(x)\xrightarrow[m\to\infty]{}\varphi_i^*(x)\quad\text{weakly in } L_2(\Omega).$$

Due to the fact that the sequence of the norms $\left\| \sum_{i=1}^{s} \delta(t - t_i^*) \otimes \varphi_i^{(m)}(x) \right\|_{W_{bd^+}^{-l}}$ is bounded (proof is similar to the proof of the boundedness of $l_{A_1(h_1)}(v)$), it is easily to prove that for an arbitrary function $v \in W_{bd^+}^{+}$

$$\left\langle \sum_{i=1}^{s} \delta(t - t_i^*) \otimes \varphi_i^{(m)}(x), v \right\rangle_{W_{bd^+}} \xrightarrow[m \to \infty]{} \left\langle A_1(t, x, h^*), v \right\rangle_{W_{bd^+}}.$$

Taking into account all above-mentioned facts, we have

$$\left\langle A_1^{(m)}, v \right\rangle_{W_{bd^+}} = \left\langle A_1^{(m)} - \sum_{i=1}^{s} \delta(t - t_i^*) \otimes \varphi_i^{(m)}(x), v \right\rangle_{W_{bd^+}} +$$

$$+ \left\langle \sum_{i=1}^{s} \delta(t - t_i^*) \otimes \varphi_i^{(m)}(x), v \right\rangle_{W_{bd^+}},$$

where $v \in W_{bd^+}^{+l}$.

Let $m \to \infty$. The first term tends to zero, as far as by the Schwarz inequality

$$\left| \left\langle A_1^{(m)} - \sum_{i=1}^{s} \delta(t - t_i^*) \otimes \varphi_i^{(m)}(x), v \right\rangle_{W_{bd^+}} \right| \le$$

$$\le \left\| A_1^{(m)} - \sum_{i=1}^{s} \delta(t - t_i^*) \otimes \varphi_i^{(m)}(x) \right\|_{W_{bd^+}^{-l}} \|v\|_{W_{bd^+}^{+l}} \xrightarrow[m \to \infty]{} 0.$$

The second term tends to $\left\langle A_1(t, x, h^*), v \right\rangle_{W_{bd^+}}$. Thus, for any function $v \in W_{bd^+}^{+l}$

$$\left\langle A_1^{(m)}, v \right\rangle_{W_{bd^+}} \xrightarrow[m \to \infty]{} \left\langle A_1^*, v \right\rangle_{W_{bd^+}},$$

i.e. the mapping $A_1(\cdot)$ is weakly continuous.

Thus, we obtain

Theorem 3. *Let a system state be determined as a solution of the problem (1.12) under the following assumptions:*

1) the performance criterion $\Phi(\cdot): H_{bd}^{+} \to R^1$ *is a functional which is weakly lower semicontinuous with respect to the system state* $u(t,x,h)$ *and below bounded;*

2) the set of admissible controls $U_{ad} \subset H$ *is bounded, closed and convex in* H*;*

3) $H = [0,T]^s \times (L_2(\Omega))^s$*;*

4) $A_1(\cdot) = \sum_{i=1}^{s} \delta(t-t_i) \otimes \varphi_i(x)$*;*

5) the estimations (1.2) and (9) are valid.

Then there exists the optimal control of the system (1.12).

Remark. *As it was shown above, there exists a sequence of controls* $\left(t_k^*, \varphi_k^*\right)$*, which weakly converges to the optimal one* $\left(\tilde{t}^*, \tilde{\varphi}^*\right)$ *in* $R^N \times L_2^N(\Omega)$*. Hence, the problem of optimization may be ill posed [88-92] in the above selected metrics, that can yields significant difficulties in computation of the optimal pulse control. This circumstance may be overcame by using the regularization of the control. Let* $\varphi_k^* \in \Phi_{ad}$ *be a convex, closed and bounded set in* $W_2^1(\Omega)$*. In this case from the sequence* $\left\{\varphi_k^*\right\}_{k=1}^{\infty}$ *we can extract a subsequence* $\left\{\varphi_{k_l}^*\right\}_{l=1}^{\infty}$*, which is weakly converges to* $\tilde{\varphi}^*$ *in* $W_2^1(\Omega)$*. Due to the fact that* Ω *is bounded and has sufficiently smooth bound, it is follows from Kondrashow theorem [27, 93] that the imbedding operator of* $W_2^1(\Omega)$ *into* $L_2(\Omega)$ *is completely continuous and, hence, the sequence* $\left\{\varphi_{k_l}^*\right\}_{l=1}^{\infty}$ *is strongly* *convergent in* $L_2(\Omega)$ *to* φ^*. *Thus, the considered problem of optimization will be well posed in* $R^N \times L_2^N(\Omega)$.

In the similar way we can prove the following

Theorem 4. *Let a system state be determined as a solution of the problem (1.12) under the following assumptions:*

1) *the performance criterion* $\Phi(\cdot): W_{bd}^{+l} \to R^1$ *is a functional which is weakly lower semicontinuous with respect to the system state* $u(t,x,h)$ *and below bounded;*
2) *the set of admissible controls* $U_{ad} \subset H$ *is bounded, closed and convex in* H;
3) $H = [0,T]^s \times (L_2(\Omega))^s$;
4) $A_1(\cdot) = \sum_{i=1}^{s} \delta(t-t_i) \otimes \varphi_i(x)$;
5) *the estimations (1.2) are valid, and the following imbeddings and inequalities hold true*

$$H_{bd^+}^+ \subset W_2^{1,0}(Q) \ \left(H_{bd}^+ \subset W_2^{1,0}(Q)\right) \text{ è } \|\cdot\|_{W_2^{1,0}(Q)} \le c\|\cdot\|_{H_{bd^+}^+}$$

$$\left(\|\cdot\|_{W_2^{1,0}(Q)} \le c\|\cdot\|_{H_{bd}^+}\right).$$

Then there exists the optimal control of system (1.12).

2. $A_2(t,x,h) = \sum_{i=1}^{s} \delta^{(k)}(t-t_i) \otimes \varphi_i(x)$,

where $\delta^{(k)}(\cdot)$ is the k-th Sobolev derivative of the $\delta(\cdot)$-function, $t, t_i \in [0,T]$, and $\varphi_i(x) \in L_2(\Omega)$.

The control is

$$h_2 = \{(t_i, \varphi_i(x))\}_{i=1}^s \in U_{ad} \subset H_2 = [0,T]^s \times (L_2(\Omega))^s,$$

where U_{ad} is bounded, closed and convex in H_2.

By $A_2(t,x,h)=\sum_{i=1}^{s}\delta^{(k)}(t-t_i)\otimes\varphi_i(x)$ we shall denote a functional, which is defined on smooth in $\overline{Q}$ functions in the following way:

$$l_{A_2(\cdot)}(v)=(-1)^k\int_{\Omega}\sum_{i=1}^{s}v_t^{(k)}(t_i,x)\varphi_i(x)d\Omega,$$

where $v_t^{(k)}(t,x)$ is the k-th derivative of the function $v(t,x)$ with respect to the variable t.

Suppose that $W_{bd^+}^{+l}\subset W_2^{k+1,0}(Q)$ $\left(W_{bd}^{+l}\subset W_2^{k+1,0}(Q)\right)$ and

$$\|\cdot\|_{W_2^{k+1,0}(Q)}\le c\|\cdot\|_{W_{bd^+}^{+l}}\quad\left(\|\cdot\|_{W_2^{k+1,0}(Q)}\le c\|\cdot\|_{W_{bd}^{+l}}\right),\tag{15}$$

where $W_2^{k+1,0}(Q)$ is the completion of the space of smooth in $\overline{Q}$ functions with respect to the norm

$$\|v\|_{W_2^{k+1,0}(Q)}=\left(\int_Q\sum_{n=0}^{k+1}\left(v_t^{(n)}\right)^2dQ\right)^{\frac{1}{2}},$$

and let $f\in W_{bd^+}^{-l}$.

Similarly to the previous cases we can prove that $F(\cdot)=f+A_2(\cdot)\in W_{bd^+}^{-l}$, i.e. the mapping $F(\cdot)=f+A_2(\cdot)$ defines a linear continuous functional on $W_{bd^+}^{+l}$, and $A_2(\cdot)$ is weakly continuous mapping, and hence, $F(\cdot)=f+A_2(\cdot)$.

Theorem 5. *Let a system state be determined as a solution of problem (1.12) under the following assumptions:*

1) the performance criterion $\Phi(\cdot):H_{bd}^{+}\to R^1$ is a functional which is weakly lower semicontinuous with respect to the system state $u(t,x,h)$ and below bounded;

2) the set of admissible controls $U_{ad} \subset H$ *is bounded, closed and convex in* H*;*

3) $H = [0,T]^s \times (L_2(\Omega))^s$*;*

4) $A_2(t,x,h) = \sum_{i=1}^{s} \delta^{(k)}(t-t_i) \otimes \varphi_i(x)$*;*

5) the estimations (1.2) and (15) are valid.

Then there exists the optimal control of the system (1.12).

In a similar way we can prove the following

Theorem 6. *Let a system state be defined as a solution of the problem (1.12) under the following assumptions:*

1) the performance criterion $\Phi(\cdot): W_{bd}^{+l} \to R^1$ *is a functional which is weakly lower semicontinuous with respect to the system state* $u(t,x,h)$ *and below bounded;*

2) the set of admissible controls $U_{ad} \subset H$ *is bounded, closed and convex in* H*;*

3) $H = [0,T]^s \times (L_2(\Omega))^s$*;*

4) $A_2(t,x,h) = \sum_{i=1}^{s} \delta^{(k)}(t-t_i) \otimes \varphi_i(x)$*;*

5) the estimations (1.2) hold true, and also the following imbedding and inequalities are valid:

$$H_{bd}^{+} \subset W_2^{k+1,0}(Q) \quad \left(H_{bd^+}^{+} \subset W_2^{k+1,0}(Q)\right)$$

and

$$\|\cdot\|_{W_2^{k+1,0}(Q)} \le c\|\cdot\|_{H_{bd}^{+}} \quad \left(\|\cdot\|_{W_2^{k+1,0}(Q)} \le c\|\cdot\|_{H_{bd^+}^{+}}\right).$$

Then there exists the optimal control of system (1.12).

In studying of the optimal control existence in the case when the right-hand side of the equation is defined by one of the mappings A_i, $i = \overline{3,5}$ we shall suppose for simplicity that the set Ω is a tube domain with respect to the variable x_1, i.e. $\Omega = [\bar{x}_1, \bar{\bar{x}}_1] \times \Omega'$.

3. $A_3(t,x,h)=\sum_{i=1}^{s}\delta(x_1-x_{1,i})\otimes\varphi_i(t,x_2,...,x_n)$,

when $x_1,x_{1,i}\in[\overline{x}_1,\overline{\overline{x}}_1]$, $\varphi_i(t,x_2,...,x_n)\in L_2((0,T)\times\Omega')$.

The control is

$$h_3=\{(x_{1,i},\varphi_i(t,x_2,x_3,\ldots,x_n))\}_{i=1}^{s}\in U_{ad}\subset H_3=$$
$$=[\overline{x}_1,\overline{\overline{x}}_1]^s\times(L_2((0,T)\times\Omega'))^s,$$

where U_{ad} is a bounded, closed and convex set in H_3.

By $A_3(t,x,h)=\sum_{i=1}^{s}\delta(x_1-x_{1,i})\otimes\varphi_i(t,x_2,...,x_n)$ we shall denote a functional, which is defined on smooth in $\overline{Q}$ functions in the following way:

$$l_{A_3(\cdot)}(v)=$$
$$=\int_{\Omega'\times[0,T]}\sum_{i=1}^{s}v(t,x_{1,i},x_2,\ldots,x_n)\varphi_i(t,x_2,\ldots,x_n)d(\Omega'\times[0,T]),$$

Suppose that $W_{bd^+}^{+l}\subset W_2^{0,1,0}(Q)$ $\left(W_{bd}^{+l}\subset W_2^{0,1,0}(Q)\right)$ and

$$\|\cdot\|_{W_2^{0,1,0}(Q)}\le c\|\cdot\|_{W_{bd^+}^{+l}}\quad\left(\|\cdot\|_{W_2^{0,1,0}(Q)}\le c\|\cdot\|_{W_{bd}^{+l}}\right), \tag{16}$$

where $W_2^{0,1,0}(Q)$ is the completion of the space of smooth in $\overline{Q}$ functions with respect to the norm

$$\|v\|_{W_2^{0,1,0}(Q)}=\left(\int_Q v^2+v_{x_1}^2\,dQ\right)^{\frac{1}{2}},$$

and let $f\in W_{bd^+}^{-l}$.

Similarly to the previous cases we can prove that $F(\cdot)=f+A_3(\cdot)\in W_{bd^+}^{-l}$. Let us prove the weak continuity of the mapping $A_3(\cdot)$.

Let $h^{(m)} \xrightarrow{W} h^*$ in H_3. We shall prove that

$$\Big\| \delta(x_1 - x_{1,i}^{(m)}) \otimes \varphi_i^{(m)}(t, x_2, \dots, x_n) - \\ - \delta(x_1 - x_{1,i}^*) \otimes \varphi_i^{(m)}(t, x_2, \dots, x_n) \Big\|_{W_{bd^+}^{-1}} \xrightarrow[m\to\infty]{} 0. \tag{17}$$

Let $v(t,x)$ is an arbitrary smooth in $\overline{Q}$ function, which satisfies the conditions (bd^+). Then similarly to the previous cases we obtain

$$\Big\langle \delta(x_1 - x_{1,i}^{(m)}) \otimes \varphi_i^{(m)}(t, x_2, \dots, x_n) - \\ - \delta(x_1 - x_{1,i}^*) \otimes \varphi_i^{(m)}(t, x_2, \dots, x_n), v \Big\rangle_{W_{bd^+}} =$$

$$= \int_0^T \int_{\Omega'} \left(v(t, x_{1,i}^{(m)}, \dots, x_n) - v(t, x_{1,i}^*, \dots, x_n) \right) \varphi_i^{(m)}(t, \dots, x_n) d\Omega' dt \le$$

$$\le \left\| \varphi_i^{(m)} \right\|_{L_2((0,T)\times\Omega')} \cdot \left| x_{1,i}^{(m)} - x_{1,i}^* \right|^{\frac{1}{2}} \|v\|_{W_{bd^+}^{+1}},$$

Taking into account the boundedness of the set of functions $\left\{ \varphi_i^{(m)} \right\}_{i=1}^s$ and the definition of the negative norm, we obtain (17). Further proof is similar to the case of $A_1(t,x,h)$.

Theorem 7. *Let a system state be determine as a solution of the problem (1.12) under the following assumptions:*

1) the performance criterion $\Phi(\cdot): H_{bd}^+ \to R^1$ *is a functional which is weakly lower semicontinuous with respect to the system state* $u(t,x,h)$ *and below bounded;*

2) the set of admissible controls $U_{ad} \subset H$ *is bounded, closed and convex in* H*;*

3) $H = [\overline{x}_1, \overline{\overline{x}}_1]^s \times L_2((0,T)\times\Omega')^s$*;*

4) $A_3(t,x,h) = \sum_{i=1}^s \delta(x_1 - x_{1,i}) \otimes \varphi_i(t, x_2, \dots, x_n)$*;*

5) the estimations (1.2) and (16) are valid.

Then there exists the optimal control of system (1.12).

In a similar way we can prove the following

Theorem 8. *Let a system state be determined as a solution of the problem (1.12) under the following assumptions:*

1) *the performance criterion* $\Phi(\cdot): W_{bd}^{+l} \to R^1$ *is a functional which is weakly lower semicontinuous with respect to the system state* $u(t,x,h)$ *and below bounded;*
2) *the set of admissible controls* $U_{ad} \subset H$ *is bounded, closed and convex in* H*;*
3) $H = [\overline{x}_1, \overline{\overline{x}}_1]^s \times L_2((0,T)\times\Omega')^s$*;*
4) $A_3(t,x,h) = \sum_{i=1}^{s} \delta(x_1 - x_{1,i}) \otimes \varphi_i(t, x_2, \ldots, x_n)$*;*
5) *the estimations (1.2) hold true, and also the following imbeddings and inequalities are valid:*

$$H_{bd^+}^{+} \subset W_2^{0,1,0}(Q) \ \left(H_{bd}^{+} \subset W_2^{0,1,0}(Q)\right) \ è \ \|\cdot\|_{W_2^{0,1,0}(Q)} \le c\|\cdot\|_{H_{bd^+}^{+}} .$$

$$\left(\|\cdot\|_{W_2^{0,1,0}(Q)} \le c\|\cdot\|_{H_{bd}^{+}}\right).$$

Then there exists the optimal control of the system (1.12).

4. $$A_4(t,x,h) = \sum_{i=1}^{s} \sum_{j=1}^{p} \delta(t - t_i) \otimes \delta(x_1 - x_{1,j}) \otimes \varphi_{ij}(x_2, \ldots, x_n),$$

where $t, t_i \in [0,T]$, $x_1, x_{1,i} \in [\overline{x}_1, \overline{\overline{x}}_1]$, $\varphi_{ij}(x_2, \ldots, x_n) \in L_2(\Omega')$.

The control is

$$h_4 = \{(t_i, x_{1,j}, \varphi_{ij}(x_2, \ldots, x_n))\}_{i,j=1}^{s,p} \in U_{ad} \subset H_4 =$$

$$= [0,T]^s \times [\overline{x}_1, \overline{\overline{x}}_1]^p \times (L_2(\Omega'))^{sp},$$

where U_{ad} is a closed and bounded set in H_4.

By

$$A_4(t,x,h)=\sum_{i=1}^{s}\sum_{j=1}^{p}\delta(t-t_i)\otimes\delta(x_1-x_{1,j})\otimes\varphi_{ij}(x_2,\ldots,x_n)$$

we shall denote a functional defined on smooth in $\overline{Q}$ functions in the following way:

$$l_{A_4(\cdot)}(v)=\int_{\Omega'}\sum_{i=1}^{s}\sum_{j=1}^{p}v(t_i,x_{1,j},x_2,\ldots,x_n)\varphi_{ij}(x_2,\ldots,x_n)d\Omega'.$$

Suppose that $W^{+l}_{bd^+}\subset W_2^{1,1,0}(Q)$ $\left(W^{+l}_{bd}\subset W_2^{1,1,0}(Q)\right)$ and

$$\|\cdot\|_{W_2^{1,1,0}(Q)}\le c\|\cdot\|_{W^{+l}_{bd^+}}\quad\left(\|\cdot\|_{W_2^{1,1,0}(Q)}\le c\|\cdot\|_{W^{+l}_{bd}}\right),\tag{18}$$

where $W_2^{1,1,0}(Q)$ is the completion of the space of smooth in $\overline{Q}$ functions with respect to the norm

$$\|v\|_{W_2^{1,1,0}(Q)}=\left(\int_Q v^2+v_t^2+v_{x_1}^2+v_{tx_1}^2\,dQ\right)^{\frac{1}{2}},$$

and let $f\in W^{-l}_{bd^+}$.

Completely similarly to the three previous cases it is possible to prove that

$$A_4(t,x,h)=$$
$$=\sum_{i=1}^{s}\sum_{j=1}^{p}\delta(t-t_i)\otimes\delta(x_1-x_{1,j})\otimes\varphi_{ij}(x_2,\ldots,x_n)\in W^{-l}_{bd^+}.$$

Let us prove that $A_4(t,x,h)$ is a weakly continuous mapping. Let $h^m\xrightarrow{W}h^*$ in H_4. We shall prove that for any i,j

$$\Big\|\delta(t-t_i^{(m)})\otimes\delta(x_1-x_{1,j}^{(m)})\otimes\varphi_{ij}^{(m)}(x_2,\ldots,x_n)-$$
$$-\delta(t-t_i^*)\otimes\delta(x_1-x_{1,j}^*)\otimes\varphi_{ij}^{(m)}(x_2,\ldots,x_n)\Big\|_{W^{-l}_{bd^+}}\xrightarrow[m\to\infty]{}0.$$

Similarly to the previous cases let us consider

$$I = \left\langle \delta(t - t_i^{(m)}) \otimes \delta(x_1 - x_{1,j}^{(m)}) \otimes \varphi_{ij}^{(m)}(\cdot) - \right.$$
$$\left. - \delta(t - t_i^*) \otimes \delta(x_1 - x_{1,j}^*) \otimes \varphi_{ij}^{(m)}(\cdot), v \right\rangle_{W_{bd^+}} = \tag{19}$$
$$= \int_{\Omega'} \left(v(t_i^{(m)}, x_{1,j}^{(m)}, x_2, \ldots, x_n) - v(t_i^*, x_{1,j}^*, x_2, \ldots, x_n) \right) \varphi_{ij}^{(m)}(\cdot) d\Omega',$$

where $v(t, x)$ is an arbitrary smooth in $\overline{Q}$ function.

Using the Newton-Leibniz formula, we obtain

$$I = \int_{\Omega'} \left(\int_{t_i^*}^{t_i^{(m)}} v_t(\eta, x_{1,j}^*, x_2, \ldots, x_n) d\eta + \right.$$
$$\left. + \int_{x_{1,j}^*}^{x_{1,j}^{(m)}} v_{x_1}(t_i^{(m)}, \xi, x_2, \ldots, x_n) d\xi \right) \varphi_{ij}^{(m)}(\cdot) d\Omega'.$$

Applying the integral Cauchy inequality and the inequality $(a+b)^2 \le 2(a^2 + b^2)$, we have

$$I \le 2^{\frac{1}{2}} \left(\int_{\Omega'} \left(\int_{t_i^*}^{t_i^{(m)}} v_t(\eta, x_{1,j}^*, x_2, \ldots, x_n) d\eta \right)^2 + \right.$$
$$\left. + \left(\int_{x_{1,j}^*}^{x_{1,j}^{(m)}} v_{x_1}(t_i^{(m)}, \xi, x_2, \ldots, x_n) d\xi \right)^2 d\Omega' \right)^{\frac{1}{2}} \left\| \varphi_{ij}^{(m)} \right\|_{L_2(\Omega')} \le \tag{20}$$
$$\le 2 \left(\int_{\Omega'} \left| \left(t_i^{(m)} - t_i^* \right) \int_{t_i^*}^{t_i^{(m)}} v_t^2(\eta, x_{1,j}^*, x_2, \ldots, x_n) d\eta \right| + \right.$$
$$\left. + \left| \left(x_{1,j}^{(m)} - x_{1,j}^* \right) \int_{x_{1,j}^*}^{x_{1,j}^{(m)}} v_{x_1}^2(t_i^{(m)}, \xi, x_2, \ldots, x_n) d\xi \right| d\Omega' \right)^{\frac{1}{2}} \cdot \left\| \varphi_{ij}^{(m)} \right\|_{L_2(\Omega')}.$$

The Sobolev imbedding theorems imply that for any function $g(x) \in W_2^1(a,b)$ the following inequality holds true [74]

$$\max_{x\in[a,b]}|g(x)| \le c\left(\int_a^b g^2 + \left(\frac{dg}{dx}\right)^2 dx\right)^{\frac{1}{2}}. \tag{21}$$

Therefore,

$$v_t^2(\eta, x_{1,j}^*, x_2, \ldots, x_n) \le$$
$$\le c\left(\int_{\bar{\bar{x}}_1}^{\bar{\bar{\bar{x}}}_1} v_t^2(\eta, \xi, x_2, \ldots, x_n) + v_{tx_1}^2(\eta, \xi, x_2, \ldots, x_n) d\xi\right), \tag{22}$$

and

$$v_{x_1}^2(t_i^{(m)}, \xi, x_2, \ldots, x_n) \le$$
$$\le c\left(\int_0^T v_{x_1}^2(\eta, \xi, x_2, \ldots, x_n) + v_{tx_1}^2(\eta, \xi, x_2, \ldots, x_n) d\eta\right). \tag{23}$$

Substituting (22) and (23) into (20), we obtain

$$I \le 2c\left(\left|t_i^{(m)} - t_i^*\right| \cdot \left|\int_{\Omega'} \int_{t_i^*}^{t_i^{(m)}} \int_{\bar{\bar{x}}_1}^{\bar{\bar{\bar{x}}}_1} v_t^2(\eta, \xi, x_2, \ldots, x_n) + \right.\right.$$
$$\left. + v_{tx_1}^2(\eta, \xi, x_2, \ldots, x_n) d\xi d\eta d\Omega'\right| + \left|x_{1,j}^{(m)} - x_{1,j}^*\right| \times$$
$$\left.\times\left|\int_{\Omega'} \int_{x_{1,j}^*}^{x_{1,j}^{(m)}} \int_0^T v_{x_1}^2(\eta, \xi, x_2, \ldots, x_n) + v_{x_1 t}^2(\eta, \xi, x_2, \ldots, x_n) d\eta d\xi d\Omega'\right|\right)^{\frac{1}{2}} \times$$
$$\times \left\|\varphi_{ij}^{(m)}\right\|_{L_2(\Omega')}.$$

Note, that as far as the sequence $\varphi_{ij}^{(m)}$ is weakly convergent, and hence, it is bounded, we have

$$I \le C\left(\left|t_i^{(m)} - t_i^*\right| \iint\limits_Q v_t^2 + v_{tx_1}^2 \, dQ + \left|x_{1,j}^{(m)} - x_{1,j}^*\right| \iint\limits_Q v_{x_1}^2 + v_{tx_1}^2 \, dQ\right)^{\frac{1}{2}} \le$$

$$\le C_1\left(\left|t_i^{(m)} - t_i^*\right| \|v\|^2_{W^{+l}_{bd^+}} + \left|x_{1,j}^{(m)} - x_{1,j}^*\right| \|v\|^2_{W^{+l}_{bd^+}}\right)^{\frac{1}{2}} =$$

$$= C_1\left(\left|t_i^{(m)} - t_i^*\right| + \left|x_{1,j}^{(m)} - x_{1,j}^*\right|\right)^{\frac{1}{2}} \|v\|_{W^{+l}_{bd^+}}.$$

Whence,

$$\left\|\delta(t - t_i^{(m)}) \otimes \delta(x_1 - x_{1,j}^{(m)}) \otimes \varphi_{ij}^{(m)}(x_2,\ldots,x_n) - \right.$$

$$\left. - \delta(t - t_i^*) \otimes \delta(x_1 - x_{1,j}^*) \otimes \varphi_{ij}^{(m)}(x_2,\ldots,x_n)\right\|_{W^{-l}_{bd^+}} \le$$

$$\le C_1\left(\left|t_i^{(m)} - t_i^*\right| + \left|x_{1,j}^{(m)} - x_{1,j}^*\right|\right)^{\frac{1}{2}} \xrightarrow[m\to\infty]{} 0.$$

Finally, we obtain

$$\lim_{m\to\infty}\left\|A_4(h^{(m)}) - \right.$$

$$\left. - \sum_{i=1}^{s}\sum_{j=1}^{p} \delta(t - t_i^*) \otimes \delta(x_1 - x_{1,j}^*) \otimes \varphi_{ij}^{(m)}(x_2,\ldots,x_n)\right\|_{W^{-l}_{bd^+}} = 0.$$

Now, let $v(t,x)$ is an arbitrary function from $W^{+l}_{bd^+}$. It is clear that the following identity holds true

$$\left\langle A_4(h^{(m)}) - A_4(h^*), v\right\rangle_{W_{bd^+}} = \left\langle A_4(h^{(m)}) - \right.$$

$$\left. - \sum_{i=1}^{s}\sum_{j=1}^{p} \delta(t - t_i^*) \otimes \delta(x_1 - x_{1,j}^*) \otimes \varphi_{ij}^{(m)}(x_2,\ldots,x_n), v\right\rangle_{W_{bd^+}} +$$

$$+ \sum_{i=1}^{s}\sum_{j=1}^{p}\left(v(t_i^*, x_{1,j}^*, \cdot), \varphi_{ij}^{(m)}(\cdot) - \varphi_{ij}^*(\cdot)\right)_{L_2(\Omega')}.$$

It follows from the previous reasoning that the first term tends to zero. Since $\varphi_{ij}^{m}(\cdot)-\varphi_{ij}^{*}(\cdot)$ weakly converges to zero in $L_2(\Omega')$, the second term tends to zero also. Thus, we have proved that the mapping $A_4(t,x,h)$ is weakly continuous.

Theorem 9. *Let a system state be determined as a solution of the problem (1.12) under the following assumptions:*

1) *the performance criterion* $\Phi(\cdot):H_{bd}^{+}\to R^1$ *is a functional which is weakly lower semicontinuous with respect to the system state* $u(t,x,h)$ *and below bounded;*
2) *the set of admissible controls* $U_{ad}\subset H$ *is bounded, closed and convex in* H;
3) $H=[0,T]^s\times[\bar{x}_1,\bar{\bar{x}}_1]^p\times(L_2(\Omega'))^{ps}$;
4) $A_4(t,x,h)=\sum_{i=1}^{s}\sum_{j=1}^{p}\delta(t-t_i)\otimes\delta(x_1-x_{1,j})\otimes\varphi_{ij}(x_2,\dots,x_n)$;
5) *the estimations (1.2) and (18) are valid.*

Then there exists the optimal control of the system (1.12).

In a similar way we can prove the following

Theorem 10. *Let a system state be determined as a solution of the problem (1.12) under the following assumptions:*

1) *the performance criterion* $\Phi(\cdot):W_{bd}^{+l}\to R^1$ *is a functional which is weakly lower semicontinuous with respect to the system state* $u(t,x,h)$ *and below bounded;*
2) *the set of admissible controls* $U_{ad}\subset H$ *is bounded, closed and convex in* H;
3) $H=[0,T]^s\times[\bar{x}_1,\bar{\bar{x}}_1]^p\times(L_2(\Omega'))^{ps}$;
4) $A_4(t,x,h)=\sum_{i=1}^{s}\sum_{j=1}^{p}\delta(t-t_i)\otimes\delta(x_1-x_{1,j})\otimes\varphi_{ij}(x_2,\dots,x_n)$;

5) the estimations (1.2) are valid, and also the following imbeddings and inequalities hold true

$$H^+_{bd^+} \subset W_2^{1,1,0}(Q) \ \left(H^+_{bd} \subset W_2^{1,1,0}(Q)\right) \text{ è } \|\cdot\|_{W_2^{1,1,0}(Q)} \le c\|\cdot\|_{H^+_{bd^+}}$$

$$\left(\|\cdot\|_{W_2^{1,1,0}(Q)} \le c\|\cdot\|_{H^+_{bd}}\right).$$

Then there exists the optimal control of the system (1.12).

5. $A_5(t,x,h) = \sum_{i=1}^{s} \delta(x_1 - a_i(t)) \otimes \varphi_i(t, x_2, \ldots, x_n)$.

The control is

$$h_5 = \{(a_i(t), \varphi_i(\cdot))\}_{i=1}^s \in U_{ad} \subset H_5 =$$
$$= \left(W_2^1(0,T)\right)^s \times \left(L_2((0,T)\times\Omega')\right)^s.$$

By $A_5(t,x,h)$ we shall denote a functional defined on smooth in $\overline{Q}$ functions in the following way:

$$l_{A_5}(v) = \int_0^T \int_{\Omega'} \sum_{i=1}^{s} v(t, a_i(t), x_2, \ldots, x_n)\,\varphi_i(t, x_2, \ldots, x_n)\,d\Omega'\,dt,$$

where $v(t,x)$ is a smooth in $\overline{Q}$ function.

Suppose that $W^{+l}_{bd^+} \subset W_2^{0,1,0}(Q)$ $\left(W^{+l}_{bd} \subset W_2^{0,1,0}(Q)\right)$ and

$$\|\cdot\|_{W_2^{0,1,0}(Q)} \le c\|\cdot\|_{W^{+l}_{bd^+}} \quad \left(\|\cdot\|_{W_2^{0,1,0}(Q)} \le c\|\cdot\|_{W^{+l}_{bd}}\right), \tag{24}$$

where $W_2^{0,1,0}(Q)$ is a completion of the space of smooth in $\overline{Q}$ functions with respect to the norm

$$\|v\|_{W_2^{0,1,0}(Q)} = \left(\int_Q v^2 + v_{x_1}^2\, dQ\right)^{\frac{1}{2}},$$

and let $f \in W^{-l}_{bd^+}$.

We shall prove that the functional $l_{A_5}(v)$ is bounded. Indeed, using the integral Cauchy inequality, we obtain

$$\left|l_{A_5}(v)\right| \le \sum_{i=1}^{s}\left(\int_0^T\int_{\Omega'}\varphi_i^2(t,x_2,\ldots,x_n)d\Omega' dt\right)^{\frac{1}{2}} \times$$

$$\times\left(\int_0^T\int_{\Omega'}v^2(t,a_i(t),x_2,\ldots,x_n)d\Omega' dt\right)^{\frac{1}{2}}.$$

Using inequalities (21) and (24), we have

$$\left|l_{A_5}(v)\right| \le c\sum_{i=1}^{s}\left\|\varphi_i\right\|_{L_2((0,T)\times\Omega')} \times$$

$$\times\left(\int_0^T\int_{\Omega'}\int_{\bar{x}_1}^{\bar{\bar{x}}_1}v^2(t,\eta,x_2,\ldots,x_n)+v_{x_1}^2(t,\eta,x_2,\ldots,x_n)d\eta d\Omega' dt\right)^{\frac{1}{2}} \le$$

$$\le c\sum_{i=1}^{s}\left\|\varphi_i\right\|_{L_2((0,T)\times\Omega')}\left\|v\right\|_{W_{bd^+}^{+l}},$$

that proves the boundedness of the functional $l_{A_5}(v)$; the linearity of $l_{A_5}(v)$ is obvious.

Let us prove that $f_{A_5}(h^{(m)}) \xrightarrow{W} f_{A_5}(h^*)$ in $W_{bd^+}^{-l}$, where $h^{(m)}$ is some weakly convergent in H_5 sequence: $h^{(m)} \xrightarrow{W} h^*$. Note, that being weakly convergent the sequence $\left\{a_i^{(m)}(t)\right\}_{m=1}^{\infty}$ is bounded in the space $W_2^1(0,T)$, and since the imbedding of the space $W_2^1(0,T)$ into $C([0,T])$ is compact, we obtain

$$a_i^{(m)}(t) \xrightarrow[m\to\infty]{} a_i^*(t)$$

with respect to the norm of the space $C([0,T])$, and, hence,

$$\sup_{t\in[0,T]}\left|a_i^{(m)}(t)-a_i^*(t)\right| \le \varepsilon_m \xrightarrow[m\to\infty]{} 0.$$

The sequence of the functions $\left\{\varphi_i^{(m)}(t,x_2,\ldots,x_n)\right\}_{m=1}^{\infty}$ weakly converges to $\varphi_i^*(t,x_2,\ldots,x_n)$ in $L_2((0,T)\times\Omega')$.

At first, let us prove that

$$\Big\|\delta(x_1 - a_i^{(m)}(t))\varphi_i^{(m)}(t,x_2,\dots,x_n) - \\ - \delta(x_1 - a_i^*(t))\varphi_i^{(m)}(t,x_2,\dots,x_n)\Big\|_{W_{bd^+}^{-1}} \xrightarrow[m\to\infty]{} 0.$$

To do this let us consider

$$I = \Big|\Big\langle \delta(x_1 - a_i^{(m)}(t))\varphi_i^{(m)}(t,x_2,\dots,x_n) - \\ - \delta(x_1 - a_i^*(t))\varphi_i^{(m)}(t,x_2,\dots,x_n), v\Big\rangle_{W_{bd^+}}\Big|,$$

for any smooth in $\overline{Q}$ function $v(t,x)$. Applying the Newton-Leibniz formula we obtain

$$I = \Big|\int_0^T\int_{\Omega'}\varphi_i^{(m)}(t,x_2,\dots,x_n)\times \\ \times\Big(v(t,a_i^{(m)}(t),x_2,\dots,x_n) - v(t,a_i^*(t),x_2,\dots,x_n)\Big)d\Omega' dt\Big| = \\ = \Big|\int_0^T\int_{\Omega'}\varphi_i^{(m)}(t,x_2,\dots,x_n)\int_{a_i^*(t)}^{a_i^{(m)}(t)} v_{x_1}(t,\eta,x_2,\dots,x_n)d\eta d\Omega' dt\Big|.$$

Applying the integral Cauchy inequality to the right-hand side, we obtain:

$$I \le \left(\int_0^T\int_{\Omega'}\left(\varphi_i^{(m)}(t,x_2,\dots,x_n)\right)^2 d\Omega' dt\right)^{\frac{1}{2}}\times \\ \times\left(\int_0^T\int_{\Omega'}\left(\int_{a_i^*(t)}^{a_i^{(m)}(t)} v_{x_1}(t,\eta,x_2,\dots,x_n)d\eta\right)^2 d\Omega' dt\right)^{\frac{1}{2}} \le \\ \le \left\|\varphi_i^{(m)}(\cdot)\right\|_{L_2(\Omega'\times(0,T))}\times$$

$$\times\left(\int_0^T\int_{\Omega'}\left|a_i^*(t)-a_i^{(m)}(t)\right|\left(\left|\int_{a_i^*(t)}^{a_i^{(m)}(t)} v_{x_1}^2(t,\eta,x_2,\dots,x_n)d\eta\right|\right)d\Omega' dt\right)^{\frac{1}{2}} \le$$

$$\le \left\|\varphi_i^{(m)}\right\|_{L_2(\Omega'\times(0,T))}\varepsilon_m^{\frac{1}{2}}\|v\|_{W_{bd^+}^{+l}}.$$

Thus, we have

$$\frac{\left|\left\langle \delta(x_1-a_i^{(m)}(t))\varphi_i^{(m)}(t,x)-\delta(x_1-a_i^*(t))\varphi_i^{(m)}(t,x),v\right\rangle_{W_{bd^+}}\right|}{\|v\|_{W_{bd^+}^{+l}}} \le$$

$$\le \varepsilon_m^{\frac{1}{2}}\left\|\varphi_i^{(m)}\right\|_{L_2(\Omega'\times(0,T))}.$$

As far as $\left\{\varphi_i^{(m)}(t,x_2,\dots,x_n)\right\}_{m=1}^{\infty}$ is weakly convergent, then the sequence of the norms $\left\|\varphi_i^m(\cdot)\right\|_{L_2(\Omega'\times(0,T))}$ is bounded. Granting that the set of smooth in $\overline{Q}$ functions $v(t,x)$ is dense in $W_{bd^+}^{+l}$, we have

$$\left\|\delta(x_1-a_i^{(m)}(t))\varphi_i^{(m)}(t,x_2,\dots,x_n)-\right.$$
$$\left.-\delta(x_1-a_i^*(t))\varphi_i^{(m)}(t,x_2,\dots,x_n)\right\|_{W_{bd^+}^{-l}} \le$$

$$\le \varepsilon_m^{\frac{1}{2}}\left\|\varphi_i^{(m)}\right\|_{L_2(\Omega'\times(0,T))} \le c\varepsilon_m^{\frac{1}{2}} \xrightarrow[m\to\infty]{} 0.$$

The last expression implies that

$$\left\|A_5(t,x,h^{(m)})-\sum_{i=1}^{s}\delta(x_1-a_i^*(t))\varphi_i^{(m)}(t,x_2,\dots,x_n)\right\|_{W_{bd^+}^{-l}} \underset{m\to\infty}{\to} 0. \quad (25)$$

To finish the proof we have to verify that for any $v(t,x)\in W_{bd^+}^{+l}$:

$$\left|\left\langle \sum_{i=1}^{s}\delta(x_1-a_i^*(t))\varphi_i^{(m)}(t,x_2,\dots,x_n)-A_5(t,x,h^*),v\right\rangle_{W_{bd^+}}\right| \underset{m\to\infty}{\to} 0. \quad (26)$$

Indeed,

$$\left|\left\langle \sum_{i=1}^{s} \delta(x_1 - a_i^*(t))\varphi_i^{(m)}(t, x_2, \ldots, x_n) - A_5(t, x, h^*), v \right\rangle_{W_{bd^+}}\right| =$$

$$= \left| \sum_{i=1}^{s} \int_0^T \int_{\Omega'} v(t, a_i^*(t), x_2, \ldots, x_n)\left(\varphi_i^{(m)} - \varphi_i^*\right) d\Omega' dt \right| \le$$

$$\le \sum_{i=1}^{s} \left| \left(v, \varphi_i^{(m)} - \varphi_i^*\right)_{L_2'(\Omega' \times (0,T))} \right| \xrightarrow[m \to \infty]{} 0,$$

as far as $\left\{\varphi_i^{(m)}(\cdot)\right\}_{m=1}^{\infty}$ weakly converges to $\varphi_i^*(\cdot)$ in $L_2\left((0,T) \times \Omega'\right)$. Thus, taking into account (25) and (26), we obtain

$$\lim_{m \to \infty} \left| \left\langle A_5(t, x, h^{(m)}) - A_5(t, x, h^*), v \right\rangle_{W_{bd^+}} \right| \le$$

$$\le \lim_{m \to \infty} \left(\left| \left\langle A_5(t, x, h^{(m)}) - \sum_{i=1}^{s} \delta(x_1 - a_i^*(t))\varphi_i^{(m)}(t, x_2, \ldots), v \right\rangle_{W_{bd^+}} \right| + \right.$$

$$\left. + \left| \left\langle \sum_{i=1}^{s} \delta(x_1 - a_i^*(t))\varphi_i^{(m)}(t, x_2, \ldots) - A_5(t, x, h^*), v \right\rangle_{W_{bd^+}} \right| \right) = 0,$$

for an arbitrary function $v(t, x) \in W_{bd^+}^{+l}$.

In other words, the mapping

$$A_5(t, x, h) = \sum_{i=1}^{s} \delta(x_1 - a_i(t)) \otimes \varphi_i(t, x_2, \ldots, x_n)$$

is weakly continuous.

Theorem 11. *Let a system state be determined as a solution of the problem (1.12) under the following assumptions:*

1) *the performance criterion* $\Phi(\cdot) : H_{bd}^+ \to R^1$ *is a functional which is weakly lower semicontinuous with respect to the system state* $u(t, x, h)$ *and below bounded;*

2) *the set of admissible controls* $U_{ad} \subset H$ *is bounded, closed and convex in* H;

3) $H = \left(W_2^1(0,T)\right)^s \times \left(L_2((0,T)\times\Omega')\right)^s$;

4) $A_5(t,x,h) = \sum_{i=1}^{s} \delta(x_1 - a_i(t)) \otimes \varphi_i(t,x_2,\ldots,x_n)$;

5) *the estimations (1.2) and (24) are valid.*

Then there exists the optimal control of the system (1.12).

Similarly we can prove the following

Theorem 12. *Let a system state be determined as a solution of the problem (1.12) under the following assumptions:*

1) *the performance criterion* $\Phi(\cdot): W_{bd}^{+l} \to R^1$ *is a functional which is weakly lower semicontinuous with respect to the system state* $u(t,x,h)$ *and below bounded;*

2) *the set of admissible controls* $U_{ad} \subset H$ *is bounded, closed and convex in* H;

3) $H = \left(W_2^1(0,T)\right)^s \times \left(L_2((0,T)\times\Omega')\right)^s$;

4) $A_5(t,x,h) = \sum_{i=1}^{s} \delta(x_1 - a_i(t)) \otimes \varphi_i(t,x_2,\ldots,x_n)$;

5) *the estimations (1.2) are valid, and also the following imbeddings and inequalities àre valid:*

$$H_{bd^+}^+ \subset W_2^{0,1,0}(Q) \ \left(H_{bd}^+ \subset W_2^{0,1,0}(Q)\right) \text{ è } \|\cdot\|_{W_2^{0,1,0}(Q)} \le c\|\cdot\|_{H_{bd^+}^+}$$
$$\left(\|\cdot\|_{W_2^{0,1,0}(Q)} \le c\|\cdot\|_{H_{bd}^+}\right).$$

Then there exists the optimal control of the system (1.12).

Note, that we may consider the problem of optimal control when the right-hand side of the state equation is a linear combination of the functionals A_i, $i = \overline{1,5}$.

Consider, for example, the problem

$$Lu = f + A_6(t,x,h), \tag{27}$$

where $A_6(t,x,h)=\sum_{i=1}^{\bar{s}}\delta(t-\bar{t}_i)\otimes\overline{\varphi}_i(x)+\sum_{i=1}^{\bar{\bar{s}}}\delta^{(1)}(t-\bar{\bar{t}}_i)\otimes\overline{\overline{\varphi}}_i(x)$.

The control is

$h_6=\{(\bar{t}_i,\bar{\bar{t}}_j,\overline{\varphi}_i(x),\overline{\overline{\varphi}}_j(x))\}_{ij=1}^{\bar{s},\bar{\bar{s}}}\in U_{ad}\subset H_6=[0,T]^{\bar{s}+\bar{\bar{s}}}\times(L_2(\Omega))^{\bar{s}+\bar{\bar{s}}}$

By

$$A_6(t,x,h)=\sum_{i=1}^{\bar{s}}\delta(t-\bar{t}_i)\otimes\overline{\varphi}_i(x)+\sum_{i=1}^{\bar{\bar{s}}}\delta^{(1)}(t-\bar{\bar{t}}_i)\otimes\overline{\overline{\varphi}}_i(x)$$

we shall denote a functional defined on smooth in $\overline{Q}$ functions in the following way:

$$l_{A_2(\cdot)}(v)=\int_\Omega\left(\sum_{i=1}^{\bar{s}}v(\bar{t}_i,x)\overline{\varphi}_i(x)\right)-\left(\sum_{i=1}^{\bar{\bar{s}}}v_t(\bar{\bar{t}}_i,x)\overline{\overline{\varphi}}_i(x)\right)d\Omega,$$

Suppose that $W_{bd^+}^{+l}\subset W_2^{2,0}(Q)$ $\left(W_{bd}^{+l}\subset W_2^{2,0}(Q)\right)$ and

$$\|\cdot\|_{W_2^{2,0}(Q)}\le c\|\cdot\|_{W_{bd^+}^{+l}}\quad\left(\|\cdot\|_{W_2^{2,0}(Q)}\le c\|\cdot\|_{W_{bd}^{+l}}\right),\tag{28}$$

and let $f\in W_{bd^+}^{-l}$.

As far as $W_2^{-1,0}(Q)\subset W_2^{-2,0}(Q)$, where $W_2^{-1,0}(Q)$, $W_2^{-2,0}(Q)$ are negative spaces constructed on the pairs $W_2^{1,0}(Q),L_2(Q)$ and $W_2^{2,0}(Q),L_2(Q)$, similarly to the cases $A_1(\cdot)$ and $A_2(\cdot)$ we shall prove that the right-hand side of (27) belongs to the space $W_2^{-2,0}(Q)$, and hence, to the $W_{bd^+}^{-l}$.

Let us prove that $A_6(t,x,h):H_6\to W_{bd^+}^{-l}$ is a weakly continuous mapping. Indeed, let $h_6^{(m)}$ is a weakly convergent in H_6 sequence $h_6^{(m)}\xrightarrow{W}h_6^*$. In is clear that

$$h_6^{(m)}=\overline{h}_6^{(m)}\times\overline{\overline{h}}_6^{(m)},$$

where $\overline{h}_6^{(m)} = (\overline{t}_i^{(m)}, \overline{\varphi}_i^{(m)})$, $\overline{\overline{h}}_6^{(m)} = (\overline{\overline{t}}_j^{(m)}, \overline{\overline{\varphi}}_j^{(m)})$ and $\overline{h}_6^{(m)} \xrightarrow{W} \overline{h}_6^*$ in H_1, $\overline{\overline{h}}_6^{(m)} \xrightarrow{W} \overline{\overline{h}}_6^*$ in H_2.

That is why from the results of this section we have that

$$A_1\left(\overline{h}_6^{(m)}\right) \xrightarrow{W} A_1\left(\overline{h}_6^*\right) \qquad \text{in } W_2^{-1,0}(Q),$$

$$A_2\left(\overline{\overline{h}}_6^{(m)}\right) \xrightarrow{W} A_2\left(\overline{\overline{h}}_6^*\right) \qquad \text{in } W_2^{-2,0}(Q).$$

Since $W_2^{2,0}(Q) \subset W_2^{1,0}(Q)$, the bilinear forms $\langle\cdot,\cdot\rangle_{W_2^{1,0}\times W_2^{-1,0}}$ and $\langle\cdot,\cdot\rangle_{W_2^{2,0}\times W_2^{-2,0}}$ coincide for the pairs from $W_2^{-1,0}\times W_2^{2,0}$, hence, $A_1\left(\overline{h}_6^{(m)}\right) \xrightarrow{W} A_1\left(\overline{h}_6^*\right)$ in $W_2^{-2,0}(Q)$.

That is why

$$A_6(t,x,h_6^{(m)}) = A_1(t,x,\overline{h}_6^{(m)}) + A_2(t,x,\overline{\overline{h}}_6^{(m)}) \xrightarrow{W} A_1(t,x,\overline{h}_6^*) +$$
$$+ A_2(t,x,\overline{\overline{h}}_6^*) = A_6(t,x,h_6^*),$$

weakly in $W_2^{-2,0}(Q)$, and hence, in $W_{bd^+}^{-l}$ also. Thus, the following theorem holds true:

Theorem 13. *Let a system state be determined as a solution of the problem (1.12) under the following assumptions:*

1) *the performance criterion* $\Phi(\cdot): H_{bd}^+ \to R^1$. *is a functional which is weakly lower semicontinuous with respect to the system state* $u(t,x,h)$ *and below bounded;*
2) *the set of admissible controls* $U_{ad} \subset H$ *is bounded, closed and convex in* H*;*
3) $H = [0,T]^{\overline{s}+\overline{\overline{s}}} \times (L_2(\Omega))^{\overline{s}+\overline{\overline{s}}}$*;*
4) $A_6(t,x,h) = \sum_{i=1}^{\overline{s}} \delta(t-\overline{t}_i)\otimes\overline{\varphi}_i(x) + \sum_{i=1}^{\overline{\overline{s}}} \delta^{(1)}(t-\overline{\overline{t}}_i)\otimes\overline{\overline{\varphi}}_i(x)$*;*
5) *the estimations (1.2) and (28) are valid.*

Then there exists the optimal control of system (1.12).

In a similar way we can prove the following

Theorem 14. *Let a system state be determined as a solution of the problem (1.12) under the following assumptions:*

1) *the performance criterion* $\Phi(\cdot): W_{bd}^{+l} \to R^1$ *is a functional which is weakly lower semicontinuous with respect to the system state* $u(t,x,h)$ *and below bounded;*
2) *the set of admissible controls* $U_{ad} \subset H$ *is bounded, closed and convex in* H*;*
3) $H = [0,T]^{\bar{s}+\bar{\bar{s}}} \times (L_2(\Omega))^{\bar{s}+\bar{\bar{s}}}$;
4) $A_6(t,x,h) = \sum_{i=1}^{\bar{s}} \delta(t-\bar{t}_i) \otimes \bar{\varphi}_i(x) + \sum_{i=1}^{\bar{\bar{s}}} \delta^{(1)}(t-\bar{\bar{t}}_i) \otimes \bar{\bar{\varphi}}_i(x)$;
5) *the estimations (1.2) hold true, and also the following imbeddings and inequalities are valid*

$$H_{bd^+}^{+} \subset W_2^{2,0}(Q) \ \left(H_{bd}^{+} \subset W_2^{2,0}(Q)\right) \text{ è } \|\cdot\|_{W_2^{2,0}(Q)} \le c\|\cdot\|_{H_{bd^+}^{+}}$$
$$\left(\|\cdot\|_{W_2^{2,0}(Q)} \le c\|\cdot\|_{H_{bd}^{+}}\right).$$

Then, there exist optimal controls of the system (1.12).

Chapter 2

GENERAL PRINCIPLES OF INVESTIGATION OF LINEAR SYSTEMS WITH GENERALIZED CONTROL

1. DIFFERENTIAL PROPERTIES OF PERFORMANCE CRITERION

Depending on the properties of the operators L and L^* it is possible to consider various controls and performance criteria. Hereinafter we shall study the differential properties of integral, quadratic with respect to the system state performance criterion for the problem of pulse optimal control and later we shall generalized the corresponding theorems for the case of the problem of the optimal control with an arbitrary right-hand side.

In the case when the following inequalities are valid for the operator L

$$\|u\|_{L_2(Q)} \le C_1 \|Lu\|_{W^{-l}_{bd^+}} \le C_2 \|u\|_{W^{+l}_{bd}}$$

$$\|u\|_{L_2(Q)} \le C_1 \|L^* u\|_{W^{-l}_{bd}} \le C_2 \|u\|_{W^{+l}_{bd^+}}$$

and the right-hand side $F(\cdot)$ of the equation (1.1.12) is of the form $F = f + A(h) = f + \sum_{i=1}^{N} \delta(t-t_i) \otimes \varphi_i(x) \in W^{-l}_{bd^+}$ (the first case considered in Section 1.1.2) we consider the performance criterion:

$$J(h) = \|u(h) - z_{ad}\|^2_{L_2(Q)}, \quad h \in U_{ad}, \tag{1}$$

where z_{ad} is a function describing the desired functioning of the system, $z_{ad} \in L_2(Q)$.

The functional (1) is defined correctly as far as Theîrem 1.1.3 guarantees that the function $u(t,x,h)$ belongs to the space $L_2(Q)$.

Suppose that $W^{+l}_{bd^+} \subset C^1([0,T];L_2(\Omega))$, and moreover this imbedding is continuous.

Theorem 1. *Provides that conditions above mentioned are satisfied, the functional (1) is differentiable by Gâteaux in the space $R^N \times L_2^N(\Omega)$ and its gradient is of the form:*

$$grad\, J(h) =$$

$$= \left(\int_\Omega v_t'(t_1,x)\varphi_1(x)d\Omega,..,\int_\Omega v_t'(t_N,x)\varphi_N(x)d\Omega, v(t_1,x),..,v(t_N,x) \right),$$

where $h = (t^*,\varphi^*) = (t_1 \ldots t_N, \varphi_1(x)\ldots\varphi_N(x)) \in H$ is a control, $v(t,x)$ is a solution of the problem

$$L^* v = 2(u - z_{ad}),\ v \in W^{+l}_{bd^+}.$$

Proof. Let $(t^*,\varphi^*) = (t_1,\ldots t_N, \varphi_1(x),\ldots\varphi_N(x))$ and $(t^* + \lambda\Delta t^*, \varphi^* + \lambda\Delta\varphi^*)$ be some elements belonging to $U_{ad} > 0$, $u(t^*,\varphi^*)$ and $u(t^* + \lambda\Delta t^*, \varphi^* + \lambda\Delta\varphi^*)$ are solutions of the boundary problem (1.1.12) corresponding to these controls. Denote

$$\Delta u = u(t^* + \lambda\Delta t^*, \varphi^* + \lambda\Delta\varphi^*) - u(t^*,\varphi^*).$$

Then the increment of the performance criterion may be represented in the following form

$$\Delta J = J(t^* + \lambda\Delta t^*, \varphi^* + \lambda\Delta\varphi^*) - J(t^*,\varphi^*) =$$

$$= \int_Q \left\{ \left[u(t^* + \lambda\Delta t^*, \varphi^* + \lambda\Delta\varphi^*) - z_{ad} \right]^2 - \left[u(t^*,\varphi^*) - z_{ad} \right]^2 \right\} dQ =$$

$$= \int_Q \Delta u \left[u(t^* + \lambda\Delta t^*, \varphi^* + \lambda\Delta\varphi^*) + u(t^*,\varphi^*) - 2z_{ad} \right]^2 dQ. \quad (2)$$

Define the adjoint state as a solution of the problem

$$L^* v = 2(u(t,\varphi) - z_{ad}).$$

It follows from the results of Chapter 1 that there exists a unique solution of this problem in the class of functions belonging to $W^{+l}_{bd^+}$. Then (2) we may rewrite in the following form

$$\Delta J = \left(\Delta u, L^* v\right)_{L_2(Q)} + \|\Delta u\|^2_{L_2(Q)}.$$

Obviously, Δu is a solution of the problem

$$L\Delta u = \sum_{i=1}^{N} \delta(t - t_i - \lambda\Delta t_i) \otimes \left(\varphi_i + \lambda\Delta\varphi_i\right) - \sum_{i=1}^{N} \delta(t - t_i) \otimes \varphi_i .$$

As a consequence of the results obtained in Chapter 1 there exists a solution of this problem which is defined as a function $\Delta u \in L_2(Q)$ such that for all $y(t,x) \in W^{+l}_{bd^+}$: $L^* y \in L_2(Q)$ (including $v(t,x)$) the following equality is true:

$$\left(\Delta u, L^* y\right)_{L_2(Q)} = \sum_{i=1}^{N} \Big[\left(y(t_i + \lambda\Delta t_i, x) - y(t_i, x), \varphi_i(x)\right)_{L_2(\Omega)} + \\ + \lambda\left(y(t_i + \lambda\Delta t_i, x), \Delta\varphi_i(x)\right)_{L_2(\Omega)} \Big].$$

We obtain

$$\Delta J = \sum_{i=1}^{N} \Big[\left(v(t_i + \lambda\Delta t_i, x) - v(t_i, x), \varphi_i(x)\right)_{L_2(\Omega)} + \\ + \lambda\left(v(t_i + \lambda\Delta t_i, x), \Delta\varphi_i(x)\right)_{L_2(\Omega)} \Big] + \|\Delta u\|^2_{L_2(Q)}. \quad (3)$$

Represent the increments $v(t_i + \lambda\Delta t_i, x) - v(t_i, x)$ from (3) as

$$v(t_i + \lambda\Delta t_i, x) - v(t_i, x) = \int_{t_i}^{t_i + \lambda\Delta t_i} v_t(t, x)\, dt .$$

We obtain

$$\Delta J = \sum_{i=1}^{N} \left(\frac{1}{\lambda\Delta t_i} \int_{t_i}^{t_i + \lambda\Delta t_i} v_t(t, x)\, dt, \varphi_i(x) \right)_{L_2(\Omega)} \lambda\Delta t_i + \\ + \lambda \sum_{i=1}^{N} \left(v(t_i, x), \Delta\varphi_i(x)\right)_{L_2(\Omega)} + \quad (4)$$

$$+\lambda\sum_{i=1}^{N}\left(\int_{t_i}^{t_i+\lambda\Delta t_i} v_t(t,x)dt,\Delta\varphi_i(x)\right)_{L_2(\Omega)}+\|\Delta u\|^2_{L_2(Q)}.$$

We shall prove that

$$\lim_{\lambda\to 0}\frac{\Delta J}{\lambda}=(J'(h),\Delta h)_H=\sum_{i=1}^{N}(v_t(t_i,x),\varphi_i(x))_{L_2(\Omega)}\Delta t_i+$$
$$+\sum_{i=1}^{N}(v(t_i,x),\Delta\varphi_i(x))_{L_2(\Omega)}, \tag{5}$$

this fact proves the theorem. Here $h=(t^*,\varphi^*)$, $(\cdot,\cdot)_H$ is an inner product in H.

To do this we shall show that $\lambda^{-1}\|\Delta u\|^2_{L_2(Q)}\xrightarrow[\lambda\to 0]{}0$. Using the inequality $\|u\|_{L_2(Q)}\le c\|F\|_{W^{-l}_{bd^+}}$ we obtain a chain of the following inequalities

$$\|\Delta u\|_{L_2(Q)}\le c\left\|\sum_{i=1}^{N}\delta(t-t_i-\lambda\Delta t_i)\otimes(\varphi_i+\lambda\Delta\varphi_i)-\right.$$
$$\left.-\sum_{i=1}^{N}\delta(t-t_i)\otimes\varphi_i\right\|_{W^{-l}_{bd^+}}\le$$
$$\le\tilde{n}\sum_{i=1}^{N}\left(\left\|(\delta(t-t_i-\lambda\Delta t_i)-\delta(t-t_i))\otimes\varphi_i\right\|_{W^{-l}_{bd^+}}+\right.$$
$$\left.+\lambda\|\delta(t-t_i-\lambda\Delta t_i)\otimes\Delta\varphi_i\|_{W^{-l}_{bd^+}}\right).$$

According to the definition of the negative norm in $W^{-l}_{bd^+}$ and continuity of the imbedding $W^{+l}_{bd^+}\subset C^1([0,T];L_2(\Omega))$ we obtain

$$\|(\delta(t-t_i-\lambda\Delta t_i)-\delta(t-t_i))\otimes\varphi_i\|_{W^{-l}_{bd^+}}=$$

$$=\sup_{y\in W^{+l}_{bd^+},y\neq 0}\frac{(y(t_i+\lambda\Delta t_i,x)-y(t_i,x),\varphi_i)_{L_2(\Omega)}}{\|y\|_{W^{+l}_{bd^+}}}\leq$$

$$\leq\lambda\Delta t_i\|\varphi_i\|_{L_2(\Omega)}\sup_{y\in W^{+l}_{bd^+},y\neq 0}\frac{\|y\|_{C^1([0,T];L_2(\Omega))}}{\|y\|_{W^{+l}_{bd^+}}}\leq\lambda\Delta t_i\|\varphi_i\|_{L_2(\Omega)}C.$$

It follows from this that $0\leq\lambda^{-1}\|\Delta u\|^2_{L_2(Q)}\leq C\lambda\xrightarrow[\lambda\to 0]{}0$.

The equality (5) implies that investigated performance criterion is differentiable by Gâteaux in $H=R^N\times L_2^N(\Omega)$ and its partial derivatives are determined by the following expressions

$$\frac{\partial J}{\partial t^*}=\left(\int_\Omega v_t'(t_1,x)\varphi_1(x)d\Omega,\ldots,\int_\Omega v_t'(t_N,x)\varphi_N(x)d\Omega\right), \tag{6}$$

$$\frac{\partial J}{\partial\varphi^*}=(v(t_1,x),\ldots,v(t_N,x)). \tag{7}$$

Thus, the theorem is proved.

Theorem 2. *The performance criterion $J(h)$ is continuously differentiable in U_{ad}.*

Proof.

$$\|\Delta J'\|_H^2=\left\|J'(t^*+\Delta t^*,\varphi^*+\Delta\varphi^*)-J'(t^*,\varphi^*)\right\|_H^2=$$

$$=\sum_{i=1}^N\left[\int_\Omega v_t'(t_i+\Delta t_i,x;t^*+\Delta t^*,\varphi^*+\Delta\varphi^*)(\varphi_i+\Delta\varphi_i)d\Omega-\right. \tag{8}$$

$$\left.-\int_\Omega v_t'(t_i,x;t^*,\varphi^*)\varphi_i d\Omega\right]^2+$$

$$+\sum_{i=1}^N\int_\Omega\left[v(t_i+\Delta t_i,x;t^*+\Delta t^*,\varphi^*+\Delta\varphi^*)-v(t_i,x;t^*,\varphi^*)\right]^2d\Omega.$$

We shall prove that when $\left\|\Delta t^*\right\|_{R^N} + \left\|\Delta\varphi^*\right\|_{L_2^N(\Omega)} \to 0$ the expression in the right-hand side (8) also tends to zero. Let us convert this expression adding and subtracting from each term in the first square brackets the following expression

$$\int_\Omega v_t'(t_i + \Delta t_i, x; t^*, \varphi^*)\varphi_i(x)d\Omega,$$

and from each term of the second sum —

$$\int_\Omega v'(t_i + \Delta t_i, x; t^*, \varphi^*)d\Omega.$$

Then, using the obvious inequalities

$$(a+b+c)^2 \le 3(a^2+b^2+c^2),$$

$$(a+b)^2 \le 2(a^2+b^2),$$

where (a,b,c) are real numbers, we obtain

$$\begin{aligned}
\left\|\Delta J'\right\|_H^2 \le 3\sum_{i=1}^N \Bigg[\int_\Omega \big(v_t'(t_i + \Delta t_i, x; t^* + \Delta t^*, \varphi^* + \Delta\varphi^*) -& \\
- v_t'(t_i + \Delta t_i, x; t^*, \varphi^*)\big)\varphi_i d\Omega\Bigg]^2 +& \\
+3\sum_{i=1}^N \left[\int_\Omega \big(v_t'(t_i + \Delta t_i, x; t^*, \varphi^*) - v_t'(t_i, x; t^*, \varphi^*)\big)\varphi_i d\Omega\right]^2 +& \\
+3\sum_{i=1}^N \left[\int_\Omega v_t'(t_i + \Delta t_i, x; t^* + \Delta t^*, \varphi^* + \Delta\varphi^*)\Delta\varphi_i d\Omega\right]^2 +& \quad (9)\\
+2\sum_{i=1}^N \int_\Omega \big[v(t_i + \Delta t_i, x; t^* + \Delta t^*, \varphi^* + \Delta\varphi^*) -& \\
- v(t_i + \Delta t_i, x; t^*, \varphi^*)\big]^2 d\Omega +& \\
+2\sum_{i=1}^N \int_\Omega \big[v(t_i + \Delta t_i, x; t^*, \varphi^*) - v(t_i, x; t^*, \varphi^*)\big]^2 d\Omega.&
\end{aligned}$$

Denote the sums in the right-hand side of (9) by I_i, $(i=\overline{1,5})$. It is required to prove that $I_i \to 0$, $i=\overline{1,5}$ as $\|\Delta t^*\|_{R^N} + \|\Delta\varphi^*\|_{L_2^N(\Omega)} \to 0$.

Lemma 1. *Let* $y(t,x)$ *is a solution of the problem*

$$L^* y = g,\ g \in L_2(Q),\ y \in W^{+l}_{bd^+},\ W^{+l}_{bd^+} \subset C^1([0,T]; L_2(\Omega))$$

Then for any $\tau \in [0,T]$ *the following inequality holds true*

$$\left\| \frac{\partial y(\tau,x)}{\partial t} \right\|_{L_2(\Omega)} \le C\|g\|_{L_2(Q)}.$$

Proof of Lemma 1 for specific operator L^* follows from the positive inequalities, which are valid as a result of the inequalities (1.1.2).

Denote

$$\Delta v(t,x) = v(t,x;t^* + \Delta t^*, \varphi^* + \Delta\varphi^*) - v(t,x;t^*,\varphi^*).$$

Then Δv is a solution of the problem

$$L^* \Delta v = 2\Delta u,\ \Delta v \in W^{+l}_{bd^+},$$

where $\Delta u = u(t,x;t^* + \Delta t^*, \varphi^* + \Delta\varphi^*) - u(t,x;t^*,\varphi^*)$.

Obviously that Δu is a solution of the problem

$$L\Delta u = \sum_{i=1}^{N} [\delta(t - t_i - \Delta t_i)(\varphi_i + \Delta\varphi_i) - \delta(t - t_i)\varphi_i].$$

Using Lemma 1 and the inequality

$$\|u\|_{L_2(Q)} \le C\|F\|_{W^{-l}_{bd^+}},$$

which validity were proved in Lemma 1.1.3, we obtain a chain of the following inequalities

$$\|\Delta v_t'(t_i + \Delta t_i, x)\|_{L_2(\Omega)} \le C\|\Delta u\|_{L_2(Q)} \le$$

$$\le C\left\| \sum_{i=1}^{N} [\delta(t - t_i - \Delta t_i)(\varphi_i + \Delta\varphi_i) - \delta(t - t_i)\varphi_i] \right\|_{W^{-l}_{bd^+}} \le \quad (10)$$

$$\leq C\left\|\sum_{i=1}^{N}[\delta(t-t_i-\Delta t_i)-\delta(t-t_i)]\varphi_i\right\|_{W_{bd^+}^{-l}}+$$

$$+C\left\|\sum_{i=1}^{N}\delta(t-t_i-\Delta t_i)\Delta\varphi_i\right\|_{W_{bd^+}^{-l}}.$$

According to the definition of the norm in $W_{bd^+}^{-l}$ and to boundedness of the set of the admissible controls U_{ad} we obtain

$$\|[\delta(t-t_i-\Delta t_i)-\delta(t-t_i)]\varphi_i\|_{W_{bd^+}^{-l}}=$$

$$=\sup_{y\neq 0,\, y\in W_{bd^+}^{+l}}\frac{\left(y(t_i+\Delta t_i,x)-y(t_i,x),\varphi_i(x)\right)_{L_2(\Omega)}}{\|y\|_{W_{bd^+}^{+l}}}\leq C|\Delta t_i|^{1/2}. \tag{11}$$

Here we have used the integral Cauchy inequality.

Similarly, estimating the i-th term of the second summand in the right-hand side (10), we obtain

$$\|\delta(t-t_i-\Delta t_i)\Delta\varphi_i\|_{W_{bd^+}^{-l}}\leq C\|\Delta\varphi_i\|_{L_2(\Omega)}. \tag{12}$$

Summing (11) and (12) with respect to i from 1 to N and substituting into (10), we obtain

$$\|\Delta v_i'(t_i+\Delta t_i,x)\|_{L_2(\Omega)}^2\leq C\left[\sum_{i=1}^{N}\left(|\Delta t_i|^{1/2}+\|\Delta\varphi_i\|_{L_2(\Omega)}\right)\right]^2\to 0, \tag{13}$$

$$\|\Delta h\|_H\to 0,$$

where $\Delta h=\left(\Delta t^*,\Delta\varphi^*\right)\in U_{ad}$.

As far as the expression in the left-hand side of (13) is the i-th term in the first sum in the right-hand side of (9) then $I_1\to 0$ with $\|\Delta h\|_H\to 0$.

The proof of the fact that $I_2\to 0$ with $\|\Delta h\|_H\to 0$ follows from $v\in C^1\left([0,T];L_2(\Omega)\right)$.

It follows from Lemma 1 and boundedness of U_{ad} that

$$\left\| \frac{\partial v(t, x; t^*, \varphi^*)}{\partial t} \right\|_{C([0,T]; L_2(\Omega))} \leq C < \infty$$

for all $(t^*, \varphi^*) \in U_{ad}$. Then applying to I_3 the Cauchy inequality, we obtain $I_3 \to 0$ as $\|\Delta h\|_H \to 0$. I_4 is estimated similarly to I_1.

The fact that $I_5 \to 0$ as $\|\Delta h\|_H \to 0$ follows from the continuity of the function $v(t, x)$ with respect to the variable t.

It follows from the proven theorem that the performance criterion $J(h)$ is continuously differentiable by Gâteaux, and hence, by Fréchet and it is possible to express the necessary conditions in the following way

$$\min_{h \in U_{ad}} \left(grad J(h^*), h - h^* \right) = \min_{h \in U_{ad}} \left\{ \sum_{i=1}^{N} \int_{\Omega} v_t'(t_i^*, x) \varphi_i^*(x) d\Omega (t_i - t_i^*) + \right.$$

$$\left. + \sum_{i=1}^{N} \int_{\Omega} v(t_i^*, x)(\varphi_i(x) - \varphi_i^*(x)) d\Omega \right\} = 0,$$

where h^* is the required solution.

In the similar way it is possible to investigate others problems of the singular optimal control.

Let us generalize the results of the previous theorems for the case of the problem of the optimal control with an arbitrary right-hand side.

Let a state of some system is described by the equation (1.1.12) and for the operator L the following inequalities hold true

$$\|u\|_{L_2(Q)} \leq C_1 \|Lu\|_{W_{bd^+}^{-l}} \leq C_2 \|u\|_{W_{bd}^{+l}},$$

$$\|v\|_{L_2(Q)} \leq C_1 \|L^* v\|_{W_{bd}^{-l}} \leq C_2 \|v\|_{W_{bd^+}^{+l}}.$$

Suppose that the performance criterion is of the following form

$$J(h) = \Phi(u(h)) = \sum_{i=1}^{p} \int_{Q} \alpha_i(t, x)(u(h) - u_i)^2 dQ, \tag{14}$$

where $u_i(t,x)$, $\alpha_i(t,x)$ are some known functions from $L_2(Q)$, $C(\overline{Q})$, respectively, and $\alpha_i(t,x) \geq \varepsilon > 0$ â $\overline{Q}$.

Theorem 3. *Let the state of the system* $u(t,x)$ *is determined as a solution of the equation (1.12) with the right-hand side* $F \in W^{-l}_{bd^+}$. *The performance criterion is of the form (14). Then, if there exists the Fréchet derivative* $F_{h^*}(\cdot): H \to W^{-l}_{bd^+}$ *of the mapping* $F(\cdot): H \to W^{-l}_{bd^+}$ *in some point* h^* *then the performance functional* $J(h)$ *is also differentiable by Fréchet in the point* h^*, *and the derivative is determined by the expression*

$$J_{h^*}(\cdot) = \left\langle F_{h^*}(\cdot), v \right\rangle_{W_{bd^+}}, \tag{15}$$

where $v(t,x)$ *is a solution of the adjoint problem*

$$L^* v = 2\sum_{i=1}^{p} \alpha_i \left(u(h^*) - u_i \right), \qquad v \in W^{+l}_{bd^+}. \tag{16}$$

Proof. It is clear that the expression $\left\langle F_{h^*}(\cdot), v \right\rangle_{W_{bd^+}}$ is correct as far as the equation (16) has a unique solution $v(t,x) \in W^{+l}_{bd^+}$ (Theorem 1.1.2) and

$$\left\langle \overline{u}, L^* v \right\rangle_{W_{bd}} = \left(2\sum_{i=1}^{p} \alpha_i (u - u_i), \overline{u} \right)_{L_2(Q)}, \tag{17}$$

for any $\overline{u} \in W^{+l}_{bd}$.

Let Δh is an increment of a control. Consider

$$J(h^* + \Delta h) - J(h^*) - \left\langle F_{h^*}(\Delta h), v \right\rangle_{W_{bd^+}} =$$

$$= \sum_{i=1}^{p} \int_Q \alpha_i(t,x) \left(u(h^* + \Delta h) - u_i \right)^2 dQ -$$

$$-\sum_{i=1}^{p}\int_{Q}\alpha_i(t,x)\left(u(h^*)-u_i\right)^2 dQ-\left\langle F_{h^*}(\Delta h),v\right\rangle_{W_{bd^+}}=$$

$$=\sum_{i=1}^{p}\int_{Q}\alpha_i\cdot\left(u(h^*+\Delta h)-u(h^*)\right)\ \left(u(h^*+\Delta h)+u(h^*)-2u_i\right)\ dQ-$$

$$-\left\langle F_{h^*}(\Delta h);v\right\rangle_{W_{bd^+}}.$$

Let $\Delta u=u(h^*+\Delta h)-u(h^*)$. Then

$$J(h^*+\Delta h)-J(h^*)-\left\langle F_{h^*}(\Delta h),v\right\rangle_{W_{bd^+}}=$$

$$=\int_{Q}\Delta u\sum_{i=1}^{p}\alpha_i(t,x)\left(\Delta u+2u(h^*)-2u_i\right)dQ-\left\langle F_{h^*}(\Delta h),v\right\rangle_{W_{bd^+}}=$$

$$=\int_{Q}\Delta u\sum_{i=1}^{p}\alpha_i(t,x)\left(2u(h^*)-2u_i\right)dQ+\int_{Q}(\Delta u)^2\sum_{i=1}^{p}\alpha_i(t,x)dQ-$$

$$-\left\langle F_{h^*}(\Delta h),v\right\rangle_{W_{bd^+}}.$$

It is obvious that Δu is a solution of the following problem:

$$L\Delta u=F(h^*+\Delta h)-F(h^*),$$

and hence, as far as $v\in W_{bd^+}^{+l}$, then

$$\left\langle\Delta u,L^*v\right\rangle_{W_{bd}}=\left\langle F(h^*+\Delta h)-F(h^*),v\right\rangle_{W_{bd^+}}.$$

Moreover, using (17) $(\overline{u}=\Delta u)$ we obtain

$$\int_{Q}\Delta u\cdot 2\sum_{i=1}^{p}\alpha_i(t,x)\left(u-u_i\right)dQ=\left\langle\Delta u,L^*v\right\rangle_{W_{bd}}=$$

$$=\left\langle F(h^*+\Delta h)-F(h^*),v\right\rangle_{W_{bd^+}}.$$

Therefore

$$\left| J(h^* + \Delta h) - J(h^*) - \left\langle F_{h^*}(\Delta h), v \right\rangle_{W_{bd^+}} \right| \le$$

$$\le \left| \left\langle F\left(h^* + \Delta h\right) - F\left(h^*\right) - F_{h^*}(\Delta h), v \right\rangle_{W_{bd^+}} \right| +$$

$$+ \left| \int_Q (\Delta u)^2 \sum_{i=1}^{p} \alpha_i(t,x) dQ \right| \le$$

$$\le \left\| F\left(h^* + \Delta h\right) - F\left(h^*\right) - F_{h^*}(\Delta h) \right\|_{W_{bd^+}^{-l}} \|v\|_{W_{bd^+}^{+l}} + c\|\Delta u\|_{L_2(Q)}^2 .$$

As far as $F_{h^*}(\cdot): H \to W_{bd^+}^{-l}$, the Fréchet derivative of the mapping $F(\cdot): H \to W_{bd^+}^{-l}$, then $\forall \varepsilon > 0\ \exists \delta_1 > 0$: it follows from the inequality $\|\Delta h\|_l < \delta_1$ that

$$\left\| F\left(h^* + \Delta h\right) - F\left(h^*\right) - F_{h^*}(\Delta h) \right\|_{W_{bd^+}^{-l}} < \frac{\varepsilon}{2\|v\|_{W_{bd^+}^{+l}}} \|\Delta h\|_l . \tag{18}$$

On the other hand, using the inequality $\|u\|_{L_2(Q)} \le c\|F\|_{W_{bd^+}^{-}}$ of Lemma 1.1.3 we obtain

$$\|\Delta u\|_{L_2(Q)}^2 \le c\left\| F\left(h^* + \Delta h\right) - F\left(h^*\right) \right\|_{W_{bd^+}^{-l}}^2 \le$$

$$\le c\left(\left\| F\left(h^* + \Delta h\right) - F\left(h^*\right) - F_{h^*}(\Delta h) \right\|_{W_{bd^+}^{-l}} + \left\| F_{h^*}(\Delta h) \right\|_{W_{bd^+}^{-l}} \right)^2 .$$

Applying the inequality $(a^2 + b^2) \le 2\left(a^2 + b^2\right)$ and (18), we obtain

$$\|\Delta u\|_{L_2(Q)}^2 \le 2c\left(\left\| F\left(h^* + \Delta h\right) - F\left(h^*\right) - F_{h^*}(\Delta h) \right\|_{W_{bd^+}^{-l}} \right)^2 +$$

$$+2c\left(\left\| F_{h^*}(\Delta h) \right\|_{W_{bd^+}^{-l}} \right)^2 \le$$

$$\leq 2c\frac{\varepsilon^2}{4\|v\|^2_{W^{+l}_{bd^+}}}\|\Delta h\|^2_l + 2c\left\|F_{h^*}\right\|^2\|\Delta h\|^2_l .$$

Let

$$\delta_2 = \frac{\varepsilon}{2\left(2c\frac{\varepsilon^2}{4\|v\|^2_{W^{+l}_{bd^+}}} + 2c\left\|F_{h^*}\right\|^2\right)}$$

If $\|\Delta h\|_l < \min\{\delta_1, \delta_2\}$ then the inequality (18) holds true and $\|\Delta u\|^2_{L_2(Q)} \leq \frac{\varepsilon}{2}\|\Delta h\|_l$. Thus, finally we obtain

$$\left| J(h^* + \Delta h) - J(h^*) - \left\langle F_{h^*}(\Delta h), v\right\rangle_{W_{bd^+}} \right| \leq \varepsilon\|\Delta h\|_l ,$$

as required.

Theorem 4. *Let the mapping* $F(\cdot): H \to W^{-l}_{bd^+}$ *has a Fréchet derivative which is continuous in the point* h^* ($\forall\varepsilon > 0\ \exists\delta > 0$: $\forall h \in U_{ad} \left\|h - h^*\right\|_l < \delta \Rightarrow \left\|F_h(\cdot) - F_{h^*}(\cdot)\right\| < \varepsilon$). *Then the derivative* $J_h(\cdot)$ *is continuous in the point* h^* *also.*

Proof. Choose an arbitrary number $\varepsilon > 0$ and let h is an arbitrary point from U_{ad}, which is close enough to h^* in order that derivative $F_h(\cdot)$ exist, and hence, by Theorem 3 the derivative $J_h(\cdot)$ exists also. Consider

$$\left|J_h(\Delta h) - J_{h^*}(\Delta h)\right| =$$

$$= \left|\left\langle F_h(\Delta h), v_h\right\rangle_{W_{bd^+}} - \left\langle F_{h^*}(\Delta h), v_{h^*}\right\rangle_{W_{bd^+}}\right| \leq$$

$$\leq \left| \left\langle F_h(\Delta h), v_h \right\rangle_{W_{bd^+}} - \left\langle F_h(\Delta h), v_{h^*} \right\rangle_{W_{bd^+}} \right| +$$

$$+ \left| \left\langle F_h(\Delta h), v_{h^*} \right\rangle_{W_{bd^+}} - \left\langle F_h(\Delta h), v_{h^*} \right\rangle_{W_{bd^+}} \right| =$$

$$= \left| \left\langle F_h(\Delta h), v_h - v_{h^*} \right\rangle_{W_{bd^+}} \right| + \left| \left\langle \left(F_h - F_{h^*}\right)(\Delta h), v_{h^*} \right\rangle_{W_{bd^+}} \right|,$$

where v_h and v_{h^*} are solutions of the problem (16) with the right-hand sides $2\sum_{i=1}^{p} \alpha_i \left(u(h) - u_i\right)$ and $2\sum_{i=1}^{p} \alpha_i \left(u(h^*) - u_i\right)$, respectively.

Since $F_h(\cdot): H \to W^{-l}_{bd^+}$ is continuous at the point h^*, there exists sufficiently small neighbourhood of the point h^* (h we shall choose namely from this neighbourhood) in which $F_h(\cdot): H \to W^{-l}_{bd^+}$ is bounded $\left(\left\|F_h(\cdot)\right\| < C_2\right)$. Let also $\left\|h - h^*\right\| < \delta$.

Applying the Schwarz inequality we obtain

$$\left|J_h(\Delta h) - J_{h^*}(\Delta h)\right| \leq \left\|F_h(\Delta h)\right\|_{W^{-l}_{bd^+}} \left\|v_h - v_{h^*}\right\|_{W^{+l}_{bd^+}} +$$

$$+ \left\|\left(F_h - F_{h^*}\right)(\Delta h)\right\|_{W^{-l}_{bd^+}} \left\|v_{h^*}\right\|_{W^{+l}_{bd^+}} \leq$$

$$\leq \left\|F_h\right\| \cdot \left\|\Delta h\right\|_l \left\|v_h - v_{h^*}\right\|_{W^{+l}_{bd^+}} + \left\|F_h - F_{h^*}\right\| \cdot \left\|\Delta h\right\|_l \left\|v_{h^*}\right\|_{W^{+l}_{bd^+}} =$$

$$= \left(C_2 \left\|v_h - v_{h^*}\right\|_{W^{+l}_{bd^+}} + \varepsilon \left\|v_{h^*}\right\|_{W^{+l}_{bd^+}} \right) \left\|\Delta h\right\|_l .$$

Note that it follows from Theorem 1.1.2 and from the proof of Lemma 1.1.3 that $\left\|v\right\|_{W^{+l}_{bd^+}} \leq c\left\|G\right\|_{L_2(Q)}$, where $v(t,x)$ is a solution of the equation (1.1.4). Therefore

$$\left|J_h(\Delta h) - J_{h^*}(\Delta h)\right| \le \left(C_2 c \left\| 2\sum_{i=1}^{p} \alpha_i(t,x)\left(u(h) - u(h^*)\right) \right\|_{L_2(Q)} + \right.$$

$$\left. + c\varepsilon \left\| 2\sum_{i=1}^{p} \alpha_i(t,x)\left(u(h^*) - u_i\right) \right\|_{L_2(Q)} \right) \|\Delta h\|_l \le$$

$$\le \left(C_2 c M \left\| u(h) - u(h^*) \right\|_{L_2(Q)} + \right.$$

$$\left. + M c\varepsilon \left(\left\| u(h^*) \right\|_{L_2(Q)} + \sum_{i=1}^{p} \|u_i\|_{L_2(Q)} \right) \right) \|\Delta h\|_l ,$$

where $2\sum_{i=1}^{p} \alpha_i(t,x) \le M$ (constant M exists by virtue of continuity, and hence by virtue of boundedness of the functions $\alpha_i(t,x)$ in $\overline{Q}$). Using the inequality $\|u\|_{L_2(Q)} \le c\|F\|_{W^{-l}_{bd^+}}$ from Lemma 1.1.3 we have

$$\left|J_h(\Delta h) - J_{h^*}(\Delta h)\right| \le \left(C_2 c M \left\| F(h) - F(h^*) \right\|_{W^{-l}_{bd^+}} + \right.$$

$$\left. + M c\varepsilon \left(\left\| F(h^*) \right\|_{W^{-l}_{bd^+}} + \sum_{i=1}^{p} \|u_i\|_{L_2(Q)} \right) \right) \|\Delta h\|_l .$$

Note that since $u_i(t,x)$ are some known functions from $L_2(Q)$ then $\sum_{i=1}^{p} \|u_i\|_{L_2(Q)} \le M_1$, and hence

$$\left|J_h(\Delta h) - J_{h^*}(\Delta h)\right| \le \left(C_2 c M \left\| F(h) - F(h^*) \right\|_{W^{-l}_{bd^+}} + \right.$$

$$\left. + M c\varepsilon \left(\left\| F(h^*) \right\|_{W^{-l}_{bd^+}} + M_1 \right) \right) \|\Delta h\|_l .$$

Let us apply to the term $\left\| F(h) - F(h^*) \right\|_{W^{-l}_{bd^+}}$ the formula of the finite increments:

$$\left\|F(h)-F(h^*)\right\|_{W^{-l}_{bd^+}} \le$$

$$\le \sup_{\theta\in[0,1]}\left\|F_{h^*+\theta(h-h^*)}(\cdot)\right\|\cdot\left\|h-h^*\right\|_l \le C_2\left\|h-h^*\right\|_l \le C_2\delta.$$

As a result we get

$$\left|J_h(\Delta h)-J_{h^*}(\Delta h)\right| \le (C_2 cMC_2\delta +$$

$$+M\varepsilon c\left(\left\|F(h^*)\right\|_{W^{-l}_{bd^+}} + M_1\right)\Big)\left\|\Delta h\right\|_l .$$

Whence choosing $\delta<\varepsilon$ we have

$$\left|J_h(\Delta h)-J_{h^*}(\Delta h)\right| \le C_3\varepsilon\left\|\Delta h\right\|_l ,$$

$$\left\|J_h(\cdot)-J_{h^*}(\cdot)\right\| \le C_3\varepsilon .$$

The theorem is proved.

Theorem 5. *Let a mapping $F(\cdot): H \to W^{-l}_{bd^+}$ has a Frèchet derivative in some bounded convex neighbourhood of a point h^*, which satisfies the Lipschitz condition with an index α, $0<\alpha\le 1$ ($\exists C_1>0$: $\forall h_1, h_2$ from the neighbourhood of the point h^* such that $\left\|F_{h_1}(\cdot)-F_{h_2}(\cdot)\right\| \le C_1\left\|h_1-h_2\right\|_l^\alpha$). Then the derivative $J_h(\cdot)$ satisfies the Lipschitz condition with the index α.*

Proof. Let us prove that it follows from the Theorem assumptions that the derivative $F_h(\cdot)$ is bounded in the neighbourhood of the point h^*. Indeed, let h is an arbitrary point from the neighbourhood of the point h^*. Then, obviously, the following inequalities hold true

$$\left\|F_h(\cdot)\right\| \le \left\|F_h(\cdot)-F_{h^*}(\cdot)\right\| + \left\|F_{h^*}(\cdot)\right\| \le C_1\left\|h-h^*\right\|_l^\alpha + \left\|F_{h^*}(\cdot)\right\| \le$$

$$\le C_1 d^\alpha + \left\|F_{h^*}(\cdot)\right\| \le C_2 ,$$

where d is the diameter of the neighbourhood of the point h^*.

Now, let h_1, h_2 be arbitrary points from this neighbourhood. Consider

$$\left|J_{h_1}(\Delta h) - J_{h_2}(\Delta h)\right| = \left|\left\langle F_{h_1}(\Delta h), v_{h_1}\right\rangle_{W_{bd^+}} - \left\langle F_{h_2}(\Delta h), v_{h_2}\right\rangle_{W_{bd^+}}\right| \leq$$

$$\leq \left|\left\langle F_{h_1}(\Delta h), v_{h_1}\right\rangle_{W_{bd^+}} - \left\langle F_{h_1}(\Delta h), v_{h_2}\right\rangle_{W_{bd^+}}\right| +$$

$$+ \left|\left\langle F_{h_1}(\Delta h), v_{h_2}\right\rangle_{W_{bd^+}} - \left\langle F_{h_2}(\Delta h), v_{h_2}\right\rangle_{W_{bd^+}}\right| =$$

$$= \left|\left\langle F_{h_1}(\Delta h), v_{h_1} - v_{h_2}\right\rangle_{W_{bd^+}}\right| + \left|\left\langle \left(F_{h_1} - F_{h_2}\right)(\Delta h), v_{h_2}\right\rangle_{W_{bd^+}}\right|,$$

where v_{h_1} and v_{h_2} are solutions of the adjoint problems (16) with the right-hand sides $2\sum_{i=1}^{p}\alpha_i\left(u(h_1) - u_i\right)$ and $2\sum_{i=1}^{p}\alpha_i\left(u(h_2) - u_i\right)$, respectively. Applying the Schwarz inequality we get

$$\left|J_{h_1}(\Delta h) - J_{h_2}(\Delta h)\right| \leq \left\|F_{h_1}(\Delta h)\right\|_{W_{bd^+}^{-l}} \left\|v_{h_1} - v_{h_2}\right\|_{W_{bd^+}^{+l}} +$$

$$+ \left\|\left(F_{h_1} - F_{h_2}\right)(\Delta h)\right\|_{W_{bd^+}^{-l}} \left\|v_{h_2}\right\|_{W_{bd^+}^{+l}} \leq$$

$$\leq \left\|F_{h_1}\right\| \cdot \left\|\Delta h\right\|_l \left\|v_{h_1} - v_{h_2}\right\|_{W_{bd^+}^{+l}} + \left\|F_{h_1} - F_{h_2}\right\| \cdot \left\|\Delta h\right\|_l \left\|v_{h_2}\right\|_{W_{bd^+}^{+l}} =$$

$$= \left(C_2\left\|v_{h_1} - v_{h_2}\right\|_{W_{bd^+}^{+l}} + C_1\left\|h_1 - h_2\right\|_l^{\alpha} \left\|v_{h_2}\right\|_{W_{bd^+}^{+l}}\right)\left\|\Delta h\right\|_l .$$

Note that it follows from the analogy of Theorem 1.1.1 and from the proof of Lemma 1.1.3 that $\|v\|_{W_{bd^+}^{+l}} \leq c\|G\|_{L_2(Q)}$, where $v(t,x)$ is a solution of the equation (1.1.4). Therefore

$$\left|J_{h_1}(\Delta h)-J_{h_2}(\Delta h)\right|\le\left(C_2c\left\|2\sum_{i=1}^{p}\alpha_i(t,x)\big(u(h_1)-u(h_2)\big)\right\|_{L_2(Q)}+\right.$$

$$\left.+C_1c\|h_1-h_2\|_l^{\alpha}\left\|2\sum_{i=1}^{p}\alpha_i(t,x)\big(u(h_2)-u_i\big)\right\|_{L_2(Q)}\right)\|\Delta h\|_l\le$$

$$\le\Big(C_2cM\|u(h_1)-u(h_2)\|_{L_2(Q)}+$$

$$+MC_1c\|h_1-h_2\|_l^{\alpha}\left(\|u(h_2)\|_{L_2(Q)}+\sum_{i=1}^{p}\|u_i\|_{L_2(Q)}\right)\Big)\|\Delta h\|_l\,,$$

where $2\sum_{i=1}^{p}\alpha_i(t,x)\le M$ in $\overline{Q}$. Using the inequality $\|u\|_{L_2(Q)}\le c\|F\|_{W_{bd^+}^{-l}}$ from Lemma 1.1.3 we obtain

$$\left|J_{h_1}(\Delta h)-J_{h_2}(\Delta h)\right|\le\Big(C_2cM\|F(h_1)-F(h_2)\|_{W_{bd^+}^{-l}}+$$

$$+MC_1c\|h_1-h_2\|_l^{\alpha}\left(\|F(h_2)\|_{W_{bd^+}^{-l}}+\sum_{i=1}^{p}\|u_i\|_{L_2(Q)}\right)\Big)\|\Delta h\|_l\,.$$

Note that as far as $u_i(t,x)$ are some known functions from $L_2(Q)$ then $\sum_{i=1}^{p}\|u_i\|_{L_2(Q)}\le M_1$. Moreover,

$$\|F(h_2)\|_{W_{bd^+}^{-l}}\le\|F(h_2)-F(h^*)\|_{W_{bd^+}^{-l}}+\|F(h^*)\|_{W_{bd^+}^{-l}}\,,$$

and applying the formula of finite increments to the first term we get

$$\|F(h_2)\|_{W_{bd^+}^{-l}}\le\sup_{\theta\in[0,1]}\left\|F_{h^*+\theta(h_2-h^*)}(\cdot)\right\|\cdot\|h_2-h^*\|_l+\|F(h^*)\|_{W_{bd^+}^{-l}}\,,$$

taking into consideration that h^*,h_2 belong to the bounded set U_{ad}, i.e. $\|h_2-h^*\|_l\le d$, we get

$$\|F(h_2)\|_{W_{bd^+}^{-l}}\le C_2d+\|F(h^*)\|_{W_{bd^+}^{-l}}\,.$$

Hence,

$$\left|J_{h_1}(\Delta h)-J_{h_2}(\Delta h)\right|\le\Big(C_2cM\|F(h_1)-F(h_2)\|_{W^{-l}_{bd^+}}+$$

$$+MC_1c\|h_1-h_2\|_l^{\alpha}\left(C_2d+\|F(h^*)\|_{W^{-l}_{bd^+}}+M_1\right)\Big)\|\Delta h\|_l\,.$$

Let us apply to the term $\|F(h_1)-F(h_2)\|_{W^{-l}_{bd^+}}$ the formula of finite increments:

$$\|F(h_1)-F(h_2)\|_{W^{-l}_{bd^+}}\le\sup_{\theta\in[0,1]}\left\|F'_{h_2+\theta(h_1-h_2)}(\cdot)\right\|\cdot\|h_1-h_2\|_l\le$$

$$\le C_2\|h_1-h_2\|_l\,.$$

As a result we obtain

$$\left|J_{h_1}(\Delta h)-J_{h_2}(\Delta h)\right|\le\Big(C_2cMC_2\|h_1-h_2\|_l^{1-\alpha}+$$

$$+MC_1c\left(C_2d+\|F(h^*)\|_{W^{-l}_{bd^+}}+M_1\right)\Big)\|h_1-h_2\|_l^{\alpha}\,\|\Delta h\|_l\,.$$

Whence,

$$\left|J_{h_1}(\Delta h)-J_{h_2}(\Delta h)\right|\le C_3\|h_1-h_2\|_l^{\alpha}\,\|\Delta h\|_l\,,$$

$$\left\|J'_{h_1}(\cdot)-J'_{h_2}(\cdot)\right\|\le C_3\|h_1-h_2\|_l^{\alpha}\,.$$

The theorem is proved.

Remark. *Granting that the mapping* $F(\cdot):H\to W^{-l}_{bd^+}$ *is of the form* $F=f+A(h)$, *and the functional* $f\in W^{-l}_{bd^+}$ *does not depend on the control* $h\in U_{ad}\subset H$, *we may rewrite the formula (15), which define the gradient, in the following form*

$$J'_{h^*}(\cdot)=\left\langle A'_{h^*}(\cdot),v\right\rangle_{W_{bd^+}},$$

where $A'_{h^*}(\cdot)$ *is a Fréchet derivative of the mapping* $A(\cdot):H\to W^{-l}_{bd^+}$, *the function* $v(t,x)$ *is chosen in a similar way as in Theorem 3. Respectively, in Theorems 3 and 4 we may require* *that the assumptions of this theorem hold true for the mapping $A(\cdot)$ instead of $F(\cdot)$*

Consider the application of the theorems of this section for the cases of the right-hand sides of specific types. We shall consider functions $F = f + A_i(t,x,h)$ ($i = \overline{1,5}$) defined in Chapter 1.

1. Let $A_1(t,x,h) = \sum_{i=1}^{s} \delta(t - t_i) \otimes \varphi_i(x)$, where $t, t_i \in [0,T]$, and $\varphi_i(x) \in L_2(\Omega)$. The control is

$$h_1 = \{(t_i, \varphi_i(x))\}_{i=1}^{s} \in U_{ad} \subset H_1 = [0,T]^s \times (L_2(\Omega))^s,$$

where U_{ad} is a convex, closed and bounded set in H_1.

Suppose that $W_{bd^+}^{+l} \subset W_2^{2,0}(Q)$ $\left(W_{bd}^{+l} \subset W_2^{2,0}(Q)\right)$ and

$$\|\cdot\|_{W_2^{2,0}(Q)} \leq c\|\cdot\|_{W_{bd^+}^{+l}} \quad \left(\|\cdot\|_{W_2^{2,0}(Q)} \leq c\|\cdot\|_{W_{bd}^{+l}}\right), \tag{9}$$

where $W_2^{2,0}(Q)$ is a completion of the space of smooth in $\overline{Q}$ functions with respect to the norm

$$\|v\|_{W_2^{2,0}(Q)} = \left(\int_Q v^2 + v_t^2 + v_{tt}^2 dQ\right)^{\frac{1}{2}},$$

and also, let $f \in W_{bd^+}^{-l}$.

Make sure that the mapping

$$A_1(t,x,h) = \sum_{i=1}^{s} \delta(t - t_i) \otimes \varphi_i(x)$$

is differentiable by Fréchet and its partial derivatives are of the form

$$\frac{\partial A_1}{\partial t_i} = -\delta^{(1)}(t - t_i) \otimes \varphi_i(x), \quad \frac{\partial A_1}{\partial \varphi_i} = \delta(t - t_i).$$

Indeed, it is clear, that the expression

$$A_{1h}(\Delta h) = -\sum_{i=1}^{s} \delta^{(1)}(t - t_i) \otimes \varphi_i(x)\Delta t_i + \sum_{i=1}^{s} \delta(t - t_i) \otimes \Delta\varphi_i$$

defines a linear continuous operator $A_{1h}(\cdot): H_1 \to W_{bd^+}^{-l}$.

Let us estimate

$$\|A_1(h+\Delta h) - A_1(h) - A_{1h}(\Delta h)\|_{W_{bd^+}^{-l}} =$$

$$= \sup_{\substack{v \in W_{bd^+}^{+l}, \\ v \neq 0}} \frac{\left|\langle A_1(h+\Delta h) - A_1(h) - A_{1h}(\Delta h), v\rangle_{W_{bd^+}}\right|}{\|v\|_{W_{bd^+}^{+l}}}.$$

Consider the numerator of this ratio. At first, let $v(t,x)$ be some smooth in $\overline{Q}$ function satisfying the boundary conditions (bd^+). Then

$$\langle A_1(h+\Delta h) - A_1(h) - A_{1h}(\Delta h), v\rangle_{W_{bd^+}} =$$

$$= \sum_{i=1}^{s} \int_{\Omega} v(t_i + \Delta t_i, x)\left(\varphi_i(x) + \Delta\varphi_i(x)\right) d\Omega -$$

$$- \sum_{i=1}^{s} \int_{\Omega} v(t_i, x)\varphi_i(x) d\Omega -$$

$$- \sum_{i=1}^{s} \int_{\Omega} v_t(t_i, x)\varphi_i(x)\Delta t_i + v(t_i, x)\Delta\varphi_i(x) d\Omega =$$

$$= \sum_{i=1}^{s} \int_{\Omega} \left(v(t_i + \Delta t_i, x) - v(t_i, x) - v_t(t_i, x)\Delta t_i\right)\varphi_i(x) +$$

$$+ \left(v(t_i + \Delta t_i, x) - v(t_i, x)\right)\Delta\varphi_i(x) d\Omega$$

Let us use the Taylor formula.

$$I = \langle A_1(h+\Delta h) - A_1(h) - A_{1h}(\Delta h), v\rangle_{W_{bd^+}} =$$

$$= \sum_{i=1}^{s} \int_{\Omega} \left(v_t(t_i + \theta_1\Delta t_i, x) - v_t(t_i, x)\right)\Delta t_i \varphi_i(x) +$$

$$+ v_t(t_i + \theta_2\Delta t_i, x)\Delta t_i \Delta\varphi_i(x) d\Omega,$$

where $\theta_1, \theta_2 \in (0,1)$.

Granting that the following inequality holds true

$$\left(\int_{\Omega} v^2(t_i,x)d\Omega\right)^{\frac{1}{2}} \le \left(\int_{Q} v^2(t,x)+v_t^2(t,x)dQ\right)^{\frac{1}{2}},$$

we get

$$I=\sum_{i=1}^{s}\int_{\Omega}\int_{t_i}^{t_i+\theta\Delta t_i} v_{tt}(\eta,x)d\eta\Delta t_i\varphi_i(x)+\int_{T}^{t_i+\theta\Delta t_i} v_{tt}(\eta,x)d\eta\Delta t_i\Delta\varphi_i(x)d\Omega.$$

Applying the integral Cauchy inequality we obtain:

$$|I|\le\sum_{i=1}^{s}\int_{\Omega}|\theta_1\Delta t_i|^{\frac{1}{2}}\left|\int_{t_i}^{t_i+\theta_1\Delta t_i} v_{tt}^2(\eta,x)d\eta\right|^{\frac{1}{2}}|\Delta t_i\varphi_i(x)|+$$

$$+T^{\frac{1}{2}}\left(\int_{0}^{T} v_{tt}^2(\eta,x)d\eta\right)^{\frac{1}{2}}|\Delta t_i\Delta\varphi_i(x)|d\Omega\le$$

$$\le\sum_{i=1}^{s}\left(|\Delta t_i|^{\frac{3}{2}}\|\varphi_i\|_{L_2(\Omega)}\|v_{tt}\|_{L_2(Q)}+c|\Delta t_i|\|\Delta\varphi_i\|_{L_2(\Omega)}\|v_{tt}\|_{L_2(Q)}\right)\le$$

$$\le\left(\sum_{i=1}^{s}|\Delta t_i|^{\frac{3}{2}}\|\varphi_i\|_{L_2(\Omega)}+c|\Delta t_i|\|\Delta\varphi_i\|_{L_2(\Omega)}\right)\|v\|_{W^{+l}_{bd^+}}.$$

Using the fact that the set of the considered smooth functions $v(t,x)$ is dense in $W^{+l}_{bd^+}$, we have

$$\|A_1(h+\Delta h)-A_1(h)-A_{1h}(\Delta h)\|_{W^{-l}_{bd^+}}\le$$

$$\le\sum_{i=1}^{s}|\Delta t_i|^{\frac{3}{2}}\|\varphi_i\|_{L_2(\Omega)}+c|\Delta t_i|\|\Delta\varphi_i\|_{L_2(\Omega)}\le$$

$$\le\left(\sum_{i=1}^{s}|\Delta t_i|^{\frac{1}{2}}\|\varphi_i\|_{L_2(\Omega)}+c|\Delta t_i|\right)\|\Delta h\|_{l_1}.$$

It is clear that $\forall\varepsilon>0\ \exists\delta>0$ such that the inequality $\|\Delta h\|_{l_1}\le\delta$ implies

$$\left(\sum_{i=1}^{s} |\Delta t_i|^{\frac{1}{2}} \|\varphi_i\|_{L_2(\Omega)} + c|\Delta t_i| \right) < \varepsilon .$$

Hence, the mapping $A_1(h)$ has a Fréchet derivative and by Theorem 3 the functional $J(h)$ in the problem of the optimal control (1.1.12) with the right-hand side $A_1(t,x,h)$ has a Fréchet derivative also. Let us prove that $A_{1h}(\cdot)$ satisfies the Lipschitz condition with the index $\alpha = \dfrac{1}{2}$.

Indeed,

$$\|A_{1h_1}(\cdot) - A_{1h_2}(\cdot)\| = \sup_{\Delta h \in \hat{I}_1} \frac{\|A_{1h_1}(\Delta h) - A_{1h_2}(\Delta h)\|_{W^{-l}_{bd^+}}}{\|\Delta h\|_{\hat{I}_1}} =$$

$$= \sup_{\Delta h \in \hat{I}_1} \sup_{v \in W^{+l}_{bd^+}} \frac{\left| \left\langle A_{1h_1}(\Delta h) - A_{1h_2}(\Delta h), v \right\rangle_{W_{bd^+}} \right|}{\|\Delta h\|_{\hat{I}_1} \|v\|_{W^{+l}_{bd^+}}}.$$

Consider the numerator of this ratio. Let $v(t,x)$ be a smooth in $\overline{Q}$ function satisfying the conditions (bd^+):

$$I = \left\langle A_{1h_1}(\Delta h) - A_{1h_2}(\Delta h), v \right\rangle_{W_{bd^+}} =$$

$$= \sum_{i=1}^{s} \Big\langle \left(-\delta^{(1)}(t - t_i^1)\varphi_i^1 + \delta^{(1)}(t - t_i^2)\varphi_i^2 \right) \Delta t_i +$$

$$+ \left(\delta(t - t_i^1) - \delta(t - t_i^2) \right) \Delta \varphi_i , v \Big\rangle_{W_{bd^+}} =$$

$$= \sum_{i=1}^{s} \int_{\Omega} \left(v_t(t_i^1, x)\varphi_i^1 - v_t(t_i^2, x)\varphi_i^2 \right) \Delta t_i +$$

$$+ \left(v(t_i^1, x) - v(t_i^2, x) \right) \Delta \varphi_i \, d\Omega =$$

$$= \sum_{i=1}^{s} \int_{\Omega} \Big(\big(v_t(t_i^1, x) - v_t(t_i^2, x)\big)\varphi_i^1 \Delta t_i + v_t(t_i^2, x)\big(\varphi_i^1 - \varphi_i^2\big)\Delta t_i +$$

$$+ \left(\int_{t_i^2}^{t_i^1} v_t(\eta, x) d\eta \right) \Delta\varphi_i \, d\Omega =$$

$$= \sum_{i=1}^{s} \int_{\Omega} \left(\int_{t_i^2}^{t_i^1} v_{tt}(\eta, x) d\eta \right) \varphi_i^1 \Delta t_i + \left(\int_{T}^{t_i^2} v_{tt}(\eta, x) d\eta \right) (\varphi_i^1 - \varphi_i^2) \Delta t_i +$$

$$+ \left(\int_{t_i^2}^{t_i^1} v_t(\eta, x) d\eta \right) \Delta\varphi_i \, d\Omega.$$

Let us use the integral Cauchy inequality:

$$|I| \le \sum_{i=1}^{s} \left(\left|t_i^1 - t_i^2\right|^{\frac{1}{2}} |\Delta t_i| \int_{\Omega} \left(\int_0^T v_{tt}^2 d\eta \right)^{\frac{1}{2}} |\varphi_i^1| d\Omega + \right.$$

$$+ T^{\frac{1}{2}} |\Delta t_i| \int_{\Omega} \left(\int_0^T v_{tt}^2 d\eta \right)^{\frac{1}{2}} |\varphi_i^1 - \varphi_i^2| d\Omega +$$

$$\left. + \left|t_i^1 - t_i^2\right|^{\frac{1}{2}} \int_{\Omega} \left(\int_0^T v_t^2 d\eta \right)^{\frac{1}{2}} |\Delta\varphi_i| d\Omega \right).$$

Again, we use the integral Cauchy inequality:

$$|I| \le \sum_{i=1}^{s} \left(\left|t_i^1 - t_i^2\right|^{\frac{1}{2}} |\Delta t_i| \, \|v_{tt}\|_{L_2(Q)} \|\varphi_i^1\|_{L_2(\Omega)} + \right.$$

$$+ T^{\frac{1}{2}} |\Delta t_i| \, \|v_{tt}\|_{L_2(Q)} \|\varphi_i^1 - \varphi_i^2\|_{L_2(\Omega)} +$$

$$\left. \left|t_i^1 - t_i^2\right|^{\frac{1}{2}} \|v_t\|_{L_2(Q)} \|\Delta\varphi_i\|_{L_2(\Omega)} \right).$$

Granting that the set of the considered functions $v(t,x)$ is dense in $W^{+l}_{bd^+}$ we get that

$$\left\| A_{1h_1}(\Delta h) - A_{1h_2}(\Delta h) \right\|_{W^{-l}_{bd^+}} \le$$

$$\le \sum_{i=1}^{s} \left(\left| t_i^1 - t_i^2 \right|^{\frac{1}{2}} \left| \Delta t_i \right| \left\| \varphi_i^1 \right\|_{L_2(\Omega)} + T^{\frac{1}{2}} \left| \Delta t_i \right| \left\| \varphi_i^1 - \varphi_i^2 \right\|_{L_2(\Omega)} + \right.$$

$$\left. + c \left| t_i^1 - t_i^2 \right|^{\frac{1}{2}} \left\| \Delta \varphi_i \right\|_{L_2(\Omega)} \right) \le$$

$$\le \left\| \Delta h \right\|_{l_1} \sum_{i=1}^{s} \left(\left| t_i^1 - t_i^2 \right|^{\frac{1}{2}} \left\| \varphi_i^1 \right\|_{L_2(\Omega)} + \sqrt{T} \left\| \varphi_i^1 - \varphi_i^2 \right\|_{L_2(\Omega)} + c \left| t_i^1 - t_i^2 \right|^{\frac{1}{2}} \right).$$

Whence,

$$\left\| A_{1h_1}(\cdot) - A_{1h_2}(\cdot) \right\| \le$$

$$\le \sum_{i=1}^{s} \left(\left| t_i^1 - t_i^2 \right|^{\frac{1}{2}} \left\| \varphi_i^1 \right\|_{L_2(\Omega)} + T^{\frac{1}{2}} \left\| \varphi_i^1 - \varphi_i^2 \right\|_{L_2(\Omega)} + c \left| t_i^1 - t_i^2 \right|^{\frac{1}{2}} \right) \le$$

$$\le \left(\sum_{i=1}^{s} \left(\left\| \varphi_i^1 \right\|_{L_2(\Omega)} + T^{\frac{1}{2}} + c \right) \right) \left\| h_1 - h_2 \right\|_{l_1}^{\frac{1}{2}} \le C \left\| h_1 - h_2 \right\|_{l_1}^{\frac{1}{2}},$$

as far as $\sum_{i=1}^{s} \left\| \varphi_i^1 \right\|_{L_2(\Omega)} \le \tilde{n} \left\| h_1 \right\|_{l_1}$, and $\left\| h_1 \right\|_{l_1}$ is bounded.

Thus, $A_{1h}(\cdot)$ satisfies the Lipschitz condition with the index $\alpha = \frac{1}{2}$. By Theorem 5 we may state that the derivative by Fréchet $J_h(\cdot)$ satisfies the Lipschitz condition with the index $\alpha = \frac{1}{2}$ and hence $J_h(\cdot)$ is continuous with respect to h. Note that we may repeat the previous reasoning concerning to the mapping $A_1(h)$ (including Theorem 3) and we can prove that

$$\left\| J_{(t_i^1,\varphi_i^1)}(\cdot) - J_{(t_i^1,\varphi_i^2)}(\cdot) \right\| \le \sum_{i=1}^{s} c \left\| \varphi_i^1 - \varphi_i^2 \right\|_{L_2(\Omega)},$$

i.e. $J_h(\cdot)$ with respect to the direction $\varphi(x)$ satisfies the Lipschitz condition (with the index 1).

2. Let $A_2(t,x,h) = \sum_{i=1}^{s} \delta^{(k)}(t-t_i) \otimes \varphi_i(x)$, where $\delta^{(k)}(\cdot)$ is the k-th Sobolev derivative of the δ-function, $t, t_i \in [0,T]$, and $\varphi_i(x) \in L_2(\Omega)$.

The control is

$$h_2 = \{(t_i, \varphi_i(x))\}_{i=1}^{s} \in U_{ad} \subset H_2 = [0,T]^s \times (L_2(\Omega))^s,$$

where U_{ad} is a convex, closed and bounded set in H_2.

Suppose that $W_{bd^+}^{+l} \subset W_2^{k+2,0}(Q)$ $\left(W_{bd}^{+l} \subset W_2^{k+2,0}(Q)\right)$ and

$$\|\cdot\|_{W_2^{k+2,0}(Q)} \le c\|\cdot\|_{W_{bd^+}^{+l}} \quad \left(\|\cdot\|_{W_2^{k+2,0}(Q)} \le c\|\cdot\|_{W_{bd}^{+l}}\right), \tag{10}$$

where $W_2^{k+2,0}(Q)$ is a completion of the space of smooth in $\overline{Q}$ functions with respect to the norm

$$\|v\|_{W_2^{k+2,0}(Q)} = \left(\int_Q \sum_{n=0}^{k+2} v_t^{(n)} dQ \right)^{\frac{1}{2}},$$

and let $f \in W_{bd^+}^{-l}$.

Similarly to the previous facts we can prove that the mapping $F(\cdot) = f + A_2(\cdot)$ has a Fréchet derivative, which satisfy the Lipschitz condition with the index $1/2$. Applying Theorems 3 and 5 we state that the performance criterion

$$J(h) = \sum_{i=1}^{p} \int_Q \alpha_i(t,x)\left(u(h) - u_i\right)^2 dQ$$

is differentiable by Fréchet, and the corresponding derivatives are defined by the expressions:

$$\frac{\partial J}{\partial t_i} = (-1)^k \int_\Omega \frac{\partial^{k+1} v(t_i, x)}{\partial t^{k+1}} \varphi_i(x) d\Omega, \quad \frac{\partial J}{\partial \varphi_i} = (-1)^k \frac{\partial^k v(t_i, x)}{\partial t^k},$$

where $v(t,x)$ is a solution of the adjoint problem

$$L^* v = 2\sum_{i=1}^{p} \alpha_i \left(u(h^*) - u_i\right).$$

The derivative $J_{h^*}(\cdot)$ satisfies the Lipschitz conditions with the index $1/2$, and with respect to the direction $\varphi(x)$ - with the index 1.

Consider the application of Theorems 3-5 of this paragraph in the case when the right-hand side of the equation (1.1.12) is defined as $A_i(t,x,h)$ ($i = \overline{3,5}$). As in previous section we suppose that the set Q is a tube with respect to the variable x_1 $(Q = [0,T] \times [\overline{x}_1, \overline{\overline{x}}_1] \times \Omega')$ and $h \in U_{ad}$, which is bounded, closed and convex set from the Hilbert space of controls H.

3. Let the right-hand side of the equation is of the form

$$A_3(t,x,h) = \sum_{i=1}^{s} \delta(x_1 - x_{1,i}) \otimes \varphi_i(t, x_2, \dots, x_n)$$

where $x_1, x_{1,i} \in [\overline{x}_1, \overline{\overline{x}}_1]$, $\varphi_i(t, x_2, \dots, x_n) \in L_2((0,T) \times \Omega')$. The control is

$$h_3 = \{(x_{1,i}, \varphi_i(t, x_2, x_3, \dots, x_n))\}_{i=1}^{s} \in U_{ad} \subset \\ \subset H_3 = [\overline{x}_1, \overline{\overline{x}}_1]^s \times (L_2((0,T) \times \Omega'))^s,$$

where U_{ad} is a convex, closed and bounded set in H_3.

Suppose that $W^{+l}_{bd^+} \subset W_2^{0,2,0}(Q)$ $\left(W^{+l}_{bd} \subset W_2^{0,2,0}(Q)\right)$ and

$$\|\cdot\|_{W_2^{0,2,0}(Q)} \le c \|\cdot\|_{W^{+l}_{bd^+}} \quad \left(\|\cdot\|_{W_2^{0,2,0}(Q)} \le c \|\cdot\|_{W^{+l}_{bd}}\right), \tag{16}$$

where $W_2^{0,2,0}(Q)$ is a completion of the space of smooth in $\overline{Q}$ functions with respect to the norm

$$\|v\|_{W_2^{0,2,0}(Q)} = \left(\int_Q v^2 + v_{x_1}^2 + v_{x_1x_1}^2 \, dQ \right)^{\frac{1}{2}},$$

and let $f \in W^{-l}_{bd^+}$.

Similarly to the case of $A_1(h)$ which was considered above it is easy to see that the mapping $A_3(\cdot): H \to W^{-l}_{bd^+}$ has a Fréchet derivative, which is defined as:

$$\frac{\partial A_3}{\partial x_{1,i}} = -\delta^{(1)}(x_1 - x_{1,i}) \otimes \varphi_i(t, x_2, \ldots, x_n),$$

$$\frac{\partial A_3}{\partial \varphi_i} = \delta(x_1 - x_{1,i}), \; i = \overline{1,s}.$$

We shall prove that $A_{3h}(\cdot)$ satisfies the Lipschitz condition with the index $1/2$. Consider

$$\left\| A_{3h_1}(\cdot) - A_{3h_2}(\cdot) \right\| = \sup_{\Delta h \in H_3} \frac{\left\| A_{3h_1}(\Delta h) - A_{3h_2}(\Delta h) \right\|_{W^{-l}_{bd^+}}}{\|\Delta h\|_{H_3}} =$$

$$= \sup_{\Delta h \in H_3} \sup_{v \in W^{+l}_{bd^+}} \frac{\left| \left\langle \grave{A}_{3h_1}(\Delta h) - \grave{A}_{3h_2}(\Delta h), v \right\rangle_{W_{bd^+}} \right|}{\|\Delta h\|_{H_3} \|v\|_{W^{+l}_{bd^+}}}.$$

Let $v(t,x)$ be a smooth in $\overline{Q}$ function from $W^{+l}_{bd^+}$. Consider the numerator of the ratio:

$$I' = \left\langle A_{3h_1}(\Delta h) - A_{3h_2}(\Delta h), v \right\rangle_{W_{bd^+}} =$$

$$= \sum_{i=1}^{s} \int_0^T \int_{\Omega'} \left(v_{x_1}(t, x_{1,i}^1, x_2, \ldots, x_n) - v_{x_1}(t, x_{1,i}^2, x_2, \ldots, x_n) \right) \varphi_i^1 \Delta x_{1,i} +$$

$$+ v_{x_1}(t, x_{1,i}^2, x_2, \ldots, x_n) \left(\varphi_i^1 - \varphi_i^2 \right) \Delta x_{1,i} +$$

$$+\left(v(t,x_{1,i}^1,x_2,\dots,x_n)-v(t,x_{1,i}^2,x_2,\dots,x_n)\right)\Delta\varphi_i\,dtd\Omega'=$$

$$=\sum_{i=1}^{s}\int_0^T\int_{\Omega'}\left(\int_{x_{1,i}^2}^{x_{1,i}^1}v_{x_1x_1}(t,\eta,x_2,\dots,x_n)d\eta\right)\varphi_i^1\Delta x_{1,i}+$$

$$+v_{x_1}(t,x_{1,i}^2,x_2,\dots,x_n)\left(\varphi_i^1-\varphi_i^2\right)\Delta x_{1,i}+$$

$$+\left(\int_{x_{1,i}^2}^{x_{1,i}^1}v_{x_1}(t,\eta,x_2,\dots,x_n)d\eta\right)\Delta\varphi_i\,dtd\Omega'.$$

Using the inequality

$$\left(\int_\Omega v^2(t_i,x)d\Omega\right)^{\frac{1}{2}}\le C\left(\int_Q v^2(t,x)+v_t^2(t,x)dQ\right)^{\frac{1}{2}}$$

and the integral Cauchy inequality we get

$$|I'|\le\sum_{i=1}^{s}\int_0^T\int_{\Omega'}\left(\int_{x_{1,i}^2}^{x_{1,i}^1}d\eta\right)^{\frac{1}{2}}\left(\int_{x_{1,i}^2}^{x_{1,i}^1}v_{x_1x_1}^2(t,\eta,x_2,\dots,x_n)d\eta\right)^{\frac{1}{2}}\left|\varphi_i^1\Delta x_{1,i}\right|+$$

$$+\left(\int_{\bar{x}_1}^{\bar{\bar{x}}_1}v_{x_1}^2(t,\eta,x_2,\dots,x_n)+v_{x_1x_1}^2(t,\eta,x_2,\dots,x_n)d\eta\right)^{\frac{1}{2}}\left|\varphi_i^1-\varphi_i^2\right|\left|\Delta x_{1,i}\right|+$$

$$+\left(\int_{x_{1,i}^2}^{x_{1,i}^1}d\eta\right)^{\frac{1}{2}}\left(\int_{x_{1,i}^2}^{x_{1,i}^1}v_{x_1}^2(t,\eta,x_2,\dots,x_n)d\eta\right)^{\frac{1}{2}}\left|\Delta\varphi_i\right|dtd\Omega'.$$

Again use the Cauchy inequality:

$$|I'|\le\sum_{i=1}^{s}\left|x_{1,i}^1-x_{1,i}^2\right|^{\frac{1}{2}}\left|\Delta x_{1,i}\right|\ \left\|v_{x_1x_1}\right\|_{L_2(Q)}\left\|\varphi_i^1\right\|_{L_2((0,T)\times\Omega')}+$$

$$+\left|\Delta x_{1,i}\right|\ \left\|\varphi_i^1-\varphi_i^2\right\|_{L_2((0,T)\times\Omega')}\left(\int_Q v_{x_1}^2+v_{x_1x_1}^2\,dQ\right)^{\frac{1}{2}}+$$

$$+\left|x_{1,i}^{1}-x_{1,i}^{2}\right|^{\frac{1}{2}}\left\|v_{x_1x_1}\right\|_{L_2(Q)}\left\|\Delta\varphi_i\right\|_{L_2((0,T)\times\Omega')}\le$$

$$\le\sum_{i=1}^{s}\left(\left|x_{1,i}^{1}-x_{1,i}^{2}\right|^{\frac{1}{2}}\left\|\varphi_i^1\right\|_{L_2((0,T)\times\Omega')}+\left\|\varphi_i^1-\varphi_i^2\right\|_{L_2((0,T)\times\Omega')}+\right.$$

$$\left.+\left|x_{1,i}^{1}-x_{1,i}^{2}\right|^{\frac{1}{2}}\right)\left\|\Delta h\right\|_{l_3}\left\|v\right\|_{W_{bd^+}^{+l}}.$$

Whence we obtain

$$\left\|A_{3h_1}(\cdot)-A_{3h_2}(\cdot)\right\|\le$$

$$\le\sum_{i=1}^{s}\left(\left|x_{1,i}^{1}-x_{1,i}^{2}\right|^{\frac{1}{2}}\left\|\varphi_i^1\right\|_{L_2((0,T)\times\Omega')}+\right.$$

$$\left.+\left\|\varphi_i^1-\varphi_i^2\right\|_{L_2((0,T)\times\Omega')}+\left|x_{1,i}^{1}-x_{1,i}^{2}\right|^{\frac{1}{2}}\right).$$

Taking into account that U_{ad} is bounded we have $\left\|\varphi_i^1\right\|_{L_2((0,T)\times\Omega')}<C$. Therefore, $A_{3h}(\cdot)$ satisfies the Lipschitz condition with the index $\frac{1}{2}$ with respect to h. By Theorems 3-5 the performance functional has a strong derivative $J_h(\cdot)$ in the region U_{ad}, which satisfy the Hölder condition with the index $\frac{1}{2}$ with respect to h and is defined by the expression $J_{h^*}(\cdot)=\left\langle A_{3h^*}(\cdot),v\right\rangle_{W_{bd^+}}$. Considering directly the performance functional $J(\cdot)$ we can prove that

$$\left\|J_{(x_{1,i}^1,\varphi_i^1)}(\cdot)-J_{(x_{1,i}^1,\varphi_i^2)}(\cdot)\right\|\le c\sum_{i=1}^{s}\left\|\varphi_i^1-\varphi_i^2\right\|_{L_2((0,T)\times\Omega')},$$

i.e. with respect to the direction φ the functional $J_h(\cdot)$ satisfies the Lipschitz condition with the index 1.

4. Let $A_4(t,x,h)=\sum_{i=1}^{s}\sum_{j=1}^{p}\delta(t-t_i)\otimes\delta(x_1-x_{1,j})\otimes\varphi_{ij}(x_2,\dots,x_n)$, where $t,t_i\in[0,T]$, $x_1,x_{1,i}\in[\overline{x}_1,\overline{\overline{x}}_1]$, $\varphi_{ij}(x_2,\dots,x_n)\in L_2(\Omega')$, $i=\overline{1,s}$, $j=\overline{1,p}$, $s,p\in N$.

The control is

$$h_4=\left\{(t_i,x_{1,j},\varphi_{ij}(x_2,\dots,x_n))\right\}_{i,j=1}^{s,p}\in U_{ad}\subset$$
$$\subset H_4=[0,T]^s\times[\overline{x}_1,\overline{\overline{x}}_1]^p\times(L_2(\Omega'))^{sp}$$

where U_{ad} is a convex, closed and bounded set in H_4.

Under

$$A_4(t,x,h)=\sum_{i=1}^{s}\sum_{j=1}^{p}\delta(t-t_i)\otimes\delta(x_1-x_{1,j})\otimes\varphi_{ij}(x_2,\dots,x_n)$$

we shall mean the following functional:

$$l_{A_4(\cdot)}(v)=\int_{\Omega'}\sum_{i=1}^{s}\sum_{j=1}^{p}v(t_i,x_{1,j},x_2,\dots,x_n)\varphi_{ij}(x_2,\dots,x_n)d\Omega',$$

where $v(t,x)$ is a smooth in $\overline{Q}'$ function.

Suppose that $W_{bd^+}^{+l}\subset W_2^3(Q)$ $\left(W_{bd}^{+l}\subset W_2^3(Q)\right)$ and

$$\|\cdot\|_{W_2^3(Q)}\le c\|\cdot\|_{W_{bd^+}^{+l}}\quad\left(\|\cdot\|_{W_2^3(Q)}\le c\|\cdot\|_{W_{bd}^{+l}}\right),\tag{18}$$

where $W_2^3(Q)$ is a completion of the space of smooth in $\overline{Q}$ functions with respect to the norm:

$$\|v\|_{W_2^3(Q)}=\left(\int_Q v^2+v_t^2+v_{x_1}^2+v_{tx_1}^2+v_{ttx_1}^2+v_{tx_1x_1}^2dQ\right)^{\frac{1}{2}},$$

and also let $f\in W_{bd^+}^{-l}$.

Similarly to the previous facts it is easy to see that $A_4(h)$ has a strong derivative:

$$\frac{\partial A_4}{\partial t_i} = -\delta^{(1)}(t - t_i) \otimes \sum_{j=1}^{p} \delta(x_1 - x_{1,j}) \otimes \varphi_{ij}(x_2, \ldots, x_n),$$

$$\frac{\partial A_4}{\partial x_{1,j}} = -\delta^{(1)}(x_1 - x_{1,j}) \otimes \sum_{i=1}^{s} \delta(t - t_i) \otimes \varphi_{ij}(x_2, \ldots, x_n),$$

$$\frac{\partial A_4}{\partial \varphi_{ij}} = \delta(t - t_i) \otimes \delta(x_1 - x_{1,j}),$$

and also that the derivative $A_{4h}(\cdot)$ satisfies the Hölder-Lipschitz condition with the index $1/2$, and the performance functional has a Fréchet derivative which also satisfies the Hölder-Lipschitz condition.

Consider the last case when the right-hand side of the equation is defined as

5. Let $A_5(t, x, h) = \sum_{i=1}^{s} \delta(x_1 - a_i(t)) \otimes \varphi_i(t, x_2, \ldots, x_n)$.

The control is

$$h_5 = \{(a_i(t), \varphi_i(\cdot))\}_{i=1}^{s} \in U_{ad} \subset$$

$$\subset H_5 = \left(W_2^1(0,T)\right)^s \times \left(L_2\left((0,T) \times \Omega'\right)\right)^s.$$

Under

$$\sum_{i=1}^{s} \delta\left(x_i - a_i(t)\right) \otimes \varphi_i\left(t, x_2, \ldots, x_n\right)$$

we shall mean the following functional

$$l_{A_5(\cdot)}(v) = \sum_{i=0}^{n} \int_0^T v\left(t, a_i(t), x_2, \ldots, x_n\right) d\Omega' \, dt,$$

where $v(t, x)$ is a smooth in $\overline{Q}$ function.

Suppose that $W_{bd^+}^{+l} \subset W_2^{0,2,0}(Q)$ $\left(W_{bd}^{+l} \subset W_2^{0,2,0}(Q)\right)$ and

$$\|\cdot\|_{W_2^{0,2,0}(Q)} \le c \|\cdot\|_{W_{bd^+}^{+l}} \quad \left(\|\cdot\|_{W_2^{0,2,0}(Q)} \le c \|\cdot\|_{W_{bd}^{+l}}\right), \tag{24}$$

where $W_2^{0,2,0}(Q)$ is a completion of the space of smooth in $\overline{Q}$ functions with respect to the norm

$$\|v\|_{W_2^{0,2,0}(Q)} = \left(\int_Q v^2 + v_{x_1}^2 + v_{x_1x_1}^2 \, dQ \right)^{\frac{1}{2}}.$$

Make sure that the mapping $A_5(\cdot)$ is strongly differentiable and the derivative is defined by the expression

$$\frac{\partial A_5}{\partial a_i} = -\delta^{(1)}(x_1 - a_i(t)) \otimes \varphi_i(t, x_2, \ldots, x_n),$$

$$\frac{\partial A_5}{\partial \varphi_i} = \delta(x_1 - a_i(t)).$$

Indeed,

$$I'' = \langle A_5(h + \Delta h) - A_5(h) - (\mathrm{grad} A_5, \Delta h), v \rangle_{W_{bd^+}} =$$

$$= \sum_{i=1}^{s} \int_0^T \int_{\Omega'} \Big(\big(v_{x_1}(t, a_i(t) + \theta \Delta a_i(t), x_2, \ldots, x_n) -$$

$$- v_{x_1}(t, a_i(t), x_2, \ldots, x_n) \big) \Delta a_i(t) \varphi_i +$$

$$+ \big(v(t, a_i(t) + \Delta a_i(t), x_2, \ldots, x_n) -$$

$$- v(t, a_i(t), x_2, \ldots, x_n) \big) \Delta \varphi_i \, d\Omega' dt.$$

Whence

$$|I''| \le \sum_{i=1}^{s} \int_0^T \int_{\Omega'} |\Delta a_i(t)|^{\frac{3}{2}} \left(\int_{\overline{x}_1}^{\overline{\overline{x}}} v_{x_1x_1}^2(t, x) d\xi \right)^{\frac{1}{2}} \varphi_i +$$

$$+ |\Delta a_i(t)|^{\frac{1}{2}} \left(\int_{\overline{x}_1}^{\overline{\overline{x}}_1} v_{x_1}^2(t, x) d\xi \right)^{\frac{1}{2}} \Delta \varphi_i \, d\Omega' dt.$$

Using the fact that $\Delta a(t) \in W_2^1(0,1)$, where $W_2^1(0,1)$ is the Sobolev space, we get $\max_{t \in [0,T]} |\Delta a(t)| \le c \|\Delta a\|_{W_2^1(0,T)}$. Therefore,

$$|I''| \le$$

$$\le C\sum_{i=1}^{s}\left(\int_0^T\int_{\Omega'}\int_{\bar{x}_1}^{\bar{\bar{x}}_1} v_{x_1x_1}^2(t,x)d\xi d\Omega' dt\right)^{\frac{1}{2}} \|\Delta a_i(t)\|_{W_2^1(0,T)}^{\frac{3}{2}} \|\varphi_i\|_{L_2((0,T)\times\Omega')} +$$

$$+ C\sum_{i=1}^{s}\left(\int_0^T\int_{\Omega'}\int_{\bar{x}_1}^{\bar{\bar{x}}_1} v_{x_1}^2(t,x)d\xi d\Omega' dt\right)^{\frac{1}{2}} \|\Delta a_i(t)\|_{W_2^1(0,T)}^{\frac{1}{2}} \|\Delta\varphi_i\|_{L_2((0,T)\times\Omega')} \le$$

$$\le C_1\|v\|_{W_{bd^+}^{+l}} \|\Delta h\|_{I_5}^{\frac{3}{2}}.$$

This fact proves that $A_5(\cdot)$ is strongly differentiable. Similarly to the previous facts we make sure that $A_{5_h}(\cdot)$ satisfies the Lipschitz condition with the index $1/2$.

2. REGULARIZATION OF CONTROL

In the previous section the questions of performance criterion gradient existence and its continuity depend upon differential properties of the right-hand side of state equation have been studied. In order to calculate the performance criterion gradient in any point, it is necessary to solve the direct and adjoint problems.

In this paragraph we shall study the optimization problem in the case when optimal control exists, but the right-hand side are not differentiable enough for calculation of performance criterion gradient [80,94]. This difficulty can be remedied by regularization of control. As in the previous sections we assume here that the operators L and L^* satisfy the inequality in the negative norms

$$\|u\|_{L_2(Q)} \le C_1\|Lu\|_{W_{bd^+}^{-l}} \le C_2\|u\|_{W_{bd}^{+l}}$$

$$\|v\|_{L_2(Q)} \le C_1\|L^*v\|_{W_{bd}^{-l}} \le C_2\|v\|_{W_{bd^+}^{+l}}$$

Consider the following averaging of the distribution F in the right-hand side:

$$f_\varepsilon(t,x) = f + \int_{-\infty}^{+\infty} \omega_\varepsilon(\tau) A(t-\tau, x) d\tau = f + A_\varepsilon(h),$$

where

$$\omega_\varepsilon(\tau) \in C_0^\infty(R^1),\ \varepsilon > 0,\ \omega_\varepsilon(t) = 0,\quad |t| \geq \varepsilon,\ \omega_\varepsilon \geq 0,$$

$$\int_{-\infty}^{+\infty} \omega_\varepsilon(\tau) d\tau = 1.$$

By the integral $\int_{-\infty}^{+\infty} \omega_\varepsilon(\tau) A(t-\tau, x) d\tau$ we mean the integral in the sense of the distribution theory.

Instead of using distribution averaging, we may consider a sequence of functions $A_\varepsilon \in L_2(Q)$: $\|f + A_\varepsilon - F\|_{W_{bd^+}^{-l}} \to 0$, when $\varepsilon \to 0$.

Consider a regularized problem of impulse optimal control.

$$Lu_\varepsilon = f + A_\varepsilon(h), \tag{1}$$

$$J_\varepsilon(h) = \int_Q [u_\varepsilon(h) - z_{ad}]^2 dQ, \tag{2}$$

$$h_\varepsilon^* = \arg\min_{h \in U_{ad}^\varepsilon} J_\varepsilon(h), \tag{3}$$

where

$$A_\varepsilon(t,x) = \int_{R^1} \omega_\varepsilon(\tau) \sum_{i=1}^{N} \delta(t - t_i - \tau) \varphi_i(x) d\tau = \sum_{i=1}^{N} \omega_\varepsilon(t - t_i) \varphi_i(x).$$

We assume that $f \in L_2(Q)$ (although all the theorems can be generalized to the case $f \in W_{bd^+}^{-l}$).

The vector $h^\varepsilon = \{t_i, \varphi_i\}_{i=\overline{1,N}} \in U_{ad}^\varepsilon \subset U_{ad}$ is a control of the optimization problem (1)-(3), where

$$U_{ad}^\varepsilon = [\varepsilon/2, T - \varepsilon/2]^N \times (\Phi_{ad})^N \subset R^N \times L_2^N(\Omega) = H_1,$$

$\Phi_{ad} \subset L_2(\Omega)$ is bounded, convex and closed set of admissible controls.

Theorem 1. *There exists an optimal control $h^{*\varepsilon}$ (generally not unique) of the optimization problem (1)-(3).*

Proof. Let $\{h_k^\varepsilon\}_{k=1}^\infty$ be the minimizing sequence of the problem (1)-(3). This means that

$$J_\varepsilon(h_k^\varepsilon) \to \inf_{h \in U_{ad}^\varepsilon} J_\varepsilon(h).$$

Since Φ_{ad} is bounded, convex, closed set and $L_2(\Omega)$ is Hilbert, the set Φ_{ad} is weakly compact in $L_2(\Omega)$. Set $[\varepsilon/2, T-\varepsilon/2]^N$ is compact in R^n. Thus, there exists a subsequence (it is denoted by $\{h_k^\varepsilon\}_{k=1}^\infty$ again) such that $t_{i,k}^{*\varepsilon} \xrightarrow[k\to\infty]{} t_i^{*\varepsilon}$ in R^1, $t_i^{*\varepsilon} \in \left[\frac{\varepsilon}{2}, T-\frac{\varepsilon}{2}\right]$, $\varphi_{ik}^{*\varepsilon} \xrightarrow[k\to\infty]{} \varphi_i^{*\varepsilon}$ weakly in $L_2(\Omega)$, $\varphi_i^{*\varepsilon} \in \Phi_{ad}$, $i = \overline{1,N}$.

Let us introduce the following notation

$$A_\varepsilon(h_k^\varepsilon) = \sum_{i=1}^N \omega_\varepsilon(t - t_{i,k}^\varepsilon)\varphi_{i,k}^\varepsilon(x).$$

We shall show that

$$A_\varepsilon(h_k^\varepsilon) \xrightarrow[k\to\infty]{} A_\varepsilon^* = \sum_{i=1}^N \omega_\varepsilon(t - t_i^{*\varepsilon})\varphi_i^{*\varepsilon}(x) \tag{4}$$

weakly in $L_2(Q)$.

For every function $y \in L_2(Q)$ and each number $i = \overline{1,N}$, we have

$$\left(A_\varepsilon(h_k^\varepsilon) - A_\varepsilon^*, y\right)_{L_2(Q)} = \left(\sum_{i=1}^N \left[\omega_\varepsilon(t - t_{i,k}^\varepsilon)\varphi_{i,k}^\varepsilon(x) - \right.\right. \tag{5}$$
$$\left.\left. - \omega_\varepsilon(t - t_i^{*\varepsilon})\varphi_{i,k}^\varepsilon(x)\right] + \omega_\varepsilon(t - t_i^{*\varepsilon})\left[\varphi_{i,k}^\varepsilon(x) - \varphi_i^{*\varepsilon}(x)\right], y\right)_{L_2(Q)}.$$

By the Schwarz inequality

$$\left|\sum_{i=1}^{N}\int_{Q}\left(\omega_{\varepsilon}(t-t_{i,k}^{\varepsilon})-\omega_{\varepsilon}(t-t_{i}^{*\varepsilon})\right)\varphi_{i,k}^{\varepsilon}(x)dQ\right|\leq$$

$$\leq\sum_{i=1}^{N}\left\|\varphi_{i,k}^{\varepsilon}\right\|_{L_2(Q)}\cdot\left\|\omega_{\varepsilon}(t-t_{i,k}^{\varepsilon})-\omega_{\varepsilon}(t-t_{i}^{*\varepsilon})\right\|_{L_2(Q)}.$$

Since $t_k^{*\varepsilon}=\left\{t_{i,k}^{\varepsilon}\right\}_{i=1}^{N}\to\left\{t_k^{*\varepsilon}\right\}_{i=1}^{N}=t^{*\varepsilon}$ in R^N, $\omega_\varepsilon\in C_0^\infty\left(R^1\right)$ and $\left\|\varphi_{i,k}^{\varepsilon}\right\|_{L_2(Q)}\leq C$, the first term of the right-hand side of (5) vanishes as $k\to\infty$. The second term of the right-hand side of (5) vanishes also, because $\varphi_k^{\varepsilon}=\left\{\varphi_{i,k}^{\varepsilon}\right\}_{i=1}^{N}\to\left\{\varphi_i^{*\varepsilon}\right\}_{i=1}^{N}=\varphi^{*\varepsilon}$ weakly in $L_2^N(\Omega)$.

Thus the formula (4) has been proved.

In accordance with the results of Section 1, a solution $u_k^{\varepsilon}\in W_{bd}^{+l}$ (in the sense of Definition 1.1.1) of the problem (1) with the right-hand side $f+A_\varepsilon\left(h_k^{\varepsilon}\right)$ exists and unique. The solution $u_k^{\varepsilon}\in W_{bd}^{+l}$ satisfies the equation

$$\left\langle u_k^{\varepsilon},L^*v\right\rangle_{W_{bd}}=\left\langle f+A_\varepsilon\left(h_k^{\varepsilon}\right),v\right\rangle_{W_{bd^+}}=\left(f+A_\varepsilon\left(h_k^{\varepsilon}\right),v\right)_{L_2(Q)},\qquad(6)$$

for all $v\in W_{bd^+}^{+l}$.

Taking into account that the set U_{ad}^{ε} is bounded, we have

$$\left\|u_k^{\dot{a}}\right\|_{W_{bd}^{+l}}\leq C\left\|f+A_{\dot{a}}\left(h_k^{\dot{a}}\right)\right\|_{L_2(Q)}\leq C<\infty.\qquad(7)$$

It follows from the inequality (7) that there exists weakly convergent to $u^{\dot{a}}\in W_{bd}^{+l}$ in W_{bd}^{+l} and so that in $L_2(Q)$ subsequence $\left\{u_{k_l}^{\varepsilon}\right\}_{l=1}^{\infty}$. Passing to the limit as $k\to\infty$ in the equality (6), we obtain

$$\left\langle u^{\varepsilon},L^*v\right\rangle_{W_{bd}}=\left\langle f+A_\varepsilon\left(h^{\varepsilon}\right),v\right\rangle_{W_{bd^+}}=\left(f+A_\varepsilon\left(h^{\varepsilon}\right),v\right)_{L_2(Q)},$$

for all $v\in W_{bd}^{+l}$.

If we take into consideration the above equality, we conclude that u^{ε} is solution of the problem (1) with the right-hand side $f + A_{\varepsilon}\left(h^{*\varepsilon}\right)$. Since square of norm in the Hilbert space is weakly lower semicontinuous, the functional (2) is weakly lower semicontinuous in domain U_{ad}^{ε}

$$J_{\varepsilon}\left(h^{\varepsilon *}\right) \leq \varliminf_{k \to \infty} J_{\varepsilon}\left(h_k^{\varepsilon}\right) = \inf_{h \in U_{ad}^{\varepsilon}} J_{\varepsilon}(h).$$

Consequently, $h^{*\varepsilon}$ is an optimal control of the problem (1)-(3).

Theorem 2. *The set of the optimal controls $\left\{h^{*\varepsilon}\right\}$ of the problems (1)-(3) contains a weakly convergent sequence $\left\{h^{*\varepsilon k}\right\}_{k=\overline{1,\infty}}$. The sequence $\left\{h^{*\varepsilon k}\right\}_{k=\overline{1,\infty}}$ weakly converges to an optimal control $h^{*} \in U_{ad}$ of the initial (non-regularized) problem, the points of impulse $t^{*\varepsilon_k}$ being strongly convergent to t^{*}.*

Proof. We shall show that at the fixed control $h \in U_{ad}^{\varepsilon}$ we have

$$\left\| u^{\varepsilon} - u \right\|_{L_2(Q)} \xrightarrow[\varepsilon \to 0]{} 0, \tag{8}$$

where u^{ε}, u – solutions of the regularized and initial problems, respectively.

For all $i = \overline{1,N}$ we have

$$\left\| \dot{u}_{d}(t - t_i)\varphi_i(x) - \ddot{a}(t - t_i)\varphi_i(x) \right\|_{W_{bd^{+}}^{-l}} =$$

$$= \sup_{v \neq 0,\, v \in W_{bd^{+}}^{+l}} \frac{\left| \left\langle \omega_{\varepsilon}(t_i - t)\varphi_i(x) - \delta(t - t_i)\varphi_i(x), v \right\rangle_{W_{bd^{+}}} \right|}{\left\| v \right\|_{W_{bd^{+}}^{+l}}}.$$

Taking into account the definition of delta function we obtain

$$\sup_{v\neq 0,\, v\in W^{+l}_{bd^+}} \frac{\left|\int_\Omega \int_{R^1} \dot u_{\mathring{d}}(t-t_i)\left[v(t,x)-v(t_i,x)\right]dt\varphi_i(x)d\Omega\right|}{\|v\|_{W^{+l}_{bd^+}}} \le$$

$$\le \sup_{v\neq 0,\, v\in W^{+l}_{bd^+}} \frac{\mathring{d}\left|\int_\Omega \varphi_i(x)d\Omega\|v\|_{W^{+l}_{bd^+}}\right|}{\|v\|_{W^{+l}_{bd^+}}} \xrightarrow[\mathring{d}\to 0]{} 0 .$$

Summing up the above inequalities over $i \neq \overline{1,N}$ and considering inequality of Lemma 1.1.3, we have (8).

By Theorem 1 for all $\varepsilon > 0$ there exists an optimal control $h^{*\varepsilon} \in U^{\varepsilon}_{ad}$ of the regularized problem. By virtue of the boundedness of the set U_{ad} the set $\left\{h^{*\varepsilon}\right\}_{\varepsilon>0}$ is weakly compact, hence, there exists a subsequence $\left\{h^{*\varepsilon_k}\right\}_{k=\overline{1,\infty}}$, which weakly converges to h^*. The statement $t^{*\varepsilon_k} \xrightarrow[k\to\infty]{} t^* \in R^N[0,T]$ becomes apparent when the equivalence between convergence in norm and weak convergence in the finite-dimensional space is considered. From the weak convergence of $\left\{h^{*\varepsilon_k}\right\}_{k=\overline{1,\infty}}$ we find, as well, that $\varphi^{*\varepsilon_k} \xrightarrow[k\to\infty]{} \varphi^* \in \left(\Phi_{ad}\right)^N$ weakly in $L_2^N(\Omega)$.

Reasoning similarly, we have

$$A^*_{\varepsilon_k} = \sum_{i=1}^N \omega_{\varepsilon_k}\left(t-t_i^{*\varepsilon_k}\right)\varphi_i^{*\varepsilon_k}(x) - \sum_{i=1}^N \omega_{\varepsilon_k}\left(t-t_i^*\right)\varphi_i^{*\varepsilon_k}(x) \xrightarrow[k\to\infty]{} 0,$$

$$\sum_{i=1}^N \omega_{\varepsilon_k}\left(t-t_i^*\right)\varphi_i^{*\varepsilon_k}(x) \xrightarrow[k\to\infty]{} \sum_{i=1}^N \omega_{\varepsilon_k}\left(t-t_i^*\right)\varphi_i^*(x),$$

$$\sum_{i=1}^N \omega_{\varepsilon_k}\left(t-t_i^*\right)\varphi_i^*(x) \xrightarrow[k\to\infty]{} \sum_{i=1}^N \delta\left(t-t_i^*\right)\otimes\varphi_i^*(x) = A\left(h^*\right)$$

weakly in $W^{-l}_{bd^+}$.

Adding up these three relations, we have that $A^*_{\varepsilon_k} \xrightarrow[k\to\infty]{} A(h^*)$ weakly in $W^{-l}_{bd^+}$.

Thus, we prove that sequence $A^*_{\varepsilon_k} = \sum_{i=1}^{N} \omega_{\varepsilon_k}\left(t - t_i^{*\varepsilon_k}\right)\varphi_i^{*\varepsilon_k}(x)$ is weakly convergent in the space $W^{-l}_{bd^+}$. Therefore, $A^*_{\varepsilon_k} = \sum_{i=1}^{N} \omega_{\varepsilon_k}\left(t - t_i^{*\varepsilon_k}\right)\varphi_i^{*\varepsilon_k}(x)$ is bounded. By applying the inequality of Lemma 1.1.3 we conclude that the sequence u^{ε_k} of solutions of the problem (1)-(3) with right-hand sides

$$f + A^*_{\varepsilon_k} = f + \sum_{i=1}^{N} \omega_{\varepsilon_k}\left(t - t_i^{*\varepsilon_k}\right)\varphi_i^{*\varepsilon_k}(x)$$

is also bounded, so that the sequence u^{ε_k} contains a weakly convergent subsequence (which is denoted by u^{ε_k} again).

Since $A^*_{\varepsilon_k} \xrightarrow[k\to\infty]{} A(h^*)$ weakly in $W^{-l}_{bd^+}$ it is easy to prove that weak limit of the subsequence u^{ε_k} in the space $L_2(Q)$ is the solution u^* of the initial optimization problem with the right-hand side $f + A^*$. From this it follows that

$$J\left(h^*\right) \le \varliminf_{k\to\infty} J_{\varepsilon_k}\left(h^{*\varepsilon_k}\right). \tag{9}$$

The statement $J_\varepsilon(h) \xrightarrow[\varepsilon\to 0]{} J(h)$ becomes apparent when it is considered that for all $h \in U^\varepsilon_{ad}$ $u^\varepsilon(h) \xrightarrow[\varepsilon\to 0]{} u(h)$ in the norm of the space $L_2(Q)$. Let h^{*1} be an optimal control of the initial optimization problem. When it is taken into account that $\inf_{h\in U^\varepsilon_{ad}} J_\varepsilon(h) = J_\varepsilon\left(h^{*\varepsilon}\right) \le J_\varepsilon\left(h^{*1}\right)$, the following relation is valid

$$\varlimsup_{k\to\infty} J_{\varepsilon_k}\left(h^{*\varepsilon_k}\right) \le J\left(h^{*1}\right). \tag{10}$$

From (9) and (10) we obtain

$$\inf_{h\in U_{ad}} J(h)\le J(h^*)\le \underline{\lim}_{k\to\infty} J_{\varepsilon_k}(h^{*\varepsilon_k})\le$$
$$\le \overline{\lim}_{k\to\infty} J_{\varepsilon_k}(h^{*\varepsilon_k})\le J(h^{*1})\le \inf_{h\in U_{ad}} J(h).$$

Therefore, $\lim_{k\to\infty} J_{\varepsilon_k}(h^{*\varepsilon_k})\le J(h^{*1})$, and h^* is the optimal control of the initial optimization problem as well as h^{*1}. Since the optimal control may be not unique, the controls h^* and h^{*1} may be different.

Remark. *If instead of* Φ_{ad} *we consider* $\Phi'_{ad}=\Phi_{ad}\cap W_2^1(\Omega)$, *then the sequence* $\varphi^{*\varepsilon_k}$ *converges to the* $\varphi^*\in L_2^N(\Omega)$ *in norm* $L_2^N(\Omega)$.

Theorem 3. *There exists a Fréchet derivative of the performance criterion* $J_\varepsilon(h)$. *The Fréchet derivative is of the following form*

$$grad\, J_\varepsilon(h)=$$
$$=\left(\left(-\int_Q \frac{\partial}{\partial t}\omega_\varepsilon(t-t_i)\varphi_i(x)v_\varepsilon(t,x)dQ\right)_{i=\overline{1,N}},\left(\int_0^T \omega_\varepsilon(t-t_i)v_\varepsilon(t_i,x)dt\right)_{i=1}^N\right),$$

where $v_\varepsilon(t,x)$ *is a solution of the adjoint problem with the right-hand side* $2(u_\varepsilon - z_{ad})$.

Proof. We denote by Δh an increment of a control h. Corresponding solution of the system (1) has an increment $\Delta u_\varepsilon(h)=u_\varepsilon(h+\Delta h)-u_\varepsilon(h)$. An increment of the performance criterion $J_\varepsilon(h)$ can be represented in the form

$$\Delta J_\varepsilon(h)=J_\varepsilon(h+\Delta h)-J_\varepsilon(h)=$$
$$=\int_Q [u_\varepsilon(h+\Delta h)-z_{ad}]^2-[u_\varepsilon(h)-z_{ad}]^2dQ=$$

$$= 2\int_Q \Delta u_\varepsilon(h)[u_\varepsilon(h) - z_{ad}]dQ + \int_Q (\Delta u_\varepsilon(h))^2 dQ.$$

Adjoint state is defined as a solution of the following problem

$$L^* v_\varepsilon = 2(u_\varepsilon(h) - z_{ad}),\ v_\varepsilon \in W^{+l}_{bd^+}.$$

As appears from Section 1.1, there exists a unique solution of this problem belonging to the space $W^{+l}_{bd^+}$. The increment $\Delta J_\varepsilon(h)$ can be transformed to the form

$$\Delta J_\varepsilon(h) = (\Delta u_\varepsilon(h), L^* v_\varepsilon)_{L_2(Q)} + \|\Delta u_\varepsilon(h)\|^2_{L_2(Q)}.$$

It is obvious that the increment $\Delta u_\varepsilon(h)$ is a solution of the following problem

$$L\Delta u_\varepsilon(h) = \sum_{i=1}^N \omega_\varepsilon(t - t_i - \Delta t_i)(\varphi_i + \Delta\varphi_i) - \sum_{i=1}^N \omega_\varepsilon(t - t_i)\varphi_i.$$

The results of Section 1 ensure the existence and uniqueness of the solution of the above problem. The solution $\Delta u_\varepsilon(h) \in W^{+l}_{bd}$ satisfies the following equation

$$(\Delta u_\varepsilon(h), L^* y)_{L_2(Q)} = \sum_{i=1}^N \Big[((\omega_\varepsilon(t - t_i - \Delta t_i) - \omega_\varepsilon(t - t_i))\varphi_i, y)_{L_2(Q)} + \\ + (\omega_\varepsilon(t - t_i - \Delta t_i)\Delta\varphi_i, y)_{L_2(Q)}\Big];$$

for all $y(t,x) \in W^{+l}_{bd^+}$, $L^* y \in L_2(Q)$.

Taking into account this relation, we have

$$\Delta J_\varepsilon(h) = \sum_{i=1}^N \int_Q (A_\varepsilon(t;t_i + \lambda\Delta t_i) - A_\varepsilon(t;t_i))v_\varepsilon(t,x)\varphi_i(x)dQ + \\ + \lambda\sum_{i=1}^N \int_Q A_\varepsilon(t;t_i + \lambda\Delta t_i)v_\varepsilon(t,x)\Delta\varphi_i(x)dQ + \int_Q |\Delta u_\varepsilon(h)|^2 dQ.$$

An increment $\omega_\varepsilon(t - t_i - \Delta t_i) - \omega_\varepsilon(t - t_i)$ can be represented in the form

$$A_{2,\varepsilon}(t, x, h) = \sum_{i=1}^{s} a_{2\varepsilon}(t, x, t_i)\varphi_i(x) .$$

Considering this equation, we have

$$\Delta J_\varepsilon(h) = \sum_{i=1}^{N}\left[-\left(\frac{d}{dt}\omega_\varepsilon(t-t_i)\varphi_i, v_\varepsilon\right)_{L_2(Q)}\Delta t_i + \right.$$
$$\left. + \left(\omega_\varepsilon(t-t_i)\Delta\varphi_i, v_\varepsilon\right)_{L_2(Q)}\right] + \left\|\Delta u_\varepsilon(h)\right\|^2_{L_2(Q)} + o(\|\Delta h\|).$$

Taking into account the estimation

$$|I| = \left|\left\langle F_{h',1,\varepsilon}(\Delta h) - F_{h'',1,\varepsilon}(\Delta h), v\right\rangle_{W_{bd^+}}\right| \le c\|\Delta h\|_{H_1}\|h'' - h'\|^{\frac{1}{2}}_{H_1}\|v\|_{W^{+l}_{bd^+}},$$

we have

$$\Delta J_\varepsilon(h) = \sum_{i=1}^{N}\left[-\left(\frac{d}{dt}\omega_\varepsilon(t-t_i)\varphi_i, v_\varepsilon\right)_{L_2(Q)}\Delta t_i + \right.$$
$$\left. + \left(\omega_\varepsilon(t-t_i)\Delta\varphi_i, v_\varepsilon\right)_{L_2(Q)}\right] + o(\|\Delta h\|),$$

which is what had to be proved.

The function $A_\varepsilon(\cdot)$ can be approximated, very useful from practical point of view, by the sum

$$\widetilde{A}_\varepsilon = \sum_{i=1}^{N} A_\varepsilon(t;t_i)\varphi_i(x),$$

where

$$A_\varepsilon(t;t_i) = \begin{cases} \dfrac{1}{\varepsilon}, & t \in \left[t_i - \dfrac{\varepsilon}{2}, t_i + \dfrac{\varepsilon}{2}\right], \\ 0, & t \notin \left[t_i - \dfrac{\varepsilon}{2}, t_i + \dfrac{\varepsilon}{2}\right]. \end{cases}$$

An analogous theorem for such way of approximation holds true.

Theorem 4. *There exists a Fréchet derivative of the performance criterion* $J_\varepsilon(h)$ *of the optimization problem with the right-hand side* $f = \widetilde{A}_\varepsilon$. *This derivative satisfies the Lipschitz condition and can be represented in the following form*

$$gradJ_\varepsilon(h) = \left(\left(\frac{1}{\varepsilon}\int_Q \varphi_i(x)\left[v_\varepsilon\left(t_i+\frac{\varepsilon}{2},x\right) - v_\varepsilon\left(t_i-\frac{\varepsilon}{2},x\right)\right]dQ\right)_{i=1}^N;\right.$$

$$\left.\left(\int_{t_i-\frac{\varepsilon}{2}}^{t_i+\frac{\varepsilon}{2}}\int_\Omega v_\varepsilon(t,x)\,d\Omega\,dt\right)_{i=1}^N\right)$$

where $v_\varepsilon(t,x)$ is a solution of the adjoint problem with the right-hand side $2(u_\varepsilon - z_{ad})$.

P r o o f. In the same manner as in the case of the previous theorem we denote by Δh an increment of control h. The corresponding solution of the system (1) has an increment $\Delta u_\varepsilon(h) = u_\varepsilon(h+\Delta h) - u_\varepsilon(h)$. An increment of the performance criterion $J_\varepsilon(h)$ can be represented in the form

$$\Delta J_\varepsilon(h) = J_\varepsilon(h+\lambda\Delta h) - J_\varepsilon(h) =$$
$$= 2\int_Q \Delta u_\varepsilon(h)[u_\varepsilon(h) - z_{ad}]dQ + \int_Q |\Delta u_\varepsilon(h)|^2 dQ.$$

As usual, we use the adjoint state and obtain

$$\Delta J_\varepsilon(h) = \sum_{i=1}^N \int_Q (A_\varepsilon(t;t_i+\lambda\Delta t_i) - A_\varepsilon(t;t_i))v_\varepsilon(t,x)\varphi_i(x)dQ +$$
$$+\lambda\sum_{i=1}^N \int_Q A_\varepsilon(t;t_i+\lambda\Delta t_i)v_\varepsilon(t,x)\Delta\varphi_i(x)dQ + \int_Q |\Delta u_\varepsilon(h)|^2 dQ. \tag{12}$$

Divide (12) by λ (12) and pass to the limit as $\lambda \to 0$

$$\lim_{\lambda\to 0} = \sum_{i=1}^N \frac{1}{\varepsilon}\int_\Omega \varphi_i(x)\left[v_\varepsilon\left(t_i+\frac{\varepsilon}{2},x\right) - v_\varepsilon\left(t_i-\frac{\varepsilon}{2},x\right)\right]d\Omega\Delta t_i +$$

$$+\sum_{i=1}^{N}\frac{1}{\varepsilon}\int_{t_i-\frac{\varepsilon}{2}}^{t_i+\frac{\varepsilon}{2}}\int_{\Omega} v_\varepsilon(t,x)dt\Delta\varphi_i d\Omega, \tag{13}$$

which is what had to be proved in the first part of the theorem. The Lipschitz condition of the performance criterion gradient follows from the a priori inequalities in the negative norms and the inequality

$$\|v\|_{C([0,T];L_2(\Omega))} \leq C\|v\|_{W^{+l}_{bd^+}}. \tag{14}$$

We shall first prove the inequality (14) for a smooth function $v \in W^{+l}_{bd^+}$. In this case we have

$$|v(\tau)| = \left|\int_T^\tau \frac{\partial v}{\partial t}dt\right| \leq T^{1/2}\left(\int_0^T\left(\frac{\partial v}{\partial t}\right)^2 dt\right)^{1/2}.$$

Square the right and left-hand sides and integrate over the domain Ω

$$\sup_{\tau\in[0,T]}\left(\int_\Omega v^2(\tau,x)d\Omega\right)^{1/2} \leq C\int_Q\left(\frac{\partial v}{\partial t}\right)^2 dQ,$$

which is what had to be proved in (14) for an arbitrary smooth function $v \in W^{+l}_{bd^+}$.

The inequality (14) for all $v \in W^{+l}_{bd^+}$ is proved by expansion by continuity.

Next, in the same manner as in Section 2.1 we generalize above-mentioned results for the systems with the right-hand side of the general form.

As in the previous case, we assume that the operators L, L^* satisfies the inequalities in the negative norms

$$\|u\|_{L_2(Q)} \leq C_1\|Lu\|_{W^{-l}_{bd^+}} \leq C_2\|u\|_{W^{+l}_{bd}},$$

$$\|v\|_{L_2(Q)} \leq C_1\|L^*v\|_{W^{-l}_{bd}} \leq C_2\|u\|_{W^{+l}_{bd^+}}$$

We also assume that $F(t,x,h) \in W^{-l}_{bd^+}$ and the functional $J(h) = \Phi(u(h))$ is defined at every function $u(t,x)$ from $L_2(Q)$. Consider the regularized problem

$$Lu_\varepsilon = F_\varepsilon(t,x,h), \tag{15}$$

where $\varepsilon > 0$, $F_\varepsilon(t,x,h) \in L_2(Q)$.

It is required to minimize the functional

$$J_\varepsilon(h) = \Phi(u_\varepsilon(h)). \tag{16}$$

Theorem 5. *Consider the optimal control problems (1.1.12) and (1)-(3). If*

1) the admissible set of controls U_{ad} is convex, closed, and bounded in the Hilbert space H;

2) maps $F_\varepsilon(h), F(h)$ satisfy the conditions:

a) $\left(h_{\varepsilon_k} \xrightarrow{W} h \text{ in } H\right) \Rightarrow \left(F_{\varepsilon_k}(h_{\varepsilon_k}) \xrightarrow{W} F(h) \text{ in } W^{-l}_{bd^+}\right)$, for an arbitrary sequence $\varepsilon_k \xrightarrow[k\to\infty]{} 0$,

b) $F_\varepsilon(h), F(h)$ is weakly continuous,

c) $\|F_\varepsilon(h) - F(h)\|_{W^{-l}_{bd^+}} \xrightarrow[\varepsilon\to 0]{} 0$ for all fixed $h \in U_{ad} \subset H$.

3) the performance criterion $\Phi(\cdot)$ is upper semicontinuous and weakly lower semicontinuous;

then there exist optimal controls h^*, h^*_ε of the problems (1.1.12) and (1)-(3), there exists a weakly convergent subsequence $h^*_{\varepsilon_i}$ and an arbitrary weakly convergent subsequence $h^*_{\varepsilon_i}$ converges to h^* weakly in H.

Proof. According to Theorem 1.2.1, the conditions 1), 2a), and 3) ensure the existence of optimal controls of the problems (1.1.12) and (1)-(3). Consider a set $\{h^*_\varepsilon\}$, $h^*_\varepsilon \in U_{ad}$. Since the set U_{ad} is bounded in the Hilbert space H, there exists a weakly convergent to

$\overline{h}$ subsequence $h^*_{\varepsilon_i}$ $\left(\varepsilon_i \xrightarrow[i\to\infty]{} 0\right)$. Consider an arbitrary weakly convergent subsequence $h^*_{\varepsilon_k} \xrightarrow[k\to\infty]{} \overline{h}$. Since the set U_{ad} is closed and convex, the element $\overline{h}$ belongs to U_{ad}. Consider the sequence of solutions $\{u_{\varepsilon_k}(h^*_{\varepsilon_k})\}$. By Lemma 1.1.3, we have

$$\left\|u_{\varepsilon_k}(h^*_{\varepsilon_k})\right\|_{L_2(Q)} \le c\left\|F_{\varepsilon_k}(h^*_{\varepsilon_k})\right\|_{W^{-l}_{bd^+}}.$$

Taking into account the condition 2a), we have that $F_{\varepsilon_k}(h^*_{\varepsilon_k}) \xrightarrow[k\to\infty]{} F(\overline{h})$ weakly in $W^{+l}_{bd^+}$. Therefore, the sequence of the norms $\left\|F_{\varepsilon_k}(h^*_{\varepsilon_k})\right\|_{W^{-l}_{bd^+}}$ is bounded, so there exists a weakly convergent subsequence $u_{\varepsilon_{k_m}}(h^*_{\varepsilon_{k_m}}) \xrightarrow{W} u$. Since $\{u_{\varepsilon_k}(h^*_{\varepsilon_k})\}$ are solutions, we have

$$\left(u_{\varepsilon_{k_m}}(h^*_{\varepsilon_{k_m}}), L^* v\right)_{L_2(Q)} = \left\langle F_{\varepsilon_{k_m}}(h^*_{\varepsilon_{k_m}}), v\right\rangle_{W_{bd^+}}, \tag{17}$$

for all $v(t,x) \in W^{+l}_{bd^+} : L^* v \in L_2(Q)$.

Passing to the limit as $m \to \infty$ in (17), we have

$$\left(u, L^* v\right)_{L_2(Q)} = \left\langle F(\overline{h}), v\right\rangle_{W_{bd^+}}. \tag{18}$$

It is now clear that the element u is a solution of the problem (1.1.12) with the right-hand side $F(\overline{h})$. But there exists an unique solution of the problem (1.1.12); therefore entire sequence $\{u_{\varepsilon_k}(h^*_{\varepsilon_k})\}$ converges weakly to $u = u(\overline{h})$. If there exists a weakly convergent to any other function $\tilde{u} \ne u(\overline{h})$ subsequence $\{u_{\varepsilon_l}(h^*_{\varepsilon_l})\}$ indeed, then we can prove in much the same way, that $\tilde{u}$ is an other solution of the problem (1.1.12) with the right-hand side $F(\overline{h})$, contrary to the uniqueness of the solution. Since the functional $\Phi(\cdot)$ is weakly lower semicontinuous, we have

$$J(\overline{h}) = \Phi(u(\overline{h})) \le \varliminf_{k\to\infty} \Phi(u_{\varepsilon_k}(h^*_{\varepsilon_k})) = \\ = \varliminf_{k\to\infty} J_{\varepsilon_k}(h^*_{\varepsilon_k}) \le \varliminf_{k\to\infty} J_{\varepsilon_k}(h), \tag{19}$$

for all $h \in U_{ad} \subset H$.

Consider the sequence $\{u_{\varepsilon_l}(h)\}$. By Lemma 1.1.3, we have

$$\left\|u_{\varepsilon_k}(h) - u(h)\right\|_{L_2(Q)} \le c\left\|F_{\varepsilon_k}(h) - F(h)\right\|_{W^{-l}_{bd^+}} \xrightarrow[k\to\infty]{} 0.$$

Therefore, $u_{\varepsilon_k}(h) \xrightarrow[k\to\infty]{} u(h)$ in the space $L_2(Q)$. Since the functional $\Phi(\cdot)$ is upper semicontinuous, it follows that

$$J(h) = \Phi(u(h)) \ge \varlimsup_{k\to\infty} \Phi(u_{\varepsilon_k}(h)) = \varlimsup_{k\to\infty} J_{\varepsilon_k}(h). \tag{20}$$

Employing the relations (19) and (20), we have

$$J(\overline{h}) \le \varliminf_{k\to\infty} J_{\varepsilon_k}(h) \le \varlimsup_{k\to\infty} J_{\varepsilon_k}(h) \le J(h),$$

for all $h \in U_{ad} \subset H$.

To put it in another way, the control $\overline{h}$ is optimal.

Consider the applications of this theorem.

1. Let there be given the right-hand side of the equation (1.1.12) in the following form:

$$F_1 = f + A_1(\cdot) = f + \sum_{i=1}^{s} \delta(t - t_i) \otimes \varphi_i(x), \tag{21}$$

where $t, t_i \in [0,T]$, $\varphi_i(x) \in L_2(\Omega)$.

The control is

$$h_1 = \{(t_i, \varphi_i(x))\}_{i=1}^{s} \in U_{ad} \subset H_1 = [0,T]^s \times (L_2(\Omega))^s,$$

where the admissible set U_{ad} is bounded, closed and convex in H_1.

Suppose that $W^{+l}_{bd^+} \subset W_2^{1,0}(Q)$ $\left(W^{+l}_{bd} \subset W_2^{1,0}(Q)\right)$ and

$$\|\cdot\|_{W_2^{1,0}(Q)} \le c\|\cdot\|_{W^{+l}_{bd^+}} \quad \left(\|\cdot\|_{W_2^{1,0}(Q)} \le c\|\cdot\|_{W^{+l}_{bd}}\right). \tag{22}$$

We also assume that $f \in W^{-l}_{bd^+}$.

Set

$$F_\varepsilon(h) = f_\varepsilon + A_{1,\varepsilon}(h), \tag{23}$$

where

$$f_\varepsilon(t,x) = \int_{R^1} \omega_\varepsilon(\tau) f(t-\tau, x)\,d\tau,$$

$$A_{1,\varepsilon}(\bar{t}, x, h) = \sum_{i=1}^{s} a_{1,\varepsilon}(t,x,t_i)\varphi_i(x),$$

$$a_{1,\varepsilon}(t,x,t_i) = \begin{cases} \dfrac{1}{2\varepsilon}, t \in [t_i - \varepsilon, t_i + \varepsilon], \\ 0, t \in R \setminus [t_i - \varepsilon, t_i + \varepsilon]. \end{cases}$$

The regularized control is

$$h^\varepsilon{}_1 = \{(t_i, \varphi_i(x))\}_{i=1}^s \in U_{ad}^\varepsilon =$$

$$= [\varepsilon, T-\varepsilon]^s \times (\Phi_{ad})^s \subset H_1 = R^s \times (L_2(\Omega))^s.$$

It is not difficult to prove that

$$\|f_\varepsilon - f\|_{W_{bd^+}^{-l}} \xrightarrow[\varepsilon\to 0]{} 0.$$

Let us verify that maps $A_1(h)$, $A_{1,\varepsilon}(h)$ satisfy the conditions 2a)-2c) of Theorem 5, from whence it follows that maps $F_1(h)$, $F_{1,\varepsilon}(h)$ satisfy the same conditions 2a)-2c).

We shall test the condition 2a). Let h_{ε_k} be an arbitrary weakly convergent sequence. Then

$$t_i^{\varepsilon_k} \xrightarrow[k\to\infty]{} t_i^* \text{ in } R^1 \ \forall i = \overline{1,s},$$

$$\varphi_i^{\varepsilon_k}(x) \xrightarrow{W} \varphi_i^* \text{ in } L_2(\Omega) \ \forall i = \overline{1,s}.$$

Let $v(t,x)$ be an arbitrary smooth function in the domain $\overline{Q}$ that satisfies the conditions (bd^+). We have

$$\left\langle A_{1,\varepsilon_k}(h_{\varepsilon_k}),v\right\rangle_{W_{bd^+}}=\left(A_{1,\varepsilon_k}(h_{\varepsilon_k}),v\right)_{L_2(Q)}=$$

$$=\sum_{i=1}^{s}\int_{\Omega}\varphi_i^{\varepsilon_k}\int_{t_i^{\varepsilon_k}-\varepsilon_k}^{t_i^{\varepsilon_k}+\varepsilon_k}\frac{1}{2\varepsilon}v(\eta,x)d\eta d\Omega.$$

By the mean value theorem, we obtain

$$\left\langle A_{1,\varepsilon_k}(h_{\varepsilon_k}),v\right\rangle_{W_{bd^+}}=\sum_{i=1}^{s}\int_{\Omega}\varphi_i^{\varepsilon_k}v(t_i^{\varepsilon_k}+\theta_{ik}\varepsilon_k,x)d\Omega,$$

where $\theta_{ik}\in[-1,1]$.

Consider the term $\left|\left\langle A_{1,\varepsilon_k}(h_{\varepsilon_k})-A_1(h^*),v\right\rangle_{W_{bd^+}}\right|$.

$$\left|\left\langle A_{1,\varepsilon_k}(h_{\varepsilon_k})-A_1(h^*),v\right\rangle_{W_{bd^+}}\right|\le$$

$$\le\sum_{i=1}^{s}\left|\int_{\Omega}\varphi_i^{\varepsilon_k}\left(v\left(t_i^{\varepsilon_k}+\theta_{ik}\varepsilon_k,x\right)-v\left(t_i^{\varepsilon_k},x\right)\right)d\Omega\right|+$$

$$+\left|\int_{\Omega}\varphi_i^{\varepsilon_k}\left(v\left(t_i^{\varepsilon_k},x\right)-v\left(t_i^*,x\right)\right)d\Omega\right|+\left|\int_{\Omega}v\left(t_i^*,x\right)\left(\varphi_i^{\varepsilon_k}-\varphi_i^*\right)d\Omega\right|\le$$

$$\le\sum_{i=1}^{s}\left|\int_{\Omega}\varphi_i^{\varepsilon_k}\int_{t_i^{\varepsilon_k}}^{t_i^{\varepsilon_k}+\theta_{ik}\varepsilon_k}v_t(\eta,x)d\eta d\Omega\right|+\left|\int_{\Omega}\varphi_i^{\varepsilon_k}\int_{t_i^*}^{t_i^{\varepsilon_k}}v_t(\eta,x)d\eta d\Omega\right|+$$

$$+\left|\int_{\Omega}v\left(t_i^*,x\right)\left(\varphi_i^{\varepsilon_k}-\varphi_i^*\right)d\Omega\right|.$$

By the Schwarz inequality, we have:

$$\left|\left\langle A_{1,\varepsilon_k}(h_{\varepsilon_k})-A_1(h^*),v\right\rangle_{W_{bd^+}}\right|\le$$

$$\le\sum_{i=1}^{s}\left|\theta_{ik}\varepsilon_k\right|^{\frac{1}{2}}\left\|\varphi_i^{\varepsilon_k}\right\|_{L_2(\Omega)}\left\|v_t\right\|_{L_2(Q)}+$$

$$+\left|t_i^{\varepsilon_k}-t_i^*\right|^{\frac{1}{2}}\left\|\varphi_i^{\varepsilon_k}\right\|_{L_2(\Omega)}\left\|v_t\right\|_{L_2(Q)}+$$
$$+\left(\left(\varphi_i^{\varepsilon_k}-\varphi_i^*\right)(\cdot),v\left(t_i^*,\cdot\right)\right)_{L_2(\Omega)}.$$

Since,

$$t_i^{\varepsilon_k}\xrightarrow[k\to\infty]{}t_i^* \text{ in } R^1 \ \ \forall i=\overline{1,s},$$
$$\varphi_i^{\varepsilon_k}(x)\xrightarrow{W}\varphi_i^* \text{ in } L_2(\Omega) \ \ \forall i=\overline{1,s},$$
$$\varepsilon_k\xrightarrow[k\to\infty]{}0 \text{ in } R^1,$$

The expression $\left|\left\langle A_{1,\varepsilon_k}(h_{\varepsilon_k})-A_1(h^*),v\right\rangle_{W_{bd^+}}\right|$ vanishes for an arbitrary smooth in $\overline{Q}$ function $v(t,x)$. This set of functions is dense in $W_{bd^+}^{+l}$.

We shall show that $\left\|A_{1,\varepsilon_k}(h_{\varepsilon_k})\right\|_{W_{bd^+}^{-l}}<C$.

Let $v(t,x)$ be an arbitrary smooth in $\overline{Q}$ function. Considering the previous reasoning and the Schwarz inequality, we have

$$\left|\left\langle A_{1,\varepsilon_k}(h_{\varepsilon_k}),v\right\rangle_{W_{bd^+}}\right|=\left|\sum_{i=1}^{s}\int_{\Omega}\varphi_i^{\varepsilon_k}v(t_i^{\varepsilon_k}+\theta_{ik}\varepsilon_k,x)d\Omega\right|\le$$
$$\le\sum_{i=1}^{s}\left\|\varphi_i^{\varepsilon_k}\right\|_{L_2(\Omega)}\left\|v(t_i^{\varepsilon_k}+\theta_{ik}\varepsilon_k,x)\right\|_{L_2(\Omega)}.$$

Applying the inequality

$$\left(\int_{\Omega}v^2(t_i,x)d\Omega\right)^{\frac{1}{2}}\le\left(\int_{Q}v^2(t,x)+v_t^2(t,x)dQ\right)^{\frac{1}{2}}, \tag{24}$$

we obtain

$$\left|\left\langle A_{1,\varepsilon_k}(h_{\varepsilon_k}),v\right\rangle_{W_{bd^+}}\right|\le\sum_{i=1}^{s}\left\|\varphi_i^{\varepsilon_k}\right\|_{L_2(\Omega)}\left(\left\|v\right\|_{L_2(Q)}^2+\left\|v_t\right\|_{L_2(Q)}^2\right)^{1/2}.$$

It follows from the weak convergence of $\varphi_i^{\varepsilon_k}(x)$ that the sequence $\varphi_i^{\varepsilon_k}(x)$ is bounded. Taking into account relation (22), we arrive at the following inequality:

$$\left|\left\langle A_{1,\varepsilon_k}(h_{\varepsilon_k}),v\right\rangle_{W_{bd^+}}\right|\le\sum_{i=1}^{s}\left\|\varphi_i^{\varepsilon_k}\right\|_{L_2(\Omega)}\left(\|v\|^2_{L_2(Q)}+\|v_t\|^2_{L_2(Q)}\right)^{1/2}\le c\|v\|_{W^{+l}_{bd^+}}.$$

Thus,

$$\frac{\left|\left\langle A_{1,\varepsilon_k}(h_{\varepsilon_k}),v\right\rangle_{W_{bd^+}}\right|}{\|v\|_{W^{+l}_{bd^+}}}\le c,$$

for all smooth in $\overline{Q}$ functions $v(t,x)$.

Considering that this set of functions $v(t,x)$ is dense in $W^{+l}_{bd^+}$, we claim that

$$\left\|A_{1,\varepsilon_k}(h_{\varepsilon_k})\right\|_{W^{-l}_{bd^+}}=\sup_{v\in W^{+l}_{bd^+},v\neq 0}\frac{\left|\left\langle A_{1,\varepsilon_k}(h_{\varepsilon_k}),v\right\rangle_{W_{bd^+}}\right|}{\|v\|_{W^{+l}_{bd^+}}}\le c.$$

The weak convergence of a sequence of functionals follows from the pointwise convergence of a sequence of functionals in the dense set and boundedness of the sequence of norms of functional. We conclude that the condition 2a) of Theorem 5 is proved.

Reasoning similarly, we convince that map $A_{1,\varepsilon}(h)$ is weakly convergent and $\left\|A_{1,\varepsilon}(h)-A_1(h)\right\|_{W^{-l}_{bd^+}}\xrightarrow[\varepsilon\to 0]{}0$ for h.

Thus, we have proved the following theorem

Theorem 6. *Consider the optimal control problems (1.1.12) and (15) with the right-hand sides of (21), (23), respectively. If*

1) *the admissible set of control U_{ad} is convex, closed, and bounded in the Hilbert space H;*

2) the performance criterion $\Phi(\cdot)$ is upper semicontinuous and weakly lower semicontinuous,

then there exist optimal controls h^*, h^*_ε of the problems (1.1.12), (21) and (15), (23), there exists a weakly convergent subsequence $h^*_{\varepsilon_i}$ and an arbitrary weakly convergent subsequence $h^*_{\varepsilon_i}$ converges to h^{*-} weakly in H.

Remark. *The performance criterion*

$$J(h) = \Phi(u(h)) = \sum_{i=1}^{p} \int_Q \alpha_i(t,x)\left(u(h) - u_i\right)^2 dQ$$

satisfies the conditions of Theorem 6.

Consider the problem of the performance criterion differentiability in the regularized optimal control problems.

The existence of an optimal control of the regularized problems follows from the results of the previous section.

We shall prove that there exists a Fréchet derivative of a map $F_\varepsilon(h) = f_\varepsilon + A_{1,\varepsilon}(h) : H \to W^{-l}_{bd^+}$

$$\frac{\partial F_{1,\varepsilon}}{\partial t_i} = \frac{1}{2\varepsilon}\varphi_i \otimes \left(\delta(t - t_i - \varepsilon) - \delta(t - t_i + \varepsilon)\right), \tag{25a}$$

$$\frac{\partial F_{1,\varepsilon}}{\partial \varphi_i} = a_{1,\varepsilon}(t, x, t_i). \tag{25b}$$

To prove this formulas, consider

$$I = \left\langle F_{1,\varepsilon}(t,x,h+\Delta h) - F_{1,\varepsilon}(t,x,h), v\right\rangle_{W_{bd^+}} =$$

$$= \sum_{i=1}^{s} \int_\Omega (\varphi_i + \Delta\varphi_i) \int_{t_i+\Delta t_i-\varepsilon}^{t_i+\Delta t_i+\varepsilon} \frac{1}{2\varepsilon} v(\eta, x)\, d\eta - \varphi_i \int_{t_i-\varepsilon}^{t_i+\varepsilon} \frac{1}{2\varepsilon} v(\eta, x)\, d\eta\, d\Omega =$$

$$= \sum_{i=1}^{s} \int_{\Omega} \varphi_i \left(\int_{t_i+\Delta t_i-\varepsilon}^{t_i+\Delta t_i+\varepsilon} \frac{1}{2\varepsilon} v(\eta, x) d\eta - \int_{t_i-\varepsilon}^{t_i+\varepsilon} \frac{1}{2\varepsilon} v(\eta, x) d\eta \right) d\Omega +$$

$$+ \sum_{i=1}^{s} \int_{\Omega} \Delta\varphi_i \int_{t_i+\Delta t_i-\varepsilon}^{t_i+\Delta t_i+\varepsilon} \frac{1}{2\varepsilon} v(\eta, x) d\eta d\Omega.$$

If a value of Δt_i is sufficiently small, the first and second integrals have intersecting intervals of integration:

$$I = \left\langle F_{1,\varepsilon}(t,x,h+\Delta h) - F_{1,\varepsilon}(t,x,h), v \right\rangle_{W_{bd^+}} =$$

$$= \sum_{i=1}^{s} \int_{\Omega} \varphi_i \left(\int_{t_i+\varepsilon}^{t_i+\Delta t_i+\varepsilon} \frac{1}{2\varepsilon} v(\eta, x) d\eta - \int_{t_i-\varepsilon}^{t_i+\Delta t_i-\varepsilon} \frac{1}{2\varepsilon} v(\eta, x) d\eta \right) d\Omega + \quad (26)$$

$$+ \sum_{i=1}^{s} \int_{\Omega} \Delta\varphi_i \int_{t_i+\Delta t_i-\varepsilon}^{t_i+\Delta t_i+\varepsilon} \frac{1}{2\varepsilon} v(\eta, x) d\eta d\Omega = I_1 + I_3 + I_3.$$

Consider every summand of the right-hand side. By the mean value theorem, we have

$$I_1 = \sum_{i=1}^{s} \int_{\Omega} \varphi_i \int_{t_i+\varepsilon}^{t_i+\Delta t_i+\varepsilon} \frac{1}{2\varepsilon} v(\eta, x) d\eta d\Omega =$$

$$= \sum_{i=1}^{s} \int_{\Omega} \varphi_i \Delta t_i \frac{1}{2\varepsilon} v(t_i + \theta_{1,i} \Delta t_i + \varepsilon, x) d\Omega =$$

$$= \sum_{i=1}^{s} \int_{\Omega} \varphi_i \Delta t_i \frac{1}{2\varepsilon} \left(v(t_i + \theta_{1,i} \Delta t_i + \varepsilon, x) - v(t_i + \varepsilon, x) \right) d\Omega +$$

$$+ \sum_{i=1}^{s} \int_{\Omega} \varphi_i \Delta t_i \frac{1}{2\varepsilon} v(t_i + \varepsilon, x) d\Omega =$$

$$= \sum_{i=1}^{s} \int_{\Omega} \varphi_i \Delta t_i \frac{1}{2\varepsilon} \left(\int_{t_i+\varepsilon}^{t_i+\theta_{1,i}\Delta t_i+\varepsilon} v_t(\eta, x) d\eta \right) d\Omega +$$

$$+ \sum_{i=1}^{s} \left\langle \frac{1}{2\varepsilon} \delta(t - t_i - \varepsilon) \otimes \varphi_i, v \right\rangle_{W_{bd^+}} \cdot \Delta t_i,$$

where $\theta_{1,i} \in [0,1]$.

We can now easily show that an order of smallness of the first integral of the right-hand side equals to $\frac{2}{3}$. By the Schwarz inequality,

$$\left| \sum_{i=1}^{s} \int_{\Omega} \varphi_i \Delta t_i \frac{1}{2\varepsilon} \left(\int_{t_i+\varepsilon}^{t_i+\theta_{1,i}\Delta t_i+\varepsilon} v_t(\eta, x) d\eta \right) d\Omega \right| \le$$

$$\le \sum_{i=1}^{s} \|\varphi_i\|_{L_2(\Omega)} \left| \Delta t_i \frac{1}{2\varepsilon} \right| \left(\int_{\Omega} \left(\int_{t_i+\varepsilon}^{t_i+\theta_{1,i}\Delta t_i+\varepsilon} v_t(\eta, x) d\eta \right)^2 d\Omega \right)^{\frac{1}{2}} \le$$

$$\le \sum_{i=1}^{s} \|\varphi_i\|_{L_2(\Omega)} \left| \Delta t_i \frac{1}{2\varepsilon} \right| \left(\int_{\Omega} \left| \int_{t_i+\varepsilon}^{t_i+\theta_{1,i}\Delta t_i+\varepsilon} d\eta \cdot \int_{t_i+\varepsilon}^{t_i+\theta_{1,i}\Delta t_i+\varepsilon} v_t^2(\eta, x) d\eta \right| d\Omega \right)^{\frac{1}{2}} \le$$

$$\le \frac{1}{2\varepsilon} \sum_{i=1}^{s} \|\varphi_i\|_{L_2(\Omega)} |\Delta t_i|^{\frac{3}{2}} \|v\|_{L_2(Q)} \le c \|\Delta h\|_{H_1}^{\frac{3}{2}} \|v\|_{W_{bd^+}^{+l}} .$$

whence

$$I_1 = \sum_{i=1}^{s} \left\langle \frac{1}{2\varepsilon} \delta(t - t_i - \varepsilon) \otimes \varphi_i, v \right\rangle_{W_{bd^+}} \cdot \Delta t_i + c \|\Delta h\|_{H_1}^{\frac{3}{2}} \|v\|_{W_{bd^+}^{+l}} .$$

Taking the analogous transformation of the second integral of the right-hand side of (26), we have

$$I_2 = \sum_{i=1}^{s} \int_{\Omega} \varphi_i \int_{t_i-\varepsilon}^{t_i+\Delta t_i-\varepsilon} \frac{1}{2\varepsilon} v(\eta, x) d\eta d\Omega =$$

$$= \sum_{i=1}^{s} \left\langle \frac{1}{2\varepsilon} \delta(t - t_i + \varepsilon) \otimes \varphi_i, v \right\rangle_{W_{bd^+}} \cdot \Delta t_i + c \|\Delta h\|_{H_1}^{\frac{3}{2}} \|v\|_{W_{bd^+}^{+l}} .$$

Finally, consider the third integral of (26)

$$I_3 = \sum_{i=1}^{s} \int_\Omega \Delta\varphi_i \int_{t_i+\Delta t_i-\varepsilon}^{t_i+\Delta t_i+\varepsilon} \frac{1}{2\varepsilon} v(\eta, x) d\eta d\Omega =$$

$$= \sum_{i=1}^{s} \int_\Omega \Delta\varphi_i \int_{t_i+\Delta t_i-\varepsilon}^{t_i-\varepsilon} \frac{1}{2\varepsilon} v(\eta, x) d\eta d\Omega + \tag{27}$$

$$+ \sum_{i=1}^{s} \int_\Omega \Delta\varphi_i \int_{t_i-\varepsilon}^{t_i+\varepsilon} \frac{1}{2\varepsilon} v(\eta, x) d\eta d\Omega +$$

$$+ \sum_{i=1}^{s} \int_\Omega \Delta\varphi_i \int_{t_i+\varepsilon}^{t_i+\Delta t_i+\varepsilon} \frac{1}{2\varepsilon} v(\eta, x) d\eta d\Omega.$$

We shall prove that the first and third integrals have the second infinitesimal order of control. By the mean value theorem and the inequality (24), we obtain

$$\left| \sum_{i=1}^{s} \int_\Omega \Delta\varphi_i \int_{t_i+\Delta t_i-\varepsilon}^{t_i-\varepsilon} \frac{1}{2\varepsilon} v(\eta, x) d\eta d\Omega \right| \le$$

$$\le \left| \sum_{i=1}^{s} \int_\Omega \Delta\varphi_i \Delta t_i \frac{1}{2\varepsilon} v(t_i + \theta_i' \Delta t_i - \varepsilon, x) d\Omega \right| \le$$

$$\le \frac{1}{2\varepsilon} \sum_{i=1}^{s} |\Delta t_i| \cdot \|\Delta\varphi_i\|_{L_2(\Omega)} \|v(t_i + \theta_i' \Delta t_i - \varepsilon, x)\|_{L_2(\Omega)} \le$$

$$\le \frac{c}{2\varepsilon} \sum_{i=1}^{s} \|\Delta h\|_{H_1}^2 \left(\|v\|_{L_2(Q)}^2 + \|v_t\|_{L_2(Q)}^2 \right)^{1/2} \le \frac{c \|\Delta h\|_{H_1}^2 \|v\|_{W_{bd^+}^{+l}}}{2\varepsilon}$$

Reasoning similarly, it is easy to make sure that the third integral has the second infinitesimal order of control. Returning to the formula (26), we obtain

$$I=\left\langle F_{1,\varepsilon}(t,x,h+\Delta h)-F_{1,\varepsilon}(t,x,h),v\right\rangle_{W_{bd^+}}=$$

$$=\sum_{i=1}^{s}\left\langle\frac{1}{2\varepsilon}\varphi_i\otimes\left(\delta\left(t-t_i-\varepsilon\right)-\delta\left(t-t_i+\varepsilon\right)\right)\cdot\Delta t_i,v\right\rangle_{W_{bd^+}}+$$

$$+\sum_{i=1}^{s}\left\langle a_{1,\varepsilon}(t,x,t_i)\Delta\varphi_i,v\right\rangle_{W_{bd^+}}+c\left\|\Delta h\right\|_{H_1}^{\frac{3}{2}}\left\|v\right\|_{W_{bd^+}^{+l}}.$$

or

$$\left\|F_{1,\varepsilon}(h+\Delta h)-F_{1,\varepsilon}(h)-F'_{h,1,\varepsilon}(\Delta h)\right\|_{W_{bd^+}^{-l}}\le c\left\|\Delta h\right\|_{H_1}^{\frac{3}{2}}.$$

as required in (25).

Consider the property of smoothness of a Fréchet derivative $F_{h,1,\varepsilon}(\cdot)$. For this purpose we study the norm

$$\left\|F_{h',1,\varepsilon}(\cdot)-F_{h'',1,\varepsilon}(\cdot)\right\|=\sup_{h\in H_1,h\neq 0}\frac{\left\|F_{h',1,\varepsilon}(\Delta h)-F_{h'',1,\varepsilon}(\Delta h)\right\|_{W_{bd^+}^{-l}}}{\left\|\Delta h\right\|_{H_1}}=$$

$$=\sup_{h\in H_1,h\neq 0}\ \sup_{v\in W_{bd^+}^{+l},v\neq 0}\frac{\left|\left\langle F_{h',1,\varepsilon}(\Delta h)-F_{h'',1,\varepsilon}(\Delta h),v\right\rangle_{W_{bd^+}}\right|}{\left\|v\right\|_{W_{bd^+}^{+l}}\left\|\Delta h\right\|_{H_1}}. \tag{28}$$

Analyse the numerator

$$I=\left\langle F_{h',1,\varepsilon}(\Delta h)-F_{h'',1,\varepsilon}(\Delta h),v\right\rangle_{W_{bd^+}}=$$

$$=\sum_{i=1}^{s}\left\langle\frac{1}{2\varepsilon}\varphi'_i\otimes\left(\delta(t-t'_i-\varepsilon)-\delta(t-t'_i+\varepsilon)\right)\cdot\Delta t_i,v\right\rangle_{W_{bd^+}}-$$

$$-\sum_{i=1}^{s}\left\langle\frac{1}{2\varepsilon}\varphi''_i\otimes\left(\delta(t-t''_i-\varepsilon)-\delta(t-t''_i+\varepsilon)\right)\cdot\Delta t_i,v\right\rangle_{W_{bd^+}}+$$

$$+\sum_{i=1}^{s}\left\langle a_{1,\varepsilon}(t,x,t'_i)\Delta\varphi_i,v\right\rangle_{W_{bd^+}}-\sum_{i=1}^{s}\left\langle a_{1,\varepsilon}(t,x,t''_i)\Delta\varphi_i,v\right\rangle_{W_{bd^+}}.$$

Rearrange the summand in the following way:

$$I = \left\langle F_{h',1,\varepsilon}(\Delta h) - F_{h'',1,\varepsilon}(\Delta h), v \right\rangle_{W_{bd^+}} =$$

$$= \sum_{i=1}^{s} \left\langle \frac{1}{2\varepsilon}\left(\varphi_i' \otimes \delta\left(t - t_i' - \varepsilon\right) - \varphi_i'' \otimes \delta\left(t - t_i'' - \varepsilon\right)\right) \cdot \Delta t_i, v \right\rangle_{W_{bd^+}} - \tag{29}$$

$$- \sum_{i=1}^{s} \left\langle \frac{1}{2\varepsilon}\left(\varphi_i' \otimes \delta\left(t - t_i' + \varepsilon\right) - \varphi_i'' \otimes \delta\left(t - t_i'' + \varepsilon\right)\right) \cdot \Delta t_i, v \right\rangle_{W_{bd^+}} +$$

$$+ \sum_{i=1}^{s} \left\langle \left(a_{1,\varepsilon}\left(t,x,t_i'\right) - a_{1,\varepsilon}\left(t,x,t_i''\right)\right) \Delta\varphi_i, v \right\rangle_{W_{bd^+}} =$$

$$= I_1 + I_2 + I_3.$$

Since the set of smooth functions $v(t,x)$ is dense in $W_{bd^+}^{+l}$, consider only smooth function $v(t,x)$. Consider all summands in the right-hand side separately:

$$I_1 = \sum_{i=1}^{s} \left\langle \frac{1}{2\varepsilon}\left(\varphi_i' \otimes \delta\left(t - t_i' - \varepsilon\right) - \varphi_i'' \otimes \delta\left(t - t_i'' - \varepsilon\right)\right) \cdot \Delta t_i, v \right\rangle_{W_{bd^+}} =$$

$$= \sum_{i=1}^{s} \left\langle \frac{1}{2\varepsilon}\left(\left(\varphi_i' - \varphi_i''\right) \otimes \delta\left(t - t_i' - \varepsilon\right)\right) \cdot \Delta t_i, v \right\rangle_{W_{bd^+}} +$$

$$+ \sum_{i=1}^{s} \left\langle \frac{1}{2\varepsilon}\left(\varphi_i'' \otimes \left(\delta\left(t - t_i' - \varepsilon\right) - \delta\left(t - t_i'' - \varepsilon\right)\right)\right) \cdot \Delta t_i, v \right\rangle_{W_{bd^+}} =$$

$$= \sum_{i=1}^{s} (2\varepsilon)^{-1} \Delta t_i \int_{\Omega} v\left(t_i' + \varepsilon, x\right)\left(\varphi_i' - \varphi_i''\right) d\Omega +$$

$$+ \sum_{i=1}^{s} (2\varepsilon)^{-1} \Delta t_i \int_{\Omega} \varphi_i''\left(v\left(t_i' + \varepsilon, x\right) - v\left(t_i'' + \varepsilon, x\right)\right) d\Omega.$$

By the Schwarz inequality and the inequality (24), we obtain

$$|I_1| \le \left| \sum_{i=1}^{s} (2\varepsilon)^{-1} \Delta t_i \int_\Omega v(t_i' + \varepsilon, x)(\varphi_i' - \varphi_i'')\, d\Omega \right| +$$

$$+ \left| \sum_{i=1}^{s} (2\varepsilon)^{-1} \Delta t_i \int_\Omega \varphi_i'' \big(v(t_i' + \varepsilon, x) - v(t_i'' + \varepsilon, x)\big)\, d\Omega \right| \le$$

$$\le c \sum_{i=1}^{s} \left(\int_\Omega (\varphi_i' - \varphi_i'')^2\, d\Omega \right)^{\frac{1}{2}} \left(\int_\Omega v^2 (t_i' + \varepsilon, x)\, d\Omega \right)^{\frac{1}{2}} \|\Delta h\|_{H_1} +$$

$$+ c \left| \sum_{i=1}^{s} \Delta t_i \int_\Omega \varphi_i'' \int_{t_i''+\varepsilon}^{t_i'+\varepsilon} v_t(\eta, x)\, d\eta\, d\Omega \right|.$$

Next, we have

$$|I_1| \le c \sum_{i=1}^{s} \|\varphi_i' - \varphi_i''\|_{L_2(\Omega)} \left(\|v\|^2_{L_2(Q)} + \|v_t\|^2_{L_2(Q)} \right)^{1/2} \|\Delta h\|_{H_1} +$$

$$+ c \sum_{i=1}^{s} \|\Delta h\|_{H_1} \|\varphi_i''\|_{L_2(\Omega)} |t_i' - t_i''|^{\frac{1}{2}} \|v_t\|_{L_2(Q)} \le$$

$$\le c \sum_{i=1}^{s} \left(\|\varphi_i' - \varphi_i''\|_{L_2(\Omega)} + \|\varphi_i''\|_{L_2(\Omega)} |t_i' - t_i''|^{\frac{1}{2}} \right) \|\Delta h\|_{H_1} \|v\|_{W^{+l}_{bd^+}} \le$$

$$\le c \|\Delta h\|_{H_1} \|v\|_{W^{+l}_{bd^+}} \|h' - h''\|^{\frac{1}{2}}_{H_1}.$$

Analogously, consider the second summand of (29)

$$|I_2| \le c \|\Delta h\|_{H_1} \|v\|_{W^{+l}_{bd^+}} \|h' - h''\|^{\frac{1}{2}}_{H_1}.$$

Examine the third summand of (29)

$$I_3 = \sum_{i=1}^{s} \left\langle \left(a_{1,\varepsilon}(t, x, t_i') - a_{1,\varepsilon}(t, x, t_i'')\right) \Delta\varphi_i, v \right\rangle_{W_{bd^+}} =$$

$$= \sum_{i=1}^{s} \int_\Omega \Delta\varphi_i \left(\int_{t_i'-\varepsilon}^{t_i''-\varepsilon} v(\eta, x)\, d\eta - \int_{t_i'+\varepsilon}^{t_i''+\varepsilon} v(\eta, x)\, d\eta \right) d\Omega.$$

By the Schwarz inequality,

$$|I_3| = \left| \sum_{i=1}^{s} \int_{\Omega} \Delta\varphi_i \left(\int_{t_i'-\varepsilon}^{t_i''-\varepsilon} v(\eta, x) d\eta - \int_{t_i'+\varepsilon}^{t_i''+\varepsilon} v(\eta, x)\, d\eta \right) d\Omega \right| \le$$

$$\le \sum_{i=1}^{s} \int_{\Omega} |\Delta\varphi_i| \cdot |t_i'' - t_i'|^{\frac{1}{2}} \times$$

$$\times \left(\left(\int_{t_i'-\varepsilon}^{t_i''-\varepsilon} v^2 (\eta, x) d\eta \right)^{\frac{1}{2}} + \left(\int_{t_i'+\varepsilon}^{t_i''+\varepsilon} v^2 (\eta, x) d\eta \right)^{\frac{1}{2}} \right) d\Omega \le$$

$$\le c \sum_{i=1}^{s} |t_i'' - t_i'|^{\frac{1}{2}} \|\Delta\varphi_i\|_{L_2(\Omega)} \|v\|_{L_2(Q)} \le c \|h'' - h'\|_{H_1}^{\frac{1}{2}} \|\Delta h\|_{H_1} \|v\|_{W_{bd^+}^{+1}}$$

Returning to (29), we obtain

$$|I| = \left| \langle F_{h',1,\varepsilon}(\Delta h) - F_{h'',1,\varepsilon}(\Delta h), v \rangle_{W_{bd^+}} \right| \le c \|\Delta h\|_{H_1} \|h'' - h'\|_{H_1}^{\frac{1}{2}} \|v\|_{W_{bd^+}^{+1}},$$

so that the equality (28) can be rewritten in the form

$$\|F_{h',1,\varepsilon}(\cdot) - F_{h'',1,\varepsilon}(\cdot)\| \le c \|h'' - h'\|_{H_1}^{\frac{1}{2}}.$$

To put it in another way, the derivative $F_{h,1,\varepsilon}(\cdot)$ satisfies the Lipschitz condition with index $\alpha = \frac{1}{2}$. By Theorem 1.5, we assert that gradient of regularized performance criterion satisfies Lipschitz condition with index $\alpha = \frac{1}{2}$ also. When performance criterion gradient is analysed directly (without Theorem 1.5), it should be seen that the performance criterion directional gradient with correspondence to direction $\varphi(x)$ satisfies the Lipschitz condition with index $\alpha = 1$.

Analogously, we could study the other right-hand sides $A_i(t, x, h)$ ($i = \overline{2,5}$) of the equation (1.1.12).

Let us show the specific cases of the regularization of the right-hand side.

2. $F_2 = f + A_2(t,x,h) = \sum_{i=1}^{s} \delta^{(1)}(t-t_i) \otimes \varphi_i(x)$

be the right-hand side of the equation, and

$$A_{2,\varepsilon}(t,x,h) = \sum_{i=1}^{s} a_{2\varepsilon}(t,x,t_i)\varphi_i(x),$$

where

$$a_{2,\varepsilon}(t,x,t_i) = \begin{cases} -\dfrac{1}{4\varepsilon^3}, \ t \in I_1 = \left[t_i + \varepsilon - \varepsilon^2, t_i + \varepsilon + \varepsilon^2\right], \\ \dfrac{1}{4\varepsilon^3}, \ \ t \in I_2 = \left[t_i - \varepsilon - \varepsilon^2, t_i - \varepsilon + \varepsilon^2\right], \\ 0, \ t \in R \setminus (I_1 \cup I_2). \end{cases}$$

Assume that $W^{+l}_{bd^+} \subset W_2^{2,0}(Q)$ $\left(W^{+l}_{bd} \subset W_2^{2,0}(Q)\right)$ and

$$\|\cdot\|_{W_2^{2,0}(Q)} \leq c\|\cdot\|_{W^{+l}_{bd^+}} \ \left(\|\cdot\|_{W_2^{2,0}(Q)} \leq c\|\cdot\|_{W^{+l}_{bd}}\right).$$

3. $A_3(t,x,h) = \sum_{i=1}^{s} \delta(x_1 - x_{1,i}) \otimes \varphi_i(t, x_2, \ldots, x_n)$

$$A_{3,\varepsilon}(t,x,h) = \sum_{i=1}^{s} a_{3,\varepsilon}(t,x,x_{1,i})\varphi_i(t,x_2,\ldots,x_n),$$

where

$$a_{3,\varepsilon}(t,x,x_{1,i}) = \begin{cases} \dfrac{1}{2\varepsilon}, \ x_1 \in I_1 = \left[x_{1,i} - \varepsilon, x_{1,i} + \varepsilon\right], \\ 0, \ t \in R \setminus I_1. \end{cases}$$

Assume that $W^{+l}_{bd^+} \subset W_2^{1,0,0}(Q)$ $\left(W^{+l}_{bd} \subset W_2^{2,0}(Q)\right)$ and

$$\|\cdot\|_{W_2^{0,1,0}(Q)} \leq c\|\cdot\|_{W^{+l}_{bd^+}} \ \left(\|\cdot\|_{W_2^{0,1,0}(Q)} \leq c\|\cdot\|_{W^{+l}_{bd}}\right).$$

4. $A_4(t,x,h) = \sum_{i=1}^{s}\sum_{j=1}^{p} \delta(t-t_i) \otimes \delta(x_1 - x_{1,j}) \otimes \varphi_{ij}(x_2,\ldots,x_n),$

$$A_{4,\varepsilon}(t,x,h)=\sum_{i=1}^{s}\sum_{j=1}^{p}a_{4,\varepsilon}(t,x,t_i,x_{1,j})\varphi_{ij}(x_2,\dots,x_n),$$

where

$$a_{4,\varepsilon}(t,x,t_i,x_{1,j})=\begin{cases}\dfrac{1}{4\varepsilon^2}, & x_1\in I_1=[x_{1,i}-\varepsilon,x_{1,i}+\varepsilon],\\ & t\in I_2=[t_i-\varepsilon,t_i+\varepsilon],\\ 0, & (x_1;t)\in R^2\setminus(I_1\times I_2).\end{cases}$$

Assume that $W^{+l}_{bd^+}\subset W_2^{1,1,0}(Q)\cdot\left(W^{+l}_{bd}\subset W_2^{1,1,0}(Q)\right)$ and

$$\|\cdot\|_{W_2^{1,1,0}(Q)}\le c\|\cdot\|_{W^{+l}_{bd^+}}\quad\left(\|\cdot\|_{W_2^{1,1,0}(Q)}\le c\|\cdot\|_{W^{+l}_{bd}}\right).$$

5. $A_5(t,x,h)=\sum_{i=1}^{s}\delta(x_1-a_i(t))\otimes\varphi_i(t,x_2,\dots,x_n)$

$$A_{5,\varepsilon}(t,x,h)=\sum_{i=1}^{s}a_{5,\varepsilon}(t,x,a_i(t))\varphi_i(t,x_2,\dots,x_n),$$

where

$$a_{5,\varepsilon}(t,x,a_i(t))=\begin{cases}\dfrac{1}{2\varepsilon}, & x_1\in I_1=[a_i(t)-\varepsilon,a_i(t)+\varepsilon]\ ,\\ 0, & \text{otherwise.}\end{cases}$$

Assume that $W^{+l}_{bd^+}\subset W_2^{0,1,0}(Q)\ \left(W^{+l}_{bd}\subset W_2^{0,1,0}(Q)\right)$ and

$$\|\cdot\|_{W_2^{0,1,0}(Q)}\le c\|\cdot\|_{W^{+l}_{bd^+}}\quad\left(\|\cdot\|_{W_2^{0,1,0}(Q)}\le c\|\cdot\|_{W^{+l}_{bd}}\right).$$

Chapter 3

NUMERICAL METHODS OF OPTIMIZATION OF LINEAR SYSTEMS WITH GENERALIZED CONTROL

1. PARAMETRIZATION OF CONTROL

In this section applying procedure of parametrization, the minimization problem of the performance criterion $J(h)$ in an infinite-dimensional space we replace with the corresponding minimization problem in a finite-dimensional space. This substitution allows us to decrease the computational complexity of the implementation of the gradient methods.

Let the system state be a solution of the following initial boundary value problem

$$Lu = \sum_{i=1}^{N} \delta(t - t_i) \otimes \varphi_i^n(x), \tag{1}$$

where $\varphi_i^n(x) = \sum_{k=1}^{n} c_{ik}\omega_k, \{\omega_k(x)\}_{k=\overline{1,\infty}}$ is an orthonormalized basis in $L_2(\Omega)$, numbers $\{c_{ik}\}_{k=\overline{1,n}}$ are such that $\varphi_i^n(x) \in H_{ad}$, and H_{ad} is closed, convex, and bounded set in $L_2(\Omega)$.

Let $t^* = \{t_i\}_{i=\overline{1,N}} \in R^N[0,T]$, and c be a $N \times n$ matrix with rows $c_i^n = (c_{i1},..,c_{in})$. We denote by $H = R^N \times (R^n)^N$ the space of control. It is required to find a minimum of the following functional

$$J(t^*, c) = \int_Q [u(t, x; t^*, c) - z_{ad}]^2 dQ \tag{2}$$

in the admissible set of control $U_{ad} = R^N[0,T] \times \left(R^n_{ad}\right)^N$, where $\left(R^n_{ad}\right)^N$ is closed, convex and bounded set in $\left(R^n\right)^N$.

Theorem 1. *There exists an optimal control* $\left(\tilde{t}^*_n, \tilde{c}_n\right) \in U_{ad}$, $J\left(\tilde{t}^*_n, \tilde{c}_n\right) \le J\left(t^*, c\right), \forall \left(t^*, c\right) \in U_{ad}$, *where* $J\left(t^*, c\right)$ *is defined in (2).*

Proof. Let $\left(t^*_k, c_k\right)$ be a minimizing sequence. Since U_{ad} is bounded, convex, closed set, the set U_{ad} is weakly compact. Thus there exists a weakly convergence subsequence (it is denoted by $\left(t^*_k, c_k\right)$ again). Since H is finite-dimensional space, the subsequence $\left(t^*_k, c_k\right)$ converges to $\left(\tilde{t}^*_n, \tilde{c}_n\right) \in U_{ad}$ strongly in H, whence for any number $1 \le i \le N$ we have $\varphi^n_{ik} \xrightarrow[k \to \infty]{} \tilde{\varphi}^n_i = \sum_{p=1}^{n} \tilde{c}_{ip} \omega_p$ in $L_2(\Omega)$. Similarly to the previous sections we can prove that $u_k \xrightarrow[k \to \infty]{} u$ in $L_2(Q)$. The fact that a norm is a continuous functional in a Hilbert space implies that functional (2) reaches its infimum in the admissible set U_{ad} (the space H is finite-dimensional).

Note that there exists a subsequence $\left(\tilde{t}^*_{n_l}, \tilde{c}_{n_l}\right)$ such that $\tilde{t}^*_{n_l} \xrightarrow[l \to \infty]{} \tilde{t}^* \in R^N[0,T]$ strongly in R^N, $\tilde{\varphi}^{n_l}_i \xrightarrow[l \to \infty]{} \tilde{\varphi}^*_i \in L_2(\Omega)$ weakly in $L_2(\Omega), i = \overline{1,N}$, where element $\left(\tilde{t}^*, \tilde{\varphi}^*\right)$ is a solution of the initial optimization problem and $\tilde{\varphi}^* = \left\{\tilde{\varphi}^*_i\right\}_{i=\overline{1,N}}$.

Theorem 2. *If* $W^{+l}_{bd^+} \subset C^1\left([0,T]; L_2(\Omega)\right)$ *(imbedding is continuous), then performance criterion* $J\left(t^*, c\right)$ *is differentiable in the space* $R^N \times \left(R^n\right)^N$ *and its gradient is of the following form*

$$grad\, J(t^*, c) =$$

$$= \left(\left(\int_\Omega v_t'(t_i, x) \sum_{p=1}^{n} c_{ik} \omega_p(x) d\Omega \right)_{i=\overline{1,N}}, \left(\int_\Omega v(t_i, x) \omega_k(x) d\Omega \right)_{i=\overline{1,N}, k=\overline{1,n}} \right),$$

where $v(t,x)$ *is a solution of the adjoint problem with the right-hand side* $2(u - z_{ad}^-)$.

P r o o f. Let (t^*, c) and $(t^* + \lambda \Delta t^*, c + \lambda \Delta c)$ are arbitrary elements from the admissible set U_{ad} and functions $u(t^*, c)$, $u(t^* + \lambda \Delta t^*, c + \lambda \Delta c)$ are corresponding states of the system (1). Denote an increment of the solution by $\Delta u = u(t^* + \lambda \Delta t^*, c + \lambda \Delta c) - u(t^*, c)$. Then an increment of the performance criterion (2) is of the following form

$$\Delta J = J(t^* + \lambda \Delta t^*, c + \lambda \Delta c) - J(t^*, c) =$$

$$= \int_Q \left[u(t^* + \lambda \Delta t^*, c + \lambda \Delta c) - z_{ad} \right]^2 - \left[u(t^*, c) - z_{ad} \right]^2 dQ = \quad (3)$$

$$= 2 \int_Q \Delta u \left[u(t^*, c) - z_{ad} \right] dQ + \int_Q |\Delta u|^2 dQ .$$

Define the adjoint state as a solution of the following equation

$$L^* v = 2\left(u(t^*, c) - z_{ad} \right).$$

It follows from Section 1 that there exists the unique solution $v \in W_{bd^+}^{+l}$ of this equation. Therefore, the equation (3) can be rewritten in the form

$$\Delta J = \left(\Delta u, L^* v \right)_{L_2(Q)} + \left\| \Delta u \right\|_{L_2(Q)}^2 .$$

It is obvious that the increment Δu is a solution of the equation

$$L \Delta u = \sum_{i=1}^{N} \delta(t - t_i - \lambda \Delta t_i) \otimes \sum_{k=1}^{n} (c_{ik} + \lambda \Delta c_{ik}) \omega_k -$$

$$- \sum_{i=1}^{N} \delta(t - t_i) \otimes \sum_{k=1}^{n} c_{ik} \omega_k .$$

It follows from Section 1 that there exists the unique solution $\Delta u \in L_2(Q)$ of this equation such that

$$\left(\Delta u, L^* y\right)_{L_2(Q)} = \sum_{i=1}^{N}\left[\left(y(t_i + \lambda\Delta t_i, x) - y(t_i, x), \sum_{k=1}^{n} c_{ik}\omega_k\right)_{L_2(\Omega)} + \right.$$

$$\left. + \lambda\left(y(t_i + \lambda\Delta t_i, x), \sum_{k=1}^{n} \Delta c_{ik}\omega_k\right)_{L_2(\Omega)}\right].$$

for any $y(t,x) \in W_{bd^+}^{+l} : L^* y \in L_2(Q)$.

Let $y = v$. Then using the previous equation, we have

$$\Delta J = \sum_{i=1}^{N}\left[\left(v(t_i + \lambda\Delta t_i, x) - v(t_i, x), \sum_{k=1}^{n} c_{ik}\omega_k\right)_{L_2(\Omega)} + \right.$$

$$\left. + \lambda\left(v(t_i + \lambda\Delta t_i, x), \sum_{k=1}^{n} \Delta c_{ik}\omega_k\right)_{L_2(\Omega)}\right] + \|\Delta u\|_{L_2(Q)}^2 . \tag{4}$$

The increment $v(t_i + \lambda\Delta t_i, x) - v(t_i, x)$ can be represented in the following form

$$v(t_i + \lambda\Delta t_i, x) - v(t_i, x) = \int_{t_i}^{t_i + \lambda\Delta t_i} v_t(t,x)\, dt .$$

Substituting this equality in (4), we obtain

$$\Delta J = \sum_{i=1}^{N}\left(\frac{1}{\Delta t_i}\int_{t_i}^{t_i+\lambda\Delta t_i} v_t(t,x)\; dt, \sum_{k=1}^{n} c_{ik}\omega_k\right)_{L_2(\Omega)} \lambda\Delta t_i +$$

$$+\lambda\sum_{i=1}^{N}\left(v(t_i, x), \sum_{k=1}^{n}\Delta c_{ik}\omega_k\right)_{L_2(\Omega)} +$$

$$+\lambda\sum_{i=1}^{N}\left(\int_{t_i}^{t_i+\lambda\Delta t_i} v_t(t,x)\; dt, \sum_{k=1}^{n}\Delta c_{ik}\omega_k\right)_{L_2(\Omega)} + \|\Delta u\|_{L_2(Q)}^2 .$$

To prove the theorem, it is sufficient to show that

$$\lim_{\lambda\to 0}\frac{\Delta J}{\lambda}=(J'(h),\Delta h)_H=\sum_{i=1}^{N}\left(v_t(t_i,x),\sum_{k=1}^{n}c_{ik}\omega_k\right)_{L_2(\Omega)}\Delta t_i+ \\ +\sum_{i=1}^{N}\sum_{k=1}^{n}(v(t_i,x),\omega_k)_{L_2(\Omega)}\Delta c_{ik}, \tag{5}$$

where $h=(t^*,c)$, $(\cdot,\cdot)_H$ is a inner product in H.

For this purpose we shall show that $\lambda^{-1}\|\Delta u\|^2_{L_2(Q)}\xrightarrow[\lambda\to 0]{}0$. By the inequality $\|u\|_{L_2(Q)}\le c\|F\|_{W^{-l}_{bd^+}}$, we obtain

$$\|\Delta u\|_{L_2(Q)}\le c\left\|\sum_{i=1}^{N}\delta(t-t_i-\lambda\Delta t_i)\otimes\sum_{k=1}^{n}(c_{ik}+\lambda\Delta c_{ik})\omega_k-\right. \\ \left.-\sum_{i=1}^{N}\delta(t-t_i)\otimes\sum_{k=1}^{n}c_{ik}\omega_k\right\|_{W^{-l}_{bd^+}}\le$$

$$\le\tilde{n}\sum_{i=1}^{N}\left(\left\|(\delta(t-t_i-\lambda\Delta t_i)-\delta(t-t_i))\otimes\sum_{k=1}^{n}c_{ik}\omega_k\right\|_{W^{-l}_{bd^+}}+\right. \\ \left.+\lambda\left\|\delta(t-t_i-\lambda\Delta t_i)\otimes\sum_{k=1}^{n}\Delta c_{ik}\omega_k\right\|_{W^{-l}_{bd^+}}\right).$$

By the definition of the norm in the space $W^{+l}_{bd^+}$ and continuity of the imbedding $W^{+l}_{bd^+}\subset C^1([0,T];L_2(\Omega))$, we conclude

$$\left\|(\delta(t-t_i-\lambda\Delta t_i)-\delta(t-t_i))\otimes\sum_{k=1}^{n}c_{ik}\omega_k\right\|_{W^{-l}_{bd^+}}=$$

$$=\sup_{y\in W^+_{bd^+},\,y\neq 0}\frac{\left(y(t_i+\lambda\Delta t_i,x)-y(t_i,x),\sum_{k=1}^{n}c_{ik}\omega_k\right)_{L_2(\Omega)}}{\|y\|_{W^+_{bd^+}}}\le$$

$$\le \lambda \Delta t_i \left\| \sum_{k=1}^{n} c_{ik} \omega_k \right\|_{L_2(\Omega)} \sup_{y \in W^+_{bd^+}, y \ne 0} \frac{\|y\|_{C^1([0,T]; L_2(\Omega))}}{\|y\|_{W^+_{bd^+}}} \le \lambda \Delta t_i C \ .$$

This proves that $0 \le \lambda^{-1} \|\Delta u\|^2_{L_2(Q)} \le C\lambda \xrightarrow[\lambda \to 0]{} 0$, as required.

It follows from the inequality (5) that the performance criterion has a Gâteaux derivative in $H = R^N \times \left(R^n\right)^N$ and its partial derivative is of the following form

$$\frac{\partial J}{\partial t_i} = \int_{\Omega} v_t'(t_i, x) \sum_{k=1}^{n} c_{ik} \omega_k(x) d\Omega \ , \ i = \overline{1, N} \ , \tag{6}$$

$$\frac{\partial J}{\partial c_{ik}} = \int_{\Omega} v(t_i, x) \omega_k(x) d\Omega \ , \ i = \overline{1, N} \ , k = \overline{1, n} \ . \tag{7}$$

Remark 1. *In the case of parametrization of other right-hand sides (A_i, $i = \overline{2,5}$) of the state equation the analogous theorems hold true.*

2. PULSE OPTIMIZATION PROBLEM

Earlier the pulse and point-pulse optimization problems were studied for some distributed systems. It was shown that a priori estimates in the negative norms allows us to prove the existence of the optimal controls of the investigated systems, to study the questions of its controllability, and to write out explicitly the gradient of the performance criterion of the original or regularized problems. In this section gradient methods are proposed for solving the problems of optimal control [95-100], which use an approximation of the performance criterion at every iteration [82, 101].

This approach is based on idea of solving the limit extremum problems [22, 102, 103].

At first, consider an optimization problem with pulse impact. Suppose that system state satisfies the following linear equation

$$Lu = f + A(h), \tag{1}$$

where L maps $D(L) \subseteq L_2(Q)$ into $R(L) \subseteq W^{-l}_{bd^+}$.

In our case the right-hand size of the equation (1) is

$$A(h) = \sum_{j=1}^{s} \ddot{a}(t - t_j) \otimes g_j(x), s \in N,$$

where $\delta(\cdot)$ is the Dirac delta-function. It is requested to find time moments of pulse impacts $h = (t_1, \ldots, t_s)$ on an admissible set $U_{ad} \subset R^s$, which minimize the following performance criterion

$$J(h) = \int_Q |u(h) - z_{ad}|^2 dQ,$$

where z_{ad} is a known element from $L_2(Q)$, U_{ad} is closed and bounded set from R^s.

Suppose that a priori estimates in the negative norm are valid:

$$\begin{aligned} \|u\|_{H^+_{bd}} &\leq C_1 \|Lu\|_{W^{-l}_{bd^+}} \leq C_2 \|u\|_{W^{+l}_{bd}}, \\ \|v\|_{H^+_{bd^+}} &\leq C_1 \|L^* v\|_{W^{-l}_{bd}} \leq C_2 \|u\|_{W^{+l}_{bd^+}}, \end{aligned} \tag{2}$$

where L^* is formally adjoint operator.

As it has been shown in the previous chapters, estimates (2) enable us to prove the existence and uniqueness theorem for the generalized solution of original and adjoint problems and the existence theorem for the optimal control h^*. If in addition the following imbedding

$$W^{+l}_{bd^+} \subset C^1([0,T]; L_2(\Omega))$$

is valid, then we can find the performance criterion gradient explicitly. Otherwise, we regularize the original problem. There exist optimal controls of regularized problems and these controls converge to the optimal controls of the original optimization problem as the parameter of regularization vanishes. The regularized problem is of the following form

$$h^*_{\text{ả}} = \arg\min_{h\in U_{ad}} J_{\text{ả}}(h), \tag{3}$$

$$J_\varepsilon(h) = \int_Q |u_\varepsilon(h) - z_{ad}|^2 dQ, \tag{4}$$

$$Lu_{\text{ả}} = F_{\text{ả}}(t,x;h),\ u_{\text{ả}} \in W^{+l}_{bd}, \tag{5}$$

$$F_{\text{ả}} = f + \int_{R^1} \text{ù}(\hat{\imath})A(t-\hat{\imath},x)d\hat{\imath}\ , \tag{6}$$

where $\omega_\varepsilon(t)\in C_0^\infty(R^1)$, $\varepsilon>0$, $\omega_\varepsilon(t)=0$ when $|t|\geq\varepsilon$, $\omega_\varepsilon(t)\geq 0$, $\int_{R^1}\omega_\varepsilon(\xi)d\xi = 1$.

By the integral (6) we mean the bilinear form in the sense of the theory of distributions. Hereinafter we shall omit ε in the regularized problem.

In order to find the optimal control we may use different gradient methods [104, 105]. It follows from the equation for gradient that to find the gradient $J'(h)$ it is necessary to solve direct and adjoint problems and to differentiate and integrate some expressions. As a rule, these procedures are implemented by numerical methods thus we have only a uniform convergent sequence of approximations $J_s'(h), s = 0,1,\ldots$ $J_s'(h)\to J'(h), s\to\infty$ instead of the exact value of $J'(h)$. Uniform convergence follows from the a priori inequalities in the negative norms, where the constants do not depend on h. The inequalities (2) enable us to consider the minimization problem for some functional, which is equivalent to solving the initial boundary value problem [62, 64, 106]. In the following sections we shall also discuss a projective numerical method for solving the initial boundary value problem.

We shall build the procedures of minimization of gradient that use the approximation $J_s'(h)$: $J_s'(h)\to J'(h)$, $s\to\infty$ on every iteration.

Note that the similar results were obtained for the problem of non-linear programming in [24, 25]. In these papers the convex functional $J(h)$ was approximated by the convex functionals $J_s(h)$, then the derivatives $J_s'(h)$ were used in the numerical method. In our case the original functional and its derivative are unknown, in addition they are non-convex.

3. ANALOGUE OF THE GRADIENT PROJECTION METHOD

In this section we shall suppose that the admissible set U_{ad} is a convex compact in R^s. We shall use the sufficient conditions for the convergence of non-linear programming algorithms [21, 107].

Lemma 1. *Suppose that a sequence of points satisfies the following conditions:*

1. $h^s \in K$ is a compact set.

2. For an arbitrary convergent subsequence $\{h^{s_k}\}_{k=\overline{0,\infty}}$ the following assumptions holds true:

a) if $\lim\limits_{k\to\infty} h^{s_k} = h^* \in \Phi^*$ then $\left\|h^{s_k+1} - h^{s_k}\right\| \xrightarrow[k\to\infty]{} 0$;

b) if $\lim\limits_{k\to\infty} h^{s_k} = h' \notin \Phi^*$ then there exists $\varepsilon_0 > 0$ such that

$$\tau_k = \min_{s>s_k}\left\{s : \left\|h^s - h^{s_k}\right\| > \varepsilon\right\} < \infty$$

for any $\varepsilon : 0 < \varepsilon \le \varepsilon_0$.

3. There exists a continuous function $W(h)$ so that the set of its values in Φ^* is at most denumerable and $W(h)$ satisfies the following inequality

$$\overline{\lim_{k\to\infty}} W\left(h^{\tau_k}\right) < \lim_{k\to\infty} W\left(h^{s_k}\right).$$

Then the sequence $W(h^s)$ converges and all the limit points of the sequence $\{h^s\}_{s=\overline{0,\infty}}$ belong to Φ^*.

Suppose that $h^0 \in U_{ad}$. Consider the following sequence of points

$$h^{s+1} = \eth_{U_{ad}}\left(h_s - \tilde{n}_s J'_s\left(h^s\right)\right),\ s = 0,1,..., \qquad (7)$$

where $J'_s(h)$ is an approximation of the performance criterion $J(h)$, $\pi_{U_{ad}}(\cdot)$ is a projective operator

$$\eth_{U_{ad}}(h) \in U_{ad};\ \left\|h - \eth_{U_{ad}}(h)\right\| \le \|ø - h\|, \forall ø \in U_{ad},$$

$\tilde{n}_s$ is a step of the algorithm.

Note that if U_{ad} is a convex polyhedral set then calculation of a value of $\pi_{U_{ad}}(\cdot)$ comes to a quadratic programming problem:

$$\eth_{U_{ad}}(h) = \arg\min_{ø \in U_{ad}} \|h - ø\|^2.$$

In the case when $U_{ad} = \left\{h : \sum_{i=1}^{n} c_i h_i \le b\right\}$ there exist algorithms which compute the value of the projective operator $\pi_{U_{ad}}(h)$ in a finite number of operations.

Theorem 1. *Suppose that the sequence $J'_s(h)$ converges to $J'(h)$ uniformly in the set U_{ad}. The set of values of the function $J(h)$ in the set*

$$\Phi^* = \left\{h^* \in U_{ad} \middle| \min_{h \in U_{ad}} \left(J'\left(h^*\right), h - h^*\right) = 0\right\},$$

is at most denumerable, and

$$\sum_{s=0}^{\infty} \tilde{n}_s = \infty,\ \tilde{n}_s \xrightarrow[s \to \infty]{} 0,\ \tilde{n}_s > 0,\ s = 0,1,...$$

Then the limit of an arbitrary convergent subsequence of (7) belongs to the set Φ^*.

Note that for the convex functions the set Φ^* is the set of minimum points of the function $J(h)$. In the other cases the set Φ^* contains all minimum points of the function $J(h)$. The function $J(h)$ that is under consideration can be non-convex because of non-linear state function dependence on the control h.

To prove the theorem we shall apply Lemma 1. It is obvious that the conditions 1 and 2a of Lemma 1 are valid. Ensure that the other conditions of the lemma are true also.

Suppose that there exists a subsequence

$$h^{s_k} \to h' \notin \Phi^*,\ k \to \infty.$$

Choose a sufficiently small $\varepsilon > 0$ such that a $4\mathring{a}$-neighbourhood of the point h' $\left(U_{4\varepsilon}(h')\right)$ and the set Φ^* are mutually disjoint sets. By the definition of the set Φ^*, there exists $\gamma > 0$ such that

$$\min_{h \in U_{ad}} \left(J'(h'), h - h'\right) \le -2\gamma < 0.$$

Granting that $J(h) \in C^1[0,T]$, for a sufficiently large k we have

$$\min_{h \in U_{ad}} \left(J'\left(h^{s_k}\right), h - h^{s_k}\right) \le -\gamma < 0.$$

Therefore, there exists $\beta > 0$ such that

$$\left(J'\left(h^{s_k}\right), \pi_{U_{ad}}\left(h^{s_k} - \rho^{s_k} J'_{s_k}\left(h^{s_k}\right)\right) - h^{s_k}\right) \le -2\beta\rho_{s_k} < 0. \tag{8}$$

Let us show that for all $k > N$ we have $\tau_k < \infty$, where the sequence τ_k was defined in Lemma 1. Suppose the contrary. Then $h^s \in U_\varepsilon\left(h^{s_k}\right)$ for all $s > s_k$, thus $h' \in U_\varepsilon\left(h_{s_k}\right)$ and $h^s \in U_{2\varepsilon}(h')$.

Define a function $W(h)$ in the following way: $W(h) = J(h)$. It is easy to see that the function $W(h)$ satisfies all conditions of Lemma 1. Thus for all $s > s_k$ we have

$$W(h^s)-W(h^{s_k})=(J'(h^{s_k}),h^s-h^{s_k})+o(\varepsilon)=$$
$$=\sum_{i=s_k}^{s-1}(J'(h^{s_k}),h^{i+1}-h^i)+o(\varepsilon). \tag{9}$$

From continuity of the function $J'(h)$, the inequality (8) and $h^i \in U_\varepsilon(h^{s_k})$, we obtain

$$W(h^s)\le W(h^{s_k})-2\beta\sum_{i=s_k}^{s-1}\rho_i+o(\varepsilon). \tag{10}$$

Approaching the limit as $s\to\infty$ in the inequality (10), we arrive at the inequality, which contradicts the boundedness of the continuous function $W(h)$ on the compact set $U_{2\varepsilon}(h')$. Therefore, $\tau_k<\infty$.

By the construction, $h^{\tau_k-1}\in U_\varepsilon(h^{s_k})$, so $h^{\tau_k}\in U_{4\varepsilon}(h')$ for sufficiently large $k\in N$ and, hence, in the case $s=\tau_k$ the inequality (10) holds true by the same reasons as in the previous case. On the other hand,

$$\varepsilon<\left\|h^{\tau_k}-h^{s_k}\right\|\le\sum_{i=s_k}^{\tau_k-1}\left\|h^{i+1}-h^i\right\|\le c\sum_{i=s_k}^{\tau_k-1}\rho_i\,,$$

whence

$$\sum_{i=s_k}^{\tau_k-1}\rho_i\ge\frac{\varepsilon}{c}. \tag{11}$$

Substituting (11) into (10), we find that

$$W(h^{\tau_k})\le W(h^{s_k})-\frac{\varepsilon\beta}{c}. \tag{12}$$

Approaching the limit as $k\to\infty$ in (12), we obtain

$$\overline{\lim_{k\to\infty}}\,W(h^{\tau_k})<\lim_{k\to\infty}W(h^{s_k}).$$

Thus, all the conditions of Theorem 1 are valid, which is what had to be proved.

Remark 1. *Since the a priori inequalities in the negative norms hold true, there exists a uniformly convergent to* $J'(h)$ *sequence* $\{J'_s(h)\}_{s=0}^{\infty}$.

Remark 2. *If the set of values of the function* $J(h)$ *in the set* Φ^* *is at most denumerable, it is possible to prove that there exists the subsequence (8), which is convergent to the set of solutions.*

Remark 3. *An analytic function satisfies the condition of countability.*

Remark 4. *There exists a function* $J(h)$ *from* C^∞ *such that the set of its value is at most denumerable in the set* Φ^*.

4. ANALOGUE OF THE CONDITIONAL GRADIENT METHOD

In the previous section applying the projective operation on the admissible set, we solved the optimization problem. Note that in some cases to execute the projective operation is very difficult. In this section we consider other approach to solving an optimization problem with constraints. Instead of projective operation the original optimization problem is substituted by a minimization problem of a linear function on the admissible set.

Suppose that the conditions of Section 2 hold true and $h^0 \in U_{ad}$, where U_{ad} is a convex, closed and bounded admissible set. The sequence of controls $h^1, h^2, \ldots$ is generated by the following algorithm:

$$h^{s+1} = h^s + \rho_s\left(\overline{h}^s - h^s\right), s = 0,1,\ldots, \tag{13}$$

$$\left(J'_s\left(h^s\right), \overline{h}^s\right) = \min_{h \in U_{ad}}\left(J'_s\left(h^s\right), h\right), \tag{14}$$

where ρ_s is a step of the algorithm, which is chosen by the rule

$$\rho_s \xrightarrow[s\to\infty]{} 0,\ \sum_{s=0}^{\infty}\rho_s = \infty,\ \rho_s \in (0,1] > 0,\ s = 0,1,\ldots$$

Theorem 2. *Suppose that the conditions of Theorem 1 are satisfied. Then the limit of an arbitrary convergent subsequence (13), (14) belongs to the set* Φ^*.

Proof. Apply Lemma 1. Conditions 1 and 2a hold true obviously. Check on the other conditions of Lemma 1. Suppose that there exists a subsequence $h^{s_k} \to h' \notin \Phi^*$, $k \to \infty$. Choose a sufficiently small $\varepsilon > 0$ such that the 4ε-neighbourhood of the point h' and the set Φ^* are mutually disjoint sets. It follows from the definition of the set Φ^* that for a sufficiently large k the following condition holds true

$$\min_{h \in U_{ad}} \left(J'\left(h^{s_k}\right), h - h^{s_k} \right) \le -4\gamma < 0 .$$

Since $J'_s(h) \to J'(h)$ as $s \to \infty$ uniformly on U_{ad}, we have

$$\left(J'_{s_k}\left(h^{s_k}\right), \bar{h}^{s_k} - h^{s_k} \right) \le \min_{h \in U_{ad}} \left(J'_{s_k}\left(h^{s_k}\right), h - h^{s_k} \right) \le -3\gamma .$$

Let us show that for all $k = 0,1,\ldots$, the value $\hat{o}_k$ defined in Lemma 1 is finite. Suppose the contrary. Then,

$$h^s \in U_\varepsilon\left(h^{s_k}\right) = \left\{ h : \left\| h - h^{s_k} \right\| \le \varepsilon \right\},\ \forall s > s_k$$

so that $h' \in U_\varepsilon\left(h^{s_k}\right)$, whence we have $h^s \in U_{2\varepsilon}\left(h'\right)$ for $s > s_k$. Suppose that $W(h) = J(h)$ and $s > s_k$. Observe that the function $J(h)$ satisfies all the conditions of Lemma 1.

Then,

$$W\left(\varphi^s\right) - W\left(\varphi^{s_k}\right) = J\left(\varphi^s\right) - J\left(\varphi^{s_k}\right) =$$
$$= \left(J'\left(\varphi^{s_k}\right) - J'_{sk}\left(\varphi^{s_k}\right), \varphi^s - \varphi^{s_k} \right) + \left(J'_{sk}\left(\varphi^{s_k}\right), \varphi^s - \varphi^{s_k} \right) + o(\varepsilon)$$

Since the first term becomes vanishingly small as $k \to \infty$ and

$$h^s - h^{s_k} = \sum_{l = s_k}^{s-1} \rho_l \left(\bar{h}^l - h^l \right),\ \forall s > s_k,$$

for a sufficiently large k we have

$$W(h^s)-W(h^{s_k})\le \sum_{l=s_k}^{s-1}\rho_l\left(J'_{s_k}(h^{s_k}),\bar{h}^l-h^l\right)+o(\varepsilon). \quad (15)$$

Taking into account continuity of the function $J'(h)$ and the inequality $\|h^{s_k}-h^l\|\le\varepsilon$ for a sufficiently small $\varepsilon>0$, we have

$$\left(J'_{s_k}(h^{s_k}),\bar{h}^l-h^l\right)<-2\gamma. \quad (16)$$

Considering inequalities (16) and (15), we obtain

$$W(h^s)\le W(h^{s_k})-\gamma\sum_{l=s_k}^{s-1}\rho_l+o(\varepsilon), \quad (17)$$

contradiction with the boundedness of the continuous function $W(h)$ as $s\to\infty$. Thus, the sequence $\hat{o}_k$ does not tend to infinity as $k\to\infty$. By the definition of the sequence $\hat{o}_k$, we have $h^{\tau_k-1}\in U_\varepsilon(h^{s_k})\subset U_{2\varepsilon}(h')$, so $h^{\tau_k}\in U_{4\varepsilon}(h')$ for sufficiently large k and, hence, in the case $s=\hat{o}_k$ the inequality (17) holds true by the same reasons as in the previous case. Also, we have

$$\mathring{a}<\|h^{\hat{o}_k}-h^{s_k}\|\le\sum_{l=s_k}^{\hat{o}_k-1}\|h^{l+1}-h^l\|\le M\sum_{l=s_k}^{\hat{o}_k-1}\tilde{n}_l,$$

whence

$$\sum_{l=s_k}^{\tau_k-1}\rho_l\ge\frac{\varepsilon}{M}.$$

Substituting this inequality into (17), we obtain

$$W(h^{\tau_k})\le W(h^{s_k})-\frac{\gamma_\varepsilon}{M}+o(\varepsilon).$$

Approaching the limit as $k\to\infty$, we find

$$\overline{\lim_{k\to\infty}}W(\varphi^{\tau_k})<\lim_{k\to\infty}W(\varphi^{s_k}).$$

Thus all the conditions of Lemma 1 are valid.

Remark. *This approach based on Lemma 1 is very convenient for proving convergence of a non-linear programming algorithm* *in the case of programmed control of the step* ρ_s. *On the other hand, it should been noted that in the case of the step selection based on the complete step condition the assumption 2a of the Lemma is not valid. If proof of the assumption 2a is difficult or impossible at all, it is useful to employ the modification of Lemma 1 [108].*

Lemma 2. *Suppose that a sequence of points* $\{h^s\}_{s=\overline{0,\infty}}$ *satisfies the following conditions:*

1. $h^s \in K$ *is a compact set,* $s = 1,2,\ldots$

2. If $\lim\limits_{k\to\infty} h^{s_k} = h' \notin \Phi^*$ *then there exists* ε_0 *such that for all* $\varepsilon : 0 < \varepsilon \le \varepsilon_0$ *we have* $\tau_k = \min\limits_{s > s_k}\left\{s : \left\|h^s - h^{s_k}\right\| > \varepsilon\right\} < \infty$;

3. There exists a continuous function $W(h)$ *such that the set of its values in the set* Φ^* *is at most denumerable and* $W(h)$ *satisfies the following inequality*

$$\overline{\lim_{k\to\infty}} W\left(h^{\tau_k}\right) < \lim_{k\to\infty} W\left(h^{s_k}\right).$$

4. If $\inf\limits_{h^* \in \Phi^*}\left\|h^{s_k} - h^*\right\| \xrightarrow[k\to\infty]{} 0$ *then*

$$W\left(h^{s_k+1}\right) - W\left(h^{s_k}\right) \xrightarrow[k\to\infty]{} 0.$$

Under these assumptions the sequence $W\left(h^s\right)$ *converges and an arbitrary limit point of the sequence* $\{h^s\}_{s=\overline{0,\infty}}$ *belongs to set* Φ^*.

Remark 2. *If the set of values of the function* $J(h)$ *in the set* Φ^* *is at most denumerable, it is possible to prove that there exists a subsequence of (7), which is convergent to the set of solutions.*

5. PROBLEMS OF JOINT OPTIMIZATION AND IDENTIFICATION

5.1 Formulation of joint optimization and identification problem

There are a lot of optimal control problems described by a model with the vector a of unknown parameters. A sequence $h^{(1)}, h^{(2)}, \ldots$ is an observation over this unknown vector a. In this case it is necessary to solve not only an optimization problem but also identification one.

There exist a lot of approaches to solving this problem. One can first solve the identification problem with a given accuracy and then find the optimal control, nevertheless this approach possess an essential disadvantages:

- unknown parameters are estimated inaccurately, so that in solving the optimization problem the errors of control may be accumulated,
- in real time problems there are not enough observations for identification of parameters of unknown vector.

The approach of joint optimization and identification is much more effective. This approach enables us to consider new optimization problems being directly related to limit extreme problems [25].

Although the above mentioned extreme problems are the problems of stochastic programming, it is necessary to develop special methods because the general methods of stochastic programming are not always effective.

The following simple example shows one aspect of necessity to develop the special method of joint optimization and identification.

It is requested to minimize positive quadratic form

$$J(h) = \langle Bh, h \rangle + \langle b, h \rangle$$

in the admissible set $U_{ad} \subset R^n$.

Assume that symmetric $n \times n$ matrix B is of the following form

$$B = \begin{bmatrix} a & 0 \\ 0 & B_1 \end{bmatrix},$$

where B_1 is the positive defined $(n-1)\times(n-1)$ matrix.

We have independent observations

$$h^1, h^2, \ldots, \ a = Mh^s, \ s = 0,1,2,\ldots, \ a > 0.$$

on an unknown parameter a.

This problem can be solved in two ways. In the first case using the some observations $h^1, h^2, \ldots, h^s$, we obtain an estimation a^s of the parameter a, and then we begin to solve the deterministic problem. In the second case solving the identification problem we are making some iteration of the basic optimization algorithm using the current value of the estimate a^s of the parameter a.

The first approach is automatically inconvenient for real time problems. In addition, this approach can become incorrect. For example, if the identically distributed random variables h^s, $s = 0,1,\ldots$ take on the negative values much more frequently than the positive ones, then the estimate of a may be negative and the function

$$J_s(h) = \langle B_s h, h \rangle + \langle b, h \rangle$$

may be non-convex. By B_s we mean

$$B_s = \begin{bmatrix} a_s & 0 \\ 0 & B_1 \end{bmatrix}.$$

In particular, this is due to

$$h^s = \begin{cases} 1000 & p = 0{,}01, \\ -1 & p = 0{,}99, \end{cases} \qquad s = 0,1,\ldots, \text{ where } p \text{ is probability,}$$

$$a = Mh^s = 9{,}01 > 0.$$

Thus, possibility to solve the optimization problem by this approach essentially depends on the accuracy of the estimation of the parameter a. Conversely, the second approach is much more convenient for real time optimization problem. The consecutive approximations of the solution of the extreme problem are based on all currently observed information about the parameter a. In this approach it is not necessary to find the vector a with high accuracy for concurrent execution of algorithms. In addition, the approach enables us to vary a method of step choice in the optimization algorithms.

In this section we consider the problem

$$F(a,x) \to \min_x, \tag{1}$$

$$x \in X, \tag{2}$$

where a is a vector of unknown parameters.

There are independent random observations $h^1, h^2, \ldots,$ such that $Mh^s = a,\ s = 1,2,\ldots,\ M\|h^s\|^2 < \infty$. The estimates $a^s,\ s = 1,2,\ldots$ can be obtained in the same way as in the papers [109, 110].

To solve the optimization problem, we apply the algorithms similar to the well-known mathematical programming algorithms [21] of stochastic approximation.

To prove the following assertion, we shall employ well-behaved sufficient conditions of convergence of stochastic programming algorithms. Following the paper [22], let us formulate the conditions.

Let X^* be a set of solutions of some optimization problem, and $x^0(\omega), x^1(\omega), \ldots,\ \omega \in \Omega$ be a random sequence of points.

Theorem 1. *Suppose that $x^s(\omega) \in M(\omega)$ for almost all ω, where $s = 0,1,\ldots$. $M(\omega)$ is a compact set. All convergent subsequences $x^{s_k}(\omega)$ are satisfy the following conditions:*

1) if $\lim_{k\to\infty} x^{s_k}(\omega) \in X^*$, then $\left\|x^{s_k+1} - x^{s_k}\right\| \xrightarrow[k\to\infty]{p=1} 0$;

2) if $\lim_{k\to\infty} x^{s_k}(\omega) \notin X^*$, $p = 1$, then there exists $\varepsilon_0(\omega) > 0$ such that for all $\varepsilon(\omega)$, $0 < \varepsilon(\omega) \le \varepsilon_0(\omega)$ a value $\hat{o}_k(\omega) < \infty$, where

$$\tau_k(\omega) = \min_{s > s_k}\left(s : \left\|x^s - x^{s_k}\right\| > \varepsilon(\omega)\right);$$

3) there exists a continuous function $W(x)$ such that the set of its values in the set X^* is at most denumerable and

$$\overline{\lim_{k\to\infty}} W\left(x^{\tau_k}(\omega)\right) < \lim_{k\to\infty} W\left(x^{s_k}(\omega)\right).$$

Under these assumptions the sequence $\left\{W\left(x^s(\omega)\right)\right\}_{s=\overline{0,\infty}}$ *is convergent and all the limit points of the sequence* $\left\{x^s(\omega)\right\}_{s=\overline{0,\infty}}$ *belong to the set* X^* *for almost all* ω.

In the case when to verify the condition 1) is difficult, it is convenient to employ other conditions of convergence.

Theorem 2. *Assume that for almost all* ω, *the sequence* $x^s(\omega)$, $s \in N$ *belongs to the compact set* $M(\omega)$. *All convergent subsequence* $x^{s_k}(\omega)$ *satisfy the following conditions:*

1) if $\lim_{k\to\infty} x^{s_k}(\omega) = x'(\omega) \notin X^*$, *then* $\exists \mathring{a}_0(\omega) > 0 : \forall \mathring{a}(\omega)$, $0 < \varepsilon(\omega) < \mathring{a}_0(\omega)$ *we have* $\hat{o}_k(\omega) < \infty$, *where* $\hat{o}_k(\omega) = \min_{s > s_k}\left(s : \left\|x^s - x^{s_k}\right\| > \mathring{a}(\omega)\right)$;

2) there exists a continuous function $W(x)$ *such that the set of its values in the set* X^* *is at most denumerable and*

$$\overline{\lim_{k\to\infty}} W\left(x^{\hat{o}_k}(\omega)\right) < \lim_{k\to\infty} W\left(x^{s_k}(\omega)\right);$$

3) if $\inf_{x^* \in X^*} \left\|x^{s_k} - x^*\right\| \xrightarrow[k\to\infty]{} 0$, *then*

$$W\left(x^{s_k+1}(\omega)\right) - W\left(x^{s_k}(\omega)\right) \xrightarrow[k\to\infty]{} 0$$

for almost all ω.

Under these assumptions the sequence $\left\{W\left(x^s(\omega)\right)\right\}_{s=\overline{0,\infty}}$ *is convergent and all limit points of the sequence* $\left\{x^s(\omega)\right\}_{s=\overline{0,\infty}}$ *belong to the set* X^* *for almost all* ω.

This theorem is proved in the same manner as the previous one.

5.2 Analogue of the generalized gradient projection method

Assume that a point x_0 belongs to the convex compact admissible set $X \subset R^n$ of the problem (1), (2). For all a the function $F(a,x)$ is convex with respect to the variable x. A sequence of points is generated by the algorithm

$$x^{s+1} = \pi_X\left(x^s - \rho_s \tilde{F}_x\left(a^s, x^s\right)\right), \tag{3}$$

where $\tilde{F}_x(a,x)$ is a generalized gradient of the function $F(a,x)$, s is a number of iteration, $x^0 \in X$ is an initial approximation, π_X is a projective operator to the set X; ρ_s is iteration step of the algorithm. The parameter a is determined by the rule

$$a^{s+1} = a^s + \delta_s\left(h^{s+1} - a^s\right),\ s = 0,1,\dots . \tag{4}$$

Theorem 3. *Let $F(a,x)$ be a continuous separately with respect to a and x,*

$$\rho_s \xrightarrow[s\to\infty]{} 0,\ \sum_{s=0}^{\infty} \rho_s = \infty,\ \delta_s \xrightarrow[s\to\infty]{} 0,\ \sum_{s=0}^{\infty} \delta_s = \infty,\ \sum_{s=0}^{\infty} \delta_s^2 < \infty.$$

Then with probability 1 the limit of an arbitrary convergent subsequence of (3) belongs to the set of the solutions X^* of the problem (1), (2), where

$$X^* = \left\{x^* \in X\colon F(a,x) - F\left(a,x^*\right) \ge 0,\ \forall x \in X\right\}.$$

Proof. Apply Theorem 1. According to the algorithm, the sequence $x^s(\omega)$ belongs to the compact set X, $s = 0,1,\dots$. Henceforth, the dependence $x^s(\omega)$ on ω shall be omitted.

The first condition of Theorem 1 is obvious

$$\rho_s \xrightarrow[s\to\infty]{} 0,\ \left|\tilde{F}_x\left(a^s, x^s\right)\right| \le M < \infty.$$

Verify the other conditions of Theorem 1. Let $\{x^{s_k}\}_{k=\overline{0,\infty}}$ be an arbitrary convergent to a point $x' \notin X^*$ subsequence. Then

$$\exists \varepsilon > 0 : \forall x \in U_{4\varepsilon}(x') = \{x \in X : \|x - x'\| \leq 4\varepsilon\},$$

$$F(a,x) - F(a,x^*) \geq \delta > 0, \ x^* \in X^*.$$

Since the function $F(a,x)$ is continuous and $a^s \xrightarrow[s\to\infty]{} a$ with probability 1, from inequality

$$F(a,x) - F(a^s,x) + F(a^s,x) - F(a^s,x^*) + F(a^s,x^*) - F(a,x^*) \geq \delta$$

for a sufficiently large s, we have

$$F(a^s,x) - F(a^s,x^*) \geq \frac{\delta}{2}.$$

From convexity of the function $F(a,x)$ we have

$$\left(\tilde{F}_x(a^s,x), x - x^*\right) \geq \frac{\delta}{2}, \ \forall x \in U_{4\varepsilon}(x'), \quad s = 0,1,\dots. \tag{5}$$

Show that $\hat{o}_k < \infty$ for all k. Suppose the contrary. Then $x^s \in U_\varepsilon(x^{s_k})$ for all $s > s_k$ so that $x' \in U_\varepsilon(x^{s_k})$. From this it follows that $x^s \in U_{2\varepsilon}(x')$ for all $s > s_k$, $k = 0,1,\dots$. Let $W(x) = \inf_{x^* \in X^*} \|x - x^*\|^2$ and $s > s_k$, then

$$W(x^{s+1}) = \inf_{x^* \in X^*} \|x^{s+1} - x^*\|^2 = \inf_{x^* \in X^*} \left\|\pi_X\left(x^s - \rho_s \tilde{F}_x(a^s,x^s)\right) - x^*\right\|^2 \leq$$

$$\leq W(x^s) - 2 \inf_{x^* \in X^*} \rho_s\left(\tilde{F}_x(a^s,x^s), x^s - x^*\right) + \rho_s^2 \left|\tilde{F}_x(a^s,x^s)\right|^2.$$

From the inequalities (5) and $\left|\tilde{F}_x(a^s,x^s)\right| \leq M$ for all $s > s_k$, we have

$$W(x^{s+1}) \leq W(x^s) - \delta\rho_s + \rho_s^2 M^2.$$

Whence, for sufficiently large k we obtain the inequality

$$W(x^{s+1}) \leq W(x^s) - \frac{1}{2}\delta\rho_s.$$

Summing up the inequality over all the values of s from s_k to $r-1$, we have

$$W(x^r) \le W(x^{s_k}) - \frac{1}{2}\delta \sum_{s=s_k}^{r-1} \rho_s. \tag{6}$$

Passing to the limit as $r \to \infty$ in inequality (6), we obtain the contradiction with the non-negativity of the function $W(x)$. Therefore, $\hat{o}_k < \infty$, $\forall\, k \in N$.

It is clear that $x^{\tau_k - 1} \in U_\varepsilon(x^{s_k}) \subset U_{2\varepsilon}(x')$, then for all sufficiently large k, $x^{\tau_k} \in U_{4\varepsilon}(x')$, and hence in the case $r = \hat{o}_k$ inequality (6) is true by the same reasons as in the previous case. On the other hand,

$$\varepsilon < \left\| x^{\tau_k} - x^{s_k} \right\| \le \sum_{s=s_k}^{\tau_k - 1} \left\| x^{s+1} - x^s \right\| \le M \sum_{s=s_k}^{\tau_k - 1} \rho_s,$$

whence $\sum_{s=s_k}^{\tau_k - 1} \rho_s \ge \frac{\varepsilon}{M}$. Substituting this inequality into (6), we find that

$$W(x^{\tau_k}) \le W(x^{s_k}) - \frac{\varepsilon\delta}{2M}. \tag{7}$$

Approaching the limit as $k \to \infty$ in the inequality (7), we obtain

$$\overline{\lim_{k \to \infty}} W(x^{\hat{o}_k}) < \lim_{k \to \infty} W(x^{s_k}).$$

Thus, all conditions of Theorem 1 hold true, which is what had to be proved.

Remark. *If in Theorem 3 the function $F(a, x)$ is continuously differentiable but non-convex, then an arbitrary convergent subsequence of the sequence (3) converges to points of a set*

$$X^* = \left\{ x^* \in X : \min_{x \in X} \left(F_x(a, x^*), x - x^* \right) = 0 \right\},$$

with probability 1.

5.3 Analogue of the conditional gradient method

In this section to solve the problem (1), (2) we consider a sequence $\{x^s\}_{s=\overline{0,\infty}}$

$$x^{s+1} = x^s + \rho_s\left(\overline{x}^s - x^s\right), \tag{8}$$

$$\left(F_x\left(a^s,x^s\right),\overline{x}^s\right) = \min_{x\in X}\left(F_x\left(a^s,x^s\right),x\right). \tag{9}$$

The parameter a is identified in the same manner as in the case discussed above

$$a^{s+1} = a^s + \delta_s\left(h^{s+1} - a^s\right), \tag{10}$$

$$Mh^s = a,\ M\left\|h^s\right\|^2 < \infty,\ s = 0,1,\ldots$$

By a solution of the problem we mean the set X^* which is defined in previous remark.

Theorem 4. *Let* $\widetilde{F}_a(x) = F(a,x)$ *be a continuously differentiable function for all* a *and* $\tilde{\tilde{F}}_x(a) = F(a,x)$ *be continuous one for all* $x \in X$. *The set of values of the function* $F(a,x)$ *in set* X^* *is at most denumerable and*

$$\rho_s \in (0,1],\ \rho_s \xrightarrow[s\to\infty]{} 0, \sum_{s=0}^{\infty}\rho_s = \infty,\ \sum_{s=0}^{\infty}\delta_s = \infty,\ \sum_{s=0}^{\infty}\delta_s^2 < \infty.$$

Then a limit of an arbitrary convergent subsequence belongs to the solution set X^* *with probability 1.*

Proof. The first condition of Theorem 1 holds true obviously. Assume that there exists a subsequence $x^{s_k} \xrightarrow[k\to\infty]{} x' \notin X^*$. Choose a sufficiently small $\varepsilon > 0$ such that the 4ε-neighbourhood of the point x' and the set X^* are mutually disjoint. By the definition of the set X^*, we obtain the inequality

$$\min_{x\in X}\left(F_x\left(a^{s_k},x^{s_k}\right),x - x^{s_k}\right) \le -2\gamma < 0. \tag{11}$$

for sufficiently large k.

Suppose the inequality $\left\|x^s - x^{s_k}\right\| < \varepsilon$ holds true for all $s > s_k$. Whence, we have $x^s \in U_{2\varepsilon}(x'),\ s > s_k$ for sufficiently large k. Thus, the inequality (11) holds true for x^s. Assume that $\hat{o}_k = \infty$. Consider the continuous function $W(x) = F(a,x)$ and $s > s_k$. Then

$$W(x^s) - W(x^{s_k}) = \left(F_x(a, x^{s_k}), x^s - x^{s_k}\right) + o(\varepsilon) =$$
$$= \left(F_x(a, x^{s_k}) - F_x(a^{s_k}, x^{s_k}), x^s - x^{s_k}\right) + \left(F_x(a^{s_k}, x^{s_k}), x^s - x^{s_k}\right) + o(\varepsilon).$$

Since $a^s \xrightarrow[s\to\infty]{p=1} a$ and

$$x^s - x^{s_k} = \sum_{l=s_k}^{s-1} \rho_l \left(\bar{x}^l - x^l\right),$$

we have

$$W(x^s) - W(x^{s_k}) = \sum_{l=s_k}^{s-1} \rho_l \left(F_x(a^{s_k}, x^{s_k}), \bar{x}^l - x^l\right) + o(\varepsilon) \qquad (12)$$

for sufficiently large k.

Taking into account continuity of the function $F_x(a,x)$ and inequality $\left\|x^l - x^{s_k}\right\| < \varepsilon$, for a sufficiently small ε we obtain

$$\left(F_x(a^{s_k}, x^{s_k}), \bar{x}^l - x^l\right) < -2\gamma. \qquad (13)$$

Considering the equalities (12) and (13), we find

$$W(x^s) \le W(x^{s_k}) - \gamma \sum_{l=s_k}^{s-1} \rho_l + o(\varepsilon) . \qquad (14)$$

Passing to the limit as $s \to \infty$, we obtain that the function $W(x)$ is unbounded, contrary to continuity of the function $W(x)$ on the compact set X. Thus, $\hat{o}_k < \infty$.

From $x^{\tau_k - 1} \in U_\varepsilon(x^{s_k}) \subset U_{2\varepsilon}(x')$, we have $x^{\tau_k} \in U_{4\varepsilon}(x')$ for sufficiently large k. Then, in the case $s = \hat{o}_k$ the inequality (14) holds true by the same reasons as in the previous case. On the other hand,

$$\varepsilon < \left\| x^{\tau_k} - x^{s_k} \right\| \le \sum_{l=s_k}^{\tau_k - 1} \left\| x^{l+1} - x^l \right\| \le M \sum_{l=s_k}^{\tau_k - 1} \rho_l ,$$

whence,

$$\sum_{l=s_k}^{\tau_k - 1} \rho_l \ge \frac{\varepsilon}{M}.$$

Substituting this inequality into (14), we find that

$$W\left(x^{\tau_k}\right) \le W\left(x^{s_k}\right) - \frac{\gamma\varepsilon}{M} + o(\varepsilon).$$

Approaching the limit as $k \to \infty$, we obtain

$$\overline{\lim_{k\to\infty}} W\left(x^{\tau_k}\right) < \lim_{k\to\infty} W\left(x^{s_k}\right).$$

Thus, all conditions of Theorem 1 hold true, which is what had to be proved.

Remark. *If the set of values of the function $F(a, x)$ in the set X^* is at most denumerable, it is possible to prove that there exists a convergent to the set of solutions subsequence of (8).*

6. GENERAL THEOREMS OF GENERALIZED OPTIMAL CONTROL

In the following sections we shall prove different a priori inequalities in the negative norms for different distributed systems, so the general optimal control theorems of the previous chapter hold true for this distributed systems. It is clear that the theorems from different section have common structure. We shall not formulate all theorems entirely in each section. We shall formulate only the changeable part of these theorems. Therefore in this section we collect only templates of the theorems of Chapters 1 and 2 in the convenient for the following sections form.

Suppose the state function satisfies the following equation

$$Lu = f + A(h), \tag{1}$$

with initial and boundary conditions (bd). (2)

From the general theorems of Section 1.2 we conclude that the following theorems hold true.

Theorem 1. *Consider the problem of optimal control (1), (2). If*

1) *performance criterion* $\Phi(\cdot): N \to R^1$ *is weakly lower semicontinuous;*
2) *the admissible set* $U_{ad} \subset H$ *is bounded, closed, convex in H;*
3) *H is a reflexive Banach space;*
4) $A(\cdot)$ *is a weakly continuous mapping of the space H into* $W^-(Q)$; $f \in W^-(Q)$;
5) *the operator* L, *the spaces* N *and* $W^-(Q)$ *are chosen from the following table*

N	Operator L	Space N	Space $W^-(Q)$
1.			

then there exists an optimal control of the system (1), (2).

Theorem 2. *Consider the system (1), (2) with the right-hand side* $F(\cdot) = f + A_i(\cdot)$, $f \in W^-(Q)$. *If the space* $W^-(Q)$ *and mapping* $A_i(\cdot)$ *are chosen from the table below then the mapping* $F(\cdot): H \to W^-(Q)$ *is weakly continuous*

N	Mapping $A_i(\cdot)$	Space $W^-(Q)$
1.		

Consider the question of the differential properties of performance criterion

$$J(h) = \Phi(u(h)) = \sum_{i=1}^{p} \int_Q \alpha_i(t,x)(u(h) - u_i)^2 \, dQ ,$$

where $u_i(t,x)$, $\alpha_i(t,x)$ are some functions from $L_2(Q)$ and $C(\overline{Q})$, respectively, and $\alpha_i(t,x) \geq \varepsilon > 0$ in $\overline{Q}$.

Taking into account the general theorem from paragraph 1.4 and inequalities in the negative norms, we have

Theorem 3. *Consider the problem (1), (2) with the right-hand side $f \in W^-(Q)$. If there exists a Fréchet derivative $f_{h^*}(\cdot): H \to W^-(Q)$ of a mapping $f(\cdot): H \to W^-(Q)$ at the certain point h^*, then at the point h^* there exists a Fréchet derivative of the performance criterion $J(h)$ in the following form*

$$J_{h^*}(\cdot) = \langle f_{h^*}(\cdot), v \rangle_{W(Q)},$$

where $v(t,x)$ is a solution of the adjoint problem

$$L^* v = 2\sum_{i=1}^{p} \alpha_i \left(u(h^*) - u_i\right),$$

bilinear form $\langle \cdot,\cdot \rangle_{W(Q)}$ is defined in the spaces $W^-(Q) \times W^+(Q)$.

If in addition Fréchet derivative $f_h(\cdot): H \to W^-(Q)$ is continuous at the point h^* or satisfies the Lipschitz condition with index α, $0 < \alpha \leq 1$ in a bounded and convex neighbourhood of the point h^*, then the gradient $J_h(\cdot)$ has the same properties (is continuous or satisfies the Lipschitz condition with index α), where operator L, spaces $W^+(Q)$ è $W^-(Q)$ are chosen form the following table

N	Operator L	Space $W^-(Q)$	Space $W^+(Q)$
1.			

Proof is followed from the theorems of 2.1 (Theorems 2.1.3-2.1.5).

Theorem 4. *Consider the problem (1), (2) with the right-hand side $f \in W^-(Q)$. There exists Fréchet a derivative*

$A_{i,h^*}(\cdot): H \to W^-(Q)$ *of the mapping* $A_i(\cdot): H \to W^-(Q)$, *if the space* $W^-(Q)$ *and the mapping* $A_i(\cdot)$ *are chosen from the following table.*

N	Space $W^-(Q)$	Map $A_i(\cdot)$
1.		

The Fréchet derivative $A_{i,h^*}(\cdot): H \to W^-(Q)$ *satisfies the Lipschitz condition with index* $\frac{1}{2}$.

Consider a regularized problem of optimal control

$$Lu_\varepsilon = F_\varepsilon(t,x,h), \tag{3}$$

where $\varepsilon > 0$, $F_\varepsilon(t,x,h) \in L_2(Q)$.

It is requested to minimize a functional

$$J_\varepsilon(h) = \Phi(u_\varepsilon(h)).$$

The functional $\Phi(\cdot)$ is defined on functions $u(t,x)$ from $L_2(Q)$.

Theorem 5. *Consider the optimal control problems (1), (2) and (3), (2). Let the following conditions hold true:*

1) *admissible set of control* U_{ad} *is convex, closed, and bounded in the Hilbert space* H;
2) *maps* $F_\varepsilon(h), F(h)$ *satisfy the condition:*
   a) $\left(h_{\varepsilon_k} \xrightarrow{W} h \text{ in } H\right) \Rightarrow \left(F_{\varepsilon_k}(h_{\varepsilon_k}) \xrightarrow{W} F(h) \text{ in } W^-(Q)\right)$, *for an arbitrary sequence* $\varepsilon_k \xrightarrow[k\to\infty]{} 0$,
   b) $F_\varepsilon(h)$, $F(h)$ *is weakly continuous,*
   c) $\|F_\varepsilon(h) - F(h)\|_{W^-(Q)} \xrightarrow[\varepsilon\to 0]{} 0$, *for all fixed* $h \in U_{ad} \subset H$;
   d) *the performance criterion* $\Phi(\cdot)$ *is upper semicontinuous and weakly lower semicontinuous.*

Under these assumptions there exist optimal controls h^*, h^*_ε *of the problems (1), (2) and (3), (2); there exists a weakly convergent subsequence* $h^*_{\varepsilon_i}$ *and an arbitrary weakly convergent subsequence* $h^*_{\varepsilon_i}$ *converges to* h^* *weakly in* H, *where the operator of initial boundary problem* L, *the space* $W^-(Q)$ *are chosen from the table*

N.	Operator L	Space $W^-(Q)$
1.		

Theorem 6. *Consider the regularized right-hand side of the equation (1), (2) from Section 2.2. If the operator* L, *the space* $W^-(Q)$, *the map* $A_i(\cdot)$ *are chosen from the following table, then conditions 2a)-2d) of previous theorem hold true for the maps* $A_i(\cdot)$, $A_{i,\varepsilon}(\cdot)$.

N	Operator L	Map $A_i(\cdot)$	Space $W^-(Q)$.
1.			

Theorem 7. *Consider the optimal control problems (3), (2) with the right-hand size* $f_\varepsilon + A_{i,\varepsilon} \in W^-(Q)$. *If the space* $W^-(Q)$, *the exponent* α, *the map* $A_{i,\varepsilon}(\cdot)$ *are chosen from the following table, then there exists a Fréchet derivative* $A_{i,\varepsilon,h^*}(\cdot): H \to W^-(Q)$ *of the map* $A_{i,\varepsilon}(\cdot): H \to W^-(Q)$, *the derivative* $A_{i,\varepsilon,h^*}(\cdot): H \to W^-(Q)$ *is defined in Section 2.2 and satisfies the Lipschitz condition with index* α *and directional derivative with respect to* $\varphi(x)$, *satisfies the Lipschitz condition with index* $\alpha = 1$.

N	Index α	Space $W^-(Q)$	Map $A_{i,\varepsilon}(\cdot)$
1.			

Chapter 4

PARABOLIC SYSTEMS

1. GENERALIZED SOLVABILITY OF PARABOLIC SYSTEMS

Consider applications of the results of Chapters 1 and 2 to investigation of systems governed by parabolic partial differential equations. Note that despite of enormous number of papers devoted to the parabolic systems there are many open problems such as the problems of singular control and the problems of coefficient control.

Let the functioning of a system is described by the following equation:

$$Lu = \frac{\partial u}{\partial t} - \sum_{i,j=1}^{n} \frac{\partial}{\partial x_i}\left(a_{ij}(x) \frac{\partial u}{\partial x_j} \right) + \sum_{i=1}^{n} b_i(x) \frac{\partial u}{\partial x_i} + c(x)u = f(t,x), \quad (1)$$

where $u(t,x)$ is a function describing the system state in a region $Q = (0,T) \times \Omega$, Ω is a bounded region of n-dimensional Euclidian space R^n with a smooth boundary $\partial\Omega$.

Let $a_{ij}(x) = a_{ji}(x)$, $b_i(x)$ be functions which are continuously differentiable in a closed region $\overline{\Omega}$, and $c(x)$ is a continuous in $\overline{\Omega}$ function such that

$$\sum_{i,j=1}^{n} a_{ij}\xi_i\xi_j \geq \alpha_A \sum_{i=1}^{n} \xi_i^2, \; c(x) \geq \sum_{i=1}^{n} \frac{\partial b_i(x)}{\partial x_i}, \; c(x) \geq 0,$$

$$\forall x \in \overline{\Omega}, \; \forall \xi_i \in R^1,$$

where the constant $\alpha_A > 0$.

Denote by W_{bd}^{+} a completion of the space of smooth functions which satisfy the initial and boundary conditions

$$u\big|_{t=0} = 0, \; u\big|_{x \in \partial\Omega} = 0 \quad (2)$$

with respect to the Sobolev norm

$$\|u\|_{W^+_{bd}} = \left(\int\limits_Q u_t^2 + \sum_{i=1}^n u_{x_i}^2 dQ \right)^{1/2},$$

$W^+_{bd^+}$ is a similar space but its smooth functions satisfy the boundary conditions of the adjoint problem

$$v|_{t=T} = 0,\ v|_{x\in\partial\Omega} = 0. \tag{3}$$

By pairs of the spaces W^+_{bd}, $L_2(Q)$ and $W^+_{bd^+}$, $L_2(Q)$ we construct negative spaces W^-_{bd} and $W^-_{bd^+}$ as a completion of the space of smooth in $\overline{Q}$ functions satisfying the conditions (2) or (3) with respect to the norm

$$\|f\|_{W^-_{bd}} = \sup_{\substack{u\in W^+_{bd},\\ u\neq 0}} \frac{\left|(f,u)_{L_2(Q)}\right|}{\|u\|_{W^+_{bd}}} \text{ or } \|f\|_{W^-_{bd^+}} = \sup_{\substack{v\in W^+_{bd^+},\\ v\neq 0}} \frac{\left|(v,f)_{L_2(Q)}\right|}{\|v\|_{W^+_{bd^+}}}$$

Let us study the properties of the differential operator (1), and also the properties of the formally adjoint operator:

$$L^*v = -\frac{\partial v}{\partial t} - \sum_{i,j=1}^n \frac{\partial}{\partial x_i}\left(a_{ij}(x)\frac{\partial v}{\partial x_j} \right) - \sum_{i=1}^n \frac{\partial}{\partial x_i}(b_i(x)v) + c(x)v. \tag{4}$$

L e m m a 1. *For any function $u(t,x)$, which is smooth in $\overline{Q}$ and satisfies conditions (2), the following estimation is valid:*

$$\|Lu\|_{W^-_{bd^+}} \le C\|u\|_{W^+_{bd}}.$$

P r o o f. The inequality is obtained as a result of applying of the formula of integration by parts and the integral Cauchy inequality to the expression $(Lu,v)_{L_2(Q)}$ which is written in the right-hand side of definition of the negative norm. Indeed, let $v(t,x)$ is a smooth in $\overline{Q}$ function satisfying conditions (3). Then

$$\left|(Lu,v)_{L_2(Q)}\right| =$$

$$= \left| \left(\frac{\partial u}{\partial t} - \sum_{i,j=1}^{n} \frac{\partial}{\partial x_i}\left(a_{ij}(x) \frac{\partial u}{\partial x_j} \right) + \sum_{i=1}^{n} b_i(x) \frac{\partial u}{\partial x_i} + c(x)u, v \right)_{L_2(Q)} \right| \le$$

$$\le I_1 + I_2 + I_3 + I_4,$$

where

$$I_1 = \left| \left(\frac{\partial u}{\partial t}, v \right)_{L_2(Q)} \right| \le \|u_t\|_{L_2(Q)} \|v\|_{L_2(Q)},$$

$$I_2 = \left| -\left(\sum_{i,j=1}^{n} \frac{\partial}{\partial x_i}\left(a_{ij}(x) \frac{\partial u}{\partial x_j} \right), v \right)_{L_2(Q)} \right| =$$

$$= \left| -\int_Q \sum_{i,j=1}^{n} \frac{\partial}{\partial x_i}\left(a_{ij}(x) \frac{\partial u}{\partial x_j} v \right) dQ + \int_Q \sum_{i,j=1}^{n} a_{ij}(x) \frac{\partial u}{\partial x_j} \frac{\partial v}{\partial x_i} dQ \right|,$$

$$I_3 = \left| \left(\sum_{i=1}^{n} b_i(x) \frac{\partial u}{\partial x_i}, v \right)_{L_2(Q)} \right| \le C \sum_{i=1}^{n} \|u_{x_i}\|_{L_2(Q)} \|v\|_{L_2(Q)},$$

$$I_4 = \left| (c(x)u, v)_{L_2(Q)} \right| \le C \|u\|_{L_2(Q)} \|v\|_{L_2(Q)}.$$

Taking into account the integral representation of the function $u(t,x)$ satisfying the initial condition $u|_{t=0} = 0$, we have

$$|u(t,x)| = \left| \int_0^t u_t(\eta, x) d\eta \right| \le$$

$$\le \left| \int_0^t d\eta \right|^{\frac{1}{2}} \left| \int_0^t u_t^2(\eta, x) d\eta \right|^{\frac{1}{2}} \le C \left(\int_0^T u_t^2(\eta, x) d\eta \right)^{\frac{1}{2}}.$$

Applying to I_1 the following inequality

$$\|u\|_{L_2(Q)} \le C \|u_t\|_{L_2(Q)}, \tag{5}$$

we have

$$I_1 \le \|u_t\|_{L_2(Q)} \|v\|_{L_2(Q)} \le C\|u_t\|_{L_2(Q)} \|v_t\|_{L_2(Q)} \le C\|u\|_{W_{bd}^+} \|v\|_{W_{bd^+}^+}.$$

Applying to the first term in I_2 the Ostrogradski-Gauss formula and taking into account the boundary conditions $v|_{x \in d\Omega} = 0$, we conclude that this integral is equal to zero. Applying the integral Cauchy inequality to the second term in I_2 and taking into account that the functions $a_{ij}(x)$ are continuous, and hence they are bounded, we get

$$I_2 = \left| \int_Q \sum_{i,j=1}^n a_{ij}(x) \frac{\partial u}{\partial x_j} \frac{\partial v}{\partial x_i} dQ \right| \le$$

$$\le C \sum_{i,j=1}^n \|u_{x_j}\|_{L_2(Q)} \|v_{x_i}\|_{L_2(Q)} \le C\|u\|_{W_{bd}^+} \|v\|_{W_{bd^+}^+}.$$

Let us apply to the third and fourth terms I_3, I_4 the inequality like (5). We get

$$I_3 \le C \sum_{i=1}^n \|u\|_{W_{bd}^+} \|v_t\|_{L_2(Q)} \le C\|u\|_{W_{bd}^+} \|v\|_{W_{bd^+}^+},$$

$$I_4 \le C\|u_t\|_{L_2(Q)} \|v_t\|_{L_2(Q)} \le C\|u\|_{W_{bd}^+} \|v\|_{W_{bd^+}^+}.$$

Finally, we have

$$\left|(Lu, v)_{L_2(Q)}\right| \le C\|u\|_{W_{bd}^+} \|v\|_{W_{bd^+}^+}.$$

Hence, the statement of lemma holds true for smooth functions $u(t,x)$ as far as the set of considered functions $v(t,x)$ is dense in $W_{bd^+}^+$, and hence the supremum in the definition of the negative norm we can take at such smooth functions.

R e m a r k. *The proved inequality allows extension of the operator L with respect to continuity on the whole space W_{bd}^+, in this case the inequality of the Lemma still valid for any function $u(t,x)$ of the space W_{bd}^+.*

Lemma 2. *For any function* $v \in W^{+}_{bd^{+}}$ *the following estimation is valid*

$$\left\|L^{*}v\right\|_{W^{-}_{bd}} \leq C\|v\|_{W^{+}_{bd^{+}}}. \tag{6}$$

The Proof of Lemma 2 is similar to the proof of Lemma 1.

Lemma 3. *For any function* $u(t,x) \in W^{+}_{bd}$ *the following estimation is valid*

$$\|u\|_{L_2(Q)} \leq C\|Lu\|_{W^{-}_{bd^{+}}}. \tag{7}$$

Proof. Consider the following auxiliary integral operator defined on smooth functions $u(t,x)$, which satisfy conditions (2):

$$v = I_t u = \int_T^t -e^{-\frac{nC_B^2}{\alpha_A}\tau} u(\tau,x)d\tau,$$

where C_B is a constant majoring $|b_i(x)|$, which exists by virtue the fact that the functions $b_i(x)$ are continuous in $\overline{\Omega}$ $\left(i=\overline{1,n}\right)$, n is the dimension of the region Ω.

It is clear that the function $v(t,x)$ satisfies conditions (3). Expressing $u(t,x)$ via $v(t,x)$ we get

$$u(t,x) = -e^{\frac{nC_B^2}{\alpha_A}t} v_t(t,x).$$

Consider

$$(Lu,v)_{L_2(Q)} = \left(u,L^{*}v\right)_{L_2(Q)} = \left(-e^{\frac{nC_B^2}{\alpha_A}t} v_t(t,x), L^{*}v\right)_{L_2(Q)} =$$

$$= I_1 + I_2 + I_3 + I_4,$$

where

$$I_1 = \int_Q e^{\frac{nC_B^2}{\alpha_A}t} v_t^2 dQ;\; I_2 = \int_Q e^{\frac{nC_B^2}{\alpha_A}t} v_t \sum_{i,j=1}^{n} \left(a_{ij}(x)v_{x_j}\right)_{x_i} dQ,$$

$$I_3 = \int_Q \sum_{i=1}^n e^{\frac{nC_B^2}{\alpha_A}t} b_i(x) v_t v_{x_i} dQ;\ I_4 = -\int_Q e^{\frac{nC_B^2}{\alpha_A}t}\left(c(x) - \sum_{i=1}^n \frac{\partial b_i(x)}{\partial x_i}\right) v_t v dQ.$$

Consider every term I_2, I_3, and I_4 separately.

$$I_2 = \int_Q e^{\frac{nC_B^2}{\alpha_A}t} v_t \sum_{i,j=1}^n \left(a_{ij}(x) v_{x_j}\right)_{x_i} dQ = \int_Q \sum_{i,j=1}^n \left(e^{\frac{nC_B^2}{\alpha_A}t} v_t a_{ij}(x) v_{x_j} \right)_{x_i} dQ -$$

$$-\int_Q \sum_{i,j=1}^n e^{\frac{nC_B^2}{\alpha_A}t} a_{ij}(x) v_{x_i t} v_{x_j} dQ.$$

Using the Ostrogradsky-Gauss formula and taking into account the condition $v_t\big|_{x\in\partial\Omega} = 0$ we can prove that the integral $\int_Q \sum_{i,j=1}^n \left(e^{\frac{nC_B^2}{\alpha_A}t} v_t a_{ij}(x) v_{x_j} \right)_{x_i} dQ$ is equal to zero. Let us apply to the second term the formula of integration by parts:

$$I_2 = -\int_Q \sum_{i,j=1}^n e^{\frac{nC_B^2}{\alpha_A}t} a_{ij}(x) v_{x_i t} v_{x_j} dQ = -\frac{1}{2}\int_Q \left(\sum_{i,j=1}^n e^{\frac{nC_B^2}{\alpha_A}t} a_{ij}(x) v_{x_i} v_{x_j} \right)_t dQ +$$

$$+\frac{1}{2}\int_Q \sum_{i,j=1}^n \frac{nC_B^2}{\alpha_A} e^{\frac{nC_B^2}{\alpha_A}t} a_{ij}(x) v_{x_i} v_{x_j} dQ.$$

Using the Ostrogradsky-Gauss formula we pass to the surface integral in the term $-\frac{1}{2}\int_Q \left(\sum_{i,j=1}^n e^{\frac{nC_B^2}{\alpha_A}t} a_{ij}(x) v_{x_i} v_{x_j} \right)_t dQ$ and take into account the condition $v\big|_{t=T} = 0$. Applying to the second term the inequality $\sum_{i,j=1}^n a_{ij}(x)\xi_i\xi_j \geq \alpha_A \sum_{i=1}^n \xi_i^2$, we have

$$I_2 = -\frac{1}{2}\int_Q \left(\sum_{i,j=1}^{n} e^{\frac{nC_B^2}{\alpha_A}t} a_{ij}(x) v_{x_i} v_{x_j} \right)_t dQ +$$

$$+\frac{1}{2}\int_Q \sum_{i,j=1}^{n} \frac{nC_B^2}{\alpha_A} e^{\frac{nC_B^2}{\alpha_A}t} a_{ij}(x) v_{x_i} v_{x_j} dQ \geq$$

$$\geq \frac{1}{2}\int_\Omega \sum_{i,j=1}^{n} a_{ij}(x) v_{x_i} v_{x_j} \Big|_{t=0} d\Omega + \frac{1}{2}\int_Q \sum_{i=1}^{n} nC_B^2 e^{\frac{nC_B^2}{\alpha_A}t} v_{x_i}^2 dQ \geq$$

$$\geq \frac{1}{2}\int_Q \sum_{i=1}^{n} nC_B^2 e^{\frac{nC_B^2}{\alpha_A}t} v_{x_i}^2 dQ .$$

Consider the term I_3.

$$I_3 = \int_Q \sum_{i=1}^{n} e^{\frac{nC_B^2}{\alpha_A}t} b_i(x) v_t v_{x_i} dQ \geq -\int_Q \sum_{i=1}^{n} C_B e^{\frac{nC_B^2}{\alpha_A}t} |v_t| \cdot |v_{x_i}| dQ .$$

Consider the last term I_4. Applying the formula of intergation by parts and taking into account the condition $c(x) \geq \sum_{i=1}^{n} \frac{\partial b_i(x)}{\partial x_i}$, we get

$$I_4 = -\int_Q e^{\frac{nC_B^2}{\alpha_A}t} \left(c(x) - \sum_{i=1}^{n} \frac{\partial b_i(x)}{\partial x_i} \right) v_t v dQ =$$

$$= -\frac{1}{2}\int_Q \left(e^{\frac{nC_B^2}{\alpha_A}t} \left(c(x) - \sum_{i=1}^{n} \frac{\partial b_i(x)}{\partial x_i} \right) v^2 \right)_t dQ +$$

$$+\frac{1}{2}\int_Q \frac{nC_B^2}{\alpha_A} e^{\frac{nC_B^2}{\alpha_A}t} \left(c(x) - \sum_{i=1}^{n} \frac{\partial b_i(x)}{\partial x_i} \right) v^2 dQ \geq$$

$$\geq \frac{1}{2}\int_\Omega \left(c(x) - \sum_{i=1}^{n} \frac{\partial b_i(x)}{\partial x_i} \right) v^2 \Big|_{t=0} d\Omega \geq 0 .$$

Finally, we obtain

$$(Lu,v)_{L_2(Q)} \geq \int_Q e^{\frac{nC_B^2}{\alpha_A}t}\left(v_t^2 + \sum_{i=1}^{n} \frac{nC_B^2}{2} v_{x_i}^2 - \sum_{i=1}^{n} C_B |v_t| \cdot |v_{x_i}|\right) dQ =$$

$$= \sum_{i=1}^{n} \int_Q \frac{1}{n} e^{\frac{nC_B^2}{\alpha_A}t}\left[\left(\frac{\sqrt{3}}{2}|v_t| - \frac{nC_B}{\sqrt{3}}|v_{x_i}|\right)^2 + \frac{1}{4}v_t^2 + \frac{n^2C_B^2}{6}v_{x_i}^2\right] dQ \geq$$

$$\geq C\int_Q \left(v_t^2 + \sum_{i=1}^{n} v_{x_i}^2\right) dQ \geq C\|v\|^2_{W^+_{bd^+}}.$$

Applying the Schwarz inequality to $(Lu,v)_{L_2(Q)}$, we obtain

$$\|Lu\|_{W^-_{bd^+}} \|v\|_{W^+_{bd^+}} \geq (Lu,v)_{L_2(Q)} \geq C\|v\|^2_{W^+_{bd^+}}.$$

Reducing the right-hand side and the left-hand side on the inequality by $\|v\|_{W^+_{bd^+}}$ and taking into account the relation between $u(t,x)$ and $v(t,x)$, we obtain

$$\|Lu\|_{W^-_{bd^+}} \geq C\|v\|_{W^+_{bd^+}} = C\|I_t u\|_{W^+_{bd^+}} \geq C\|u\|_{L_2(Q)}.$$

The validity of the inequality (7) on the whole space W^+_{bd} we prove by passing to the limit.

Lemma 4. *For any functions $v(t,x) \in W^+_{bd^+}$ the following inequality holds true:*

$$\|v\|_{L_2(Q)} \leq C\|L^* v\|_{W^-_{bd}}.$$

The Proof of Lemma 4 is carried out in a similar way as for Lemma 3. The auxiliary operator is of the form

$$u = I_t v = \int_0^t e^{\frac{2nC_B^2}{\alpha_A}\tau} v(\tau,x)\,d\tau.$$

Theorem 1. *For any function $f \in L_2(Q)$ there exists a unique solution of the problem (1), (2) in the sense of Definition 1.1.1.*

Theorem 2. *For any functional $f \in W^{-}_{bd^{+}}$, then a unique solution of the problem (1), (2) exists in the sense of Definition 1.1.4.*

2. ANALOGUE OF GALERKIN METHOD

Let us construct the method of numerical solving the parabolic equation with a generalized right-hand side.

Consider the following equation

$$Lu = \frac{\partial u}{\partial t} - \sum_{i,j=1}^{n} \frac{\partial}{\partial x_i}\left(a_{ij}(x) \frac{\partial u}{\partial x_j} \right) + \sum_{i=1}^{n} b_i(x) \frac{\partial u}{\partial x_i} + c(x) u = f(t,x), \quad (1)$$

$$u\big|_{t=0} = 0 \,,\; u\big|_{x \in \partial\Omega} = 0 \,, \quad (2)$$

where $u(t,x)$ is a system state defined on the set

$$Q = (0,T) \times \Omega \,,\; \Omega \subset R^n \,,$$

$$a_{ij}(x) = a_{ji}(x) \in C^1(\overline{\Omega}),\; b_i(x) \in C^1(\overline{\Omega}),\; c(x) \in C(\overline{\Omega}),$$

$$\sum_{i,j=1}^{n} a_{ij}(x) \xi_i \xi_j \ge \alpha_A \sum_{i=1}^{n} \xi_i^2 \,,\; \forall \xi_i \in R^1 \,,\; c(x) \ge 0 \,,\; |b_i(x)| < C_B \,, \quad (3)$$

$$c(x) \ge \sum_{i=1}^{n} \frac{\partial b_i(x)}{\partial x_i} .$$

Let the system state is described by the initial-boundary value problem (1), (2), $f(t,x) \in L_2(Q)$.

We shall look for the approximate solution in the following form

$$u_N(t,x) = \sum_{i=1}^{N} g_i(t) \omega_i(x), \quad (4)$$

where $\omega_i(x)$ is a orthonormal basis in $L_2(\Omega)$, which consists of functions from the space $C_0^2(\overline{\Omega})$, and the functions $g_i(t)$ are selected as a solutions of the Cauchy problem for the system of N linear ordinary equations:

$$\sum_{k=1}^{N}\left(\frac{dg_k}{dt}\omega_k+g_k\left(-\sum_{i,j=1}^{n}\frac{\partial}{\partial x_i}\left(a_{ij}\frac{\partial\omega_k}{\partial x_j}\right)+\right.\right.$$

$$\left.\left.+\sum_{i=1}^{n}b_i\frac{\partial\omega_k}{\partial x_i}+c\omega_k\right),\omega_l\right)_{L_2(\Omega)}=(f,\omega_l)_{L_2(\Omega)}, \tag{5}$$

$$g_i(0)=0,\ i=\overline{1,N},\ l=\overline{1,N}.$$

Lemma 1. *The following inequality is valid*

$$\|u_N\|_{W_{bd}^+}\le C\|f\|_{L_2(Q)}.$$

Proof. Multiplying both parts of expression (5) by $e^{-\frac{2nC_B^2}{\alpha_A}t}\dfrac{dg_l}{dt}$ and summing with respect to l from 1 to N and integrating with respect to t from 0 to T, we obtain

$$\left(Lu_N,e^{-\frac{2nC_B^2}{\alpha_A}t}\frac{\partial u_N}{\partial t}\right)_{L_2(Q)}=\left(e^{-\frac{2nC_B^2}{\alpha_A}t}\frac{\partial u_N}{\partial t},\frac{\partial u_N}{\partial t}\right)_{L_2(Q)}-$$

$$-\left(e^{-\frac{2nC_B^2}{\alpha_A}t}\frac{\partial u_N}{\partial t},\sum_{i,j=1}^{n}\frac{\partial}{\partial x_i}\left(a_{ij}\frac{\partial u_N}{\partial x_j}\right)\right)_{L_2(Q)}+ \tag{6}$$

$$+\left(e^{-\frac{2nC_B^2}{\alpha_A}t}\frac{\partial u_N}{\partial t},\sum_{i=1}^{n}b_i\frac{\partial u_N}{\partial x_i}\right)_{L_2(Q)}+\left(e^{-\frac{2nC_B^2}{\alpha_A}t}\frac{\partial u_N}{\partial t},cu_N\right)_{L_2(Q)}=$$

$$= I_1 + I_2 + I_3 + I_4 = \left(e^{-\frac{2nC_B^2 t}{\alpha_A}} \frac{\partial u_N}{\partial t}, f \right)_{L_2(Q)}$$

Integrating by parts and using the initial and boundary conditions, we get the following expressions

$$I_1 = \left(e^{-\frac{2nC_B^2 t}{\alpha_A}} \frac{\partial u_N}{\partial t}, \frac{\partial u_N}{\partial t} \right)_{L_2(Q)},$$

$$I_2 = -\left(e^{-\frac{2nC_B^2 t}{\alpha_A}} \frac{\partial u_N}{\partial t}, \sum_{i,j=1}^{n} \frac{\partial}{\partial x_i}\left(a_{ij} \frac{\partial u_N}{\partial x_j} \right) \right)_{L_2(Q)} =$$

$$= -\int_Q e^{-\frac{2nC_B^2 t}{\alpha_A}} \sum_{i,j=1}^{n} \frac{\partial}{\partial x_i}\left(\frac{\partial u_N}{\partial t} a_{ij} \frac{\partial u_N}{\partial x_j} \right) dQ + \int_Q \sum_{i,j=1}^{n} e^{-\frac{2nC_B^2 t}{\alpha_A}} a_{ij} \frac{\partial^2 u_N}{\partial x_i \partial t} \frac{\partial u_N}{\partial x_j} dQ.$$

Using the Ostrogradsky-Gauss formula we obtain that the first integral in the right-hand side equals to zero $\left(\frac{\partial u_N}{\partial t} \Big|_{x \in \partial\Omega} = 0 \right)$. Let us apply to the second integral the integration by parts. Thus, we have

$$I_2 = \frac{1}{2} \int_Q \sum_{i,j=1}^{n} \frac{\partial}{\partial t} \left(e^{-\frac{2nC_B^2 t}{\alpha_A}} a_{ij} \frac{\partial u_N}{\partial x_i} \frac{\partial u_N}{\partial x_j} \right) dQ +$$

$$+ \frac{1}{2} \int_Q \sum_{i,j=1}^{n} \frac{2nC_B^2}{\alpha_A} e^{-\frac{2nC_B^2 t}{\alpha_A}} a_{ij} \frac{\partial u_N}{\partial x_i} \frac{\partial u_N}{\partial x_j} dQ.$$

Applying the Ostrogradsky-Gauss formula again and taking into account the coercitivity condition: $\sum_{i,j=1}^{n} a_{ij} \xi_i \xi_j \geq \alpha_A \sum_{i=1}^{n} \xi_i^2$, $\forall x \in \overline{\Omega}$, $\forall \xi_i \in R^1$, we obtain

$$I_2 = \frac{1}{2}\int_\Omega \sum_{i,j=1}^n e^{-\frac{2nC_B^2}{\alpha_A}T} a_{ij} \frac{\partial u_N}{\partial x_i}\frac{\partial u_N}{\partial x_j}\Big|_{t=T} d\Omega +$$

$$+\frac{1}{2}\int_Q \sum_{i,j=1}^n \frac{2nC_B^2}{\alpha_A} e^{-\frac{2nC_B^2}{\alpha_A}t} a_{ij} \frac{\partial u_N}{\partial x_i}\frac{\partial u_N}{\partial x_j} dQ \ge \int_Q nC_B^2 e^{-\frac{2nC_B^2}{\alpha_A}t} \sum_{i=1}^n \left(\frac{\partial u_N}{\partial x_i}\right)^2 dQ.$$

The term I_3 we estimate in the following way.

$$I_3 = \left(e^{-\frac{2nC_B^2}{\alpha_A}t} \frac{\partial u_N}{\partial t}, \sum_{i=1}^n b_i \frac{\partial u_N}{\partial x_i} \right)_{L_2(Q)} \ge$$

$$\ge -\int_Q \sum_{i=1}^n C_B e^{-\frac{2nC_B^2}{\alpha_A}t} \left|\frac{\partial u_N}{\partial t}\right| \cdot \left|\frac{\partial u_N}{\partial x_i}\right| dQ.$$

Let us apply to the term I_4 the formula of integration by parts.

$$I_4 = \left(e^{-\frac{2nC_B^2}{\alpha_A}t} \frac{\partial u_N}{\partial t}, cu_N \right)_{L_2(Q)} \ge \frac{1}{2}\int_Q \frac{2nC_B^2}{\alpha_A} e^{-\frac{2nC_B^2}{\alpha_A}t} cu_N^2 dQ \ge 0.$$

Summing the obtained expressions, we have

$$\left(e^{-\frac{2nC_B^2}{\alpha_A}t} \frac{\partial u_N}{\partial t}, Lu_N \right)_{L_2(Q)} \ge \int_Q \left(e^{-\frac{2nC_B^2}{\alpha_A}t} \left(\frac{\partial u_N}{\partial t}\right)^2 + \right.$$

$$\left. + nC_B^2 e^{-\frac{2nC_B^2}{\alpha_A}t} \sum_{i=1}^n \left(\frac{\partial u_N}{\partial x_i}\right)^2 - C_B e^{-\frac{2nC_B^2}{\alpha_A}t} \sum_{i=1}^n \left|\frac{\partial u_N}{\partial t}\right| \cdot \left|\frac{\partial u_N}{\partial x_i}\right| \right) dQ =$$

$$= \int_Q \frac{1}{2} e^{-\frac{2nC_B^2}{\alpha_A}t} \left(\frac{\partial u_N}{\partial t}\right)^2 + e^{-\frac{2nC_B^2}{\alpha_A}t} \sum_{i=1}^n \frac{1}{2n}\left(\left|\frac{\partial u_N}{\partial t}\right| - C_B n \left|\frac{\partial u_N}{\partial x_i}\right| \right)^2 +$$

$$+e^{-\frac{2nC_B^2}{\alpha_A}t}\frac{nC_B^2}{2}\sum_{i=1}^{n}\left(\frac{\partial u_N}{\partial x_i}\right)^2 dQ \geq \int_Q \frac{1}{2}e^{-\frac{2nC_B^2}{\alpha_A}t}\left(\frac{\partial u_N}{\partial t}\right)^2 +$$

$$+e^{-\frac{2nC_B^2}{\alpha_A}t}\frac{nC_B^2}{2}\sum_{i=1}^{n}\left(\frac{\partial u_N}{\partial x_i}\right)^2 dQ \geq$$

$$\geq C\int_Q \left(\frac{\partial u_N}{\partial t}\right)^2 + \sum_{i=1}^{n}\left(\frac{\partial u_N}{\partial x_i}\right)^2 dQ \geq C\|u_N\|^2_{W_{bd}^+} \ .$$

Comparing the obtained expressions with (6) and using the integral Cauchy inequality, we obtain

$$\|f\|_{L_2(Q)}\|u_N\|_{W_{bd}^+} \geq \|f\|_{L_2(Q)}\left\|e^{-\frac{2nC_B^2}{\alpha_A}t}\frac{\partial u_N}{\partial t}\right\|_{L_2(Q)} \geq$$

$$\geq \left(f, e^{-\frac{2nC_B^2}{\alpha_A}t}\frac{\partial u_N}{\partial t}\right)_{L_2(Q)} \geq C\|u_N\|^2_{W_{bd}^+}.$$

Reducing both parts by $\|u_N\|_{W_{bd}^+}$, we finish to prove Lemma 1.

Theorem 1. *Let $f \in L_2(Q)$. The sequence of approximations (4) converges to the solution $u(t,x)$ of the problem (1), (2) in the sense of Definition 1.1.1 with respect to the norm of the space $L_2(Q)$.*

Proof. By Lemma 1, we can extract the weakly convergent subsequence $\{u_{N_k}\}_{k=1}^{\infty}$ from the bounded sequence $\{u_N\}_{N=1}^{\infty}$. Let $\{u_{N_k}\}_{k=1}^{\infty}$ be weakly convergent to $\hat{u} \in W_{bd}^+$ By the Banach-Sax theorem, we can extract from it a subsequence $\{u_{N_{k_i}}\}_{i=1}^{\infty}$ such that $\hat{u}_\nu \overset{def}{=} \frac{1}{\nu}\sum_{i=1}^{\nu} u_{N_{k_i}}$ strongly converges in W_{bd}^+ to the same function $\hat{u}$, i.e.

$$\left\|\hat{u}_\nu - \hat{u}\right\|_{W^+_{bd}} \xrightarrow[\nu\to\infty]{} 0.$$

By virtue of the compactness of the imbedding $W^+_{bd} \subset L_2(Q)$, it is strongly convergent in $L_2(Q)$.

Taking into account the inequality of Lemma 2.1.1

$$\|Lu\|_{W^-_{bd^+}} \le C\|u\|_{W^+_{bd}}$$

and the linearity of the operator L, we get

$$\left\|L\hat{u}_i - L\hat{u}_j\right\|_{W^-_{bd^+}} \le C\left\|\hat{u}_i - \hat{u}_j\right\|_{W^+_{bd}}.$$

By the fact that sequence $\{\hat{u}_\nu\}_{\nu=1}^{\infty}$ is fundamental, we have

$$\left\|L\hat{u}_i - L\hat{u}_j\right\|_{W^-_{bd^+}} \xrightarrow[i,j\to\infty]{} 0.$$

As far as the space $W^-_{bd^+}$ is complete the fundamental sequence $\{L\hat{u}_\nu\}_{\nu=1}^{\infty}$ has a limit $\hat{f} \in W^-_{bd^+}$, i.e.

$$\left\|L\hat{u}_\nu - \hat{f}\right\|_{W^-_{bd^+}} \xrightarrow[\nu\to\infty]{} 0.$$

Remark. *Element $\hat{f} \in W^-_{bd^+}$ is identically defined by $\hat{u} \in W^+_{bd}$ and does not depend on the sequence $u_m \in W^+_{bd}$: $\left\|u_m - \hat{u}\right\|_{W^+_{bd}} \xrightarrow[m\to\infty]{} 0$.*

Let us prove that $\hat{f} = f$ in the sense of the equality in the space $W^-_{bd^+}$. Let us multiply both parts of the equality (5) on an arbitrary smooth function $\varphi_l(t)$, which satisfies the condition $\varphi_l(0) = 0$ and integrate it with respect to t from 0 to T. Denoting

$$\varphi_l \omega_l = \psi_l,$$

we obtain

$$(\psi_l, Lu_N)_{L_2(Q)} = (\psi_l, f)_{L_2(Q)}, \; l = \overline{1,N}.$$

On the basis of the definition of the function $\hat{u}_\nu$ the following equalities hold true

$$(\psi_l, f)_{L_2(Q)} = (\psi_l, Lu_N)_{L_2(Q)} = (\psi_l, L\hat{u}_N)_{L_2(Q)}, \ \left(l = \overline{1, N_{k_1}}\right). \quad (7)$$

On the other hand, by the Schwarz inequality we have

$$\left|\left(\psi_l, L\hat{u}_N - \hat{f}\right)_{L_2(Q)}\right| \le \|\psi_l\|_{W^+_{bd^+}} \left\|L\hat{u}_N - \hat{f}\right\|_{W^-_{bd^+}}.$$

By virtue of the above reasoning, the right-hand side of this inequality tends to zero as $N \to \infty$. That is why it follows from (7) that

$$\left(\psi_l, \hat{f}\right)_{L_2(Q)} = (\psi_l, f)_{L_2(Q)}, \ l = \overline{1, N_{k_1}}. \quad (8)$$

We can make the number N_{k_1} be arbitrarily large (dropping the necessary of the first terms of the sequence u_N and repeating the analogous reasoning). In view of the fact that the set of functions $\{\psi_l\}_{l=1}^{\infty}$ is total in $W^+_{bd^+}$, we have $\hat{f} = f$ in $L_2(Q)$, whence it is easy to prove that $\hat{u}$ is the solution of the problem (1), (2) in the sense of definition 1.1.1.

By virtue of the uniqueness of the solution of the problem (1), (2), all sequence $\{u_N\}_{N=1}^{\infty}$ converges to $\hat{u}$ with respect to the norm $L_2(Q)$.

Let us consider the case when the right-hand side of the equation (1) is an element of the negative Hilbert space $W^-_{bd^+}$. Granting the density of the space $L_2(Q)$ in $W^-_{bd^+}$, let us select a sequence $f_p \in L_2(Q): \|f - f_p\|_{W^-_{bd^+}} \xrightarrow[p\to\infty]{} 0$ (we can do it applying the well-known procedure of averaging).

Consider the problem (1), (2) with the right-hand side $f_p(t,x)$. We shall look for the approximate solution of this problem in the form

$$u_{N,p} = \sum_{i=1}^{N} g_{i,p}(t)\omega_i(x), \tag{9}$$

where $\omega_i(x)$ is above defined orthonormal basis in $L_2(Q)$, and $g_{i,p}(t)$ is determined from the solution of the Cauchy problem for the system of linear ordinary differential equations:

$$\frac{dg_{l,p}}{dt} + \sum_{k=1}^{N} g_{k,p}\left(\left(\sum_{i,j=1}^{n}\left(-\frac{\partial}{\partial x_i}\left(a_{ij}\frac{\partial}{\partial x_j}\right)\right) + \sum_{i=1}^{n} b_i \frac{\partial}{\partial x_i} + c\right)\omega_k, \omega_l\right)_{L_2(\Omega)} =$$

$$= \left(f_p, \omega_l\right)_{L_2(\Omega)},$$

$$g_{i,p}(0) = 0,\ i = \overline{1,N},\ l = \overline{1,N}. \tag{10}$$

As far as the estimations in the negative norms are valid for the operator L passing to the limit $p \to \infty$ we shall obtain that the following statement holds true

Theorem 2. *Let $f \in W^-_{bd^+}$. Then the sequence $\{u_{N(p),p}(t,x)\}_{p=1}^{\infty}$, obtained by the method (9), (10) with the help of the special choosing of $N = N(p)$, which necessarily exists, converges to the generalized solution from the space $L_2(Q)$ (in the sense of Definition 1.1.4) of the problem (1), (2) in the following sense*

$$\left\| u_{N(p),p} - u \right\|_{L_2(Q)} \xrightarrow[p\to\infty]{} 0.$$

Proof. By Theorem 1, sequences of the approximations $\{u_{N,p}(t,x)\}_{N=1}^{\infty}$ converge to the generalized solutions of the problems $Lu_p = f_p$ with respect to the norm $L_2(Q)$, i.e.

$$\left\| u_{N,p} - u_p \right\|_{L_2(Q)} \xrightarrow[N\to\infty]{} 0.$$

Let us show that the sequence $\{u_p(t,x)\}_{p=1}^{\infty}$ is fundamental in the space $L_2(Q)$. We have

$$\|u_p - u_l\|_{L_2(Q)} \le \|u_p - u_{N,p}\|_{L_2(Q)} + \|u_{N,p} - u_{N,l}\|_{L_2(Q)} + \|u_{N,l} - u_l\|_{L_2(Q)}.$$

Next,

$$\|u_p - u_l\|_{L_2(Q)} \le \|u_p - u_{N,p}\|_{L_2(Q)} + C\|f_p - f_l\|_{W^-_{bd^+}} + \|u_{N,l} - u_l\|_{L_2(Q)}.$$

Here we used the estimation of Lemma 1 and the triangle inequality.

Let us pass to the limit with respect to $N \to \infty$. We shall obtain that $\|u_p - u_l\|_{L_2(Q)} \le C\|f_p - f_l\|_{W^-_{bd^+}} \xrightarrow[p,l\to\infty]{} 0$. Hence, there exists a function $u \in L_2(Q)$, such that $\|u - u_p\|_{L_2(Q)} \xrightarrow[p\to\infty]{} 0$. It follows from the proof of Theorem 1.1.3 that the function $u \in L_2(Q)$ is a generalized solution of the problem (1), (2) in the sense of Definition 1.1.4 with the right-hand side $f(t,x) \in W^-_{bd^+}$.

Consider $\|u_{N,p} - u_p\|_{L_2(Q)}$. On the basis of Theorem 1 we can choose $N = N(p)$ such that $\|u_{N(p),p} - u_p\|_{L_2(Q)} < \delta_p$, where $\{\delta_p\}_{p=1}^{\infty}$ is a sequence of positive numbers converging to zero.

Hence,

$$\|u_{N(p),p} - u\|_{L_2(Q)} \le \|u_{N(p),p} - u_p\|_{L_2(Q)} + \|u_p - u\|_{L_2(Q)} \le$$
$$\le \delta_p + \|u_p - u\|_{L_2(Q)} \xrightarrow[p\to\infty]{} 0.$$

The theorem is proved.

Thus, in the case when the right-hand side belongs to the negative space $W^-_{bd^+}$, it is necessary to solve the Cauchy problem (10) and to co-ordinate the parameters N and p in a special manner.

3. PULSE OPTIMAL CONTROL OF PARABOLIC SYSTEMS

Let us consider the problems connected with the pulse optimal control in the case when the function of the system state is a solution of the Dirichlet initial-boundary value problem for parabolic equation:

$$Lu = f + A(h), \tag{1}$$

$$u\big|_{t=0} = 0,\ u\big|_{x\in\partial\Omega} = 0, \tag{2}$$

All notations correspond to Chapter 1, where we proved the inequalities in the negative norms for the parabolic operator.

Using the templates of the theorems mentioned in Chapter 3, let us write the tables for this theorems.

Table 1.

N	Operator	Space N	Space $W^-(Q)$
1.	L	$L_2(Q)$	$W^-_{bd^+}$
2.	L	W^+_{bd}	$L_2(Q)$

Table 2.

N	Operator $A_i(\cdot)$	Space $W^-(Q)$
1.	$A_1(\cdot)$	W^+_{bd}
2.	$A_3(\cdot)$	W^+_{bd}
3.	$A_5(\cdot)$	W^+_{bd}

Table 3.

N	Operator L	Space $W^-(Q)$	Space $W^+(Q)$
1.	L	$W^-_{bd^+}$	$W^+_{bd^+}$

Table 4.

N	Space $W^-(Q)$	Mapping $A_i(\cdot)$
1.		

Table 5.

N	Operator L	Space $W^-(Q)$
1.	L	$W^-_{bd^+}$

Table 6.

N	Operator L	Mapping $A_i(\cdot)$	Space $W^-(Q)$
1.	L	$A_1(\cdot)$	$W^-_{bd^+}$
2.	L	$A_3(\cdot)$	$W^-_{bd^+}$
3.	L	$A_5(\cdot)$	$W^-_{bd^+}$

Table 7.

N	Index α	Space $W^-(Q)$	Mapping $A_{i,\varepsilon}(\cdot)$
1.	1/2	$W^-_{bd^+}$	$A_{1,\varepsilon}(\cdot)$
2.	1/2	$W^-_{bd^+}$	$A_{3,\varepsilon}(\cdot)$
3.	1/2	$W^-_{bd^+}$	$A_{5,\varepsilon}(\cdot)$

Let us investigate the problem of the existence of the optimal control of coefficients of the equation.

Let the system state is described by the following equation:

$$Lu = \frac{\partial u}{\partial t} - \sum_{i,j=1}^{N} \frac{\partial}{\partial x_i}\left(a_{ij}(x;h_1)\frac{\partial u}{\partial x_j} \right) + \sum_{i=1}^{N} b_i(x;h_2)\frac{\partial u}{\partial x_i} + c(x;h_3)u = \tag{3}$$
$$= f(x;h_4),$$
$$u\big|_{t=0} = 0; \quad u\big|_{x\in\partial\Omega} = 0,$$

where $u(t,x)$ is a function describing the system state in the domain $Q=(0,T)\times\Omega$, $\Omega=[x_1',x_2'']\times\Omega'$ is a bounded domain in n-dimensioned Euclidian space R^n with smooth boundary $\partial\Omega$, which is a tube with respect to the variable x_1. The control of the system is carried out with the help of control impacts $h=\left(h_{ij}^{(1)},h_i^{(2)},h^{(3)},h^{(4)}\right)$. The functional $\Phi(u(h))=J(h)$, which is to be minimized on the set U_{ad} of admissible controls, is defined on the solutions of the problem (3) and is weakly lower semi-continuous with respect to the system state $u(t,x)$.

Let $a_{ij}(x)=a_{ji}(x)$, $b_i(x)$ be continuously differentiable function in the closed domain $\overline{\Omega}$, and $c(x)$ be continuous function in $\overline{\Omega}$. We shall assume that $h_{ij}^{(1)}$, $h_i^{(2)}$, $h^{(3)}$ are some numerical parameters and for all values of this parameters

$$\sum_{i,j=1}^{n} a_{ij}\xi_i\xi_j \geq \alpha_A \sum_{i=1}^{n}\xi_i^2, \quad \forall\xi_i\in R^1,$$
$$c(x) \geq \sum_{i=1}^{n}\frac{\partial b_i(x)}{\partial x_i}, \tag{4}$$
$$c(x)\geq 0,$$

where the constant $\alpha_A > 0$.

Let us consider the specific case of the control impact which arise in practice and investigate the problem of existence of the optimal controls.

Let the following conditions are satisfied:

$$a_{ij}\left(x;h_{ij}^{(1)}\right)=a_{ij}(x)h_{ij}^{(1)},\ h_{ij}^{(1)}\in\left[a_{ij}^{(1)},b_{ij}^{(1)}\right]\subset R,\ \left(a_{ij}^{(1)}>0\right),$$
$$b_i\left(x;h_i^{(2)}\right)=b_i(x)h_i^{(2)},\ h_i^{(2)}\in\left[a_i^{(2)},b_i^{(2)}\right]\subset R,\ \left(a_i^{(2)}>0\right),\qquad(5)$$
$$c\left(x;h^{(3)}\right)=c(x)h^{(3)},\ h^{(3)}\in\left[a^{(3)},b^{(3)}\right]\subset R,\ \left(a^{(3)}>0\right),$$
$$f\left(t,x,h^{(4)}\right)=\left(\sum_{k=1}^{m}h_k^{(4)}\psi_k\left(t,x_2,\ldots x_n\right)\right)\otimes\delta\left(x_1-\xi_k\right),$$

where the vector $h_{ij}^{(1)},\ h_i^{(2)},\ h^{(3)},\ h_k^{(4)}$ is a control.

Then the following statement holds true.

Theorem 10. *Let the system state is determined as a solution of the problem (3), (2) and the following conditions are satisfied:*

1) the performance criterion $\Phi(\cdot):L_2(Q)\to R^1$ *is a weakly lower semi-continuous with respect to the system state* $u(t,x;\varphi)$ *functional;*

2) the set of admissible controls $U_{ad}\subset H$ *is bounded, closed and convex in H;*

3) H is a reflexive Banach space;

4) f *is a weakly continuous operator mapping H into* $W_{bd^+}^-$.

Under these assumptions there exists an optimal control of system (3), (2).

Proof. Note that by virtue of the relations (4), (5) all statements proved in the Section 1 for the parabolic operator are valid for differential operators (3). Moreover, the constant C in the estimations with respect to the negative norms (Lemmas 1.1-1.4) does not depend on the control $h_{ij}^{(1)}, h_i^{(2)}, h^{(3)}$. Hence, the proof of Theorem 1 is completely similar to the proof of the general theorem about existence of optimal control (Theorem 1.2.1)

Remark. *As far as the operator f is non-linear then the functional $J(h)$ may be non-convex and the optimal control may be non-unique.*

Let the functioning of the system is described by the equation (3) with the conditions (4), (5).

Let us investigate the differential properties of the performance criterion

$$J(h) = \Phi(u(h)) = \sum_{i=1}^{p} \int_Q \alpha_i(t,x)(u(h) - u_i)^2 \, dQ\,, \qquad (6)$$

where $u_i(t,x)$, $\alpha_i(t,x)$ are some unknown functions from $L_2(Q)$ and $C(\overline{Q})$, respectively, and $\alpha_i(t,x) \geq \varepsilon > 0$ in $\overline{Q}$ for the case of the optimal control of the system coefficients.

Let us give to the control h an increment $\lambda\Delta h$:

$$\Delta J(h) = J(h + \lambda\Delta h) - J(h) =$$

$$= \sum_{i=1}^{p} \int_Q \alpha_i(t,x) \Delta u (u(h + \lambda\Delta h) + u(h) - 2u_i) dQ =$$

$$= 2\sum_{i=1}^{p} \int_Q \alpha_i(t,x) \Delta u (u(h) - 2u_i) dQ + \sum_{i=1}^{p} \int_Q \alpha_i(t,x) |\Delta u|^2 \, dQ\,.$$

where $\Delta u = u(h + \lambda\Delta h) - u(h)$.

Let us introduce the adjoint state as a solution of the following problem

$$L^* v = 2\sum_{i=1}^{p} \int_Q \alpha_i(t,x)(u(h) - u_i) dQ\,, \qquad (7)$$

$$v|_{t=T} = 0\,,\ v|_{x\in\partial\Omega} = 0\,.$$

Then we may write the increment of the performance criterion in the following form:

$$\Delta J(h) = (\Delta u, L^* v)_{L_2(Q)} + \sum_{i=1}^{p} \int_Q \alpha_i(t,x) |\Delta u|^2 \, dQ\,.$$

As far as

$$L_{h+\lambda\Delta h}(u(h+\lambda\Delta h))-L_h(u(h))=f\left(h^{(4)}+\lambda\Delta h^{(4)}\right)-f\left(h^{(4)}\right),$$

then, obviously, Δu satisfies the equation

$$\frac{\partial\Delta u}{\partial t}-\sum_{i,j=1}^{N}\frac{\partial}{\partial x_i}\left(a_{ij}(x)h_{ij}^{(1)}\frac{\partial\Delta u}{\partial x_j}\right)+\sum_{i=1}^{N}b_i(x)h_i^{(2)}\frac{\partial\Delta u}{\partial x_i}+c(x)h_i^{(3)}\Delta u=$$

$$=\sum_{i,j=1}^{N}\lambda\frac{\partial}{\partial x_i}\left(a_{ij}(x)\Delta h_{ij}^{(1)}\frac{\partial u(h+\lambda\Delta h)}{\partial x_j}\right)-$$

$$-\sum_{i=1}^{N}\lambda b_i(x)\Delta h_i^{(2)}\frac{\partial u(h+\lambda\Delta h)}{\partial x_i}-\lambda c(x)\Delta h_i^{(3)}u(h+\lambda\Delta h)+$$

$$+\lambda\sum_{k=1}^{m}\Delta h_k^{(4)}\psi_k(t,x_2,\dots,x_n)\otimes\delta(x_1-\xi_k).$$

It follows from the results of Section 1.1 that the solution of this problem exists, it is unique and it is determined as a function $\Delta u\in L_2(Q)$ such that for any $y(t,x)\in W^{+}_{bd^{+}}$: $L^{*}y\in L_2(Q)$ (including $v(t,x)$), the following equality holds true:

$$\left(\Delta u,L^{*}y\right)_{L_2(Q)}=\int_Q\left(-\sum_{i,j=1}^{N}\lambda a_{ij}(x)\Delta h_{ij}^{(1)}\frac{\partial u(h+\lambda\Delta h)}{\partial x_j}\frac{\partial y}{\partial x_i}-\right.$$

$$\left.-\sum_{i=1}^{N}\lambda b_i(x)\Delta h_i^{(2)}\frac{\partial u(h+\lambda\Delta h)}{\partial x_i}y-\lambda c(x)\Delta h_i^{(3)}u(h+\lambda\Delta h)y\right)dQ+$$

$$+\int_{[0,T]\times\Omega'}\lambda\sum_{k=1}^{m}\Delta h_k^{(4)}\psi_k(t,x_2,\dots,x_n)y(t,\xi_k,x_2,\dots,x_n)dtd\Omega'.$$

Thus,

$$\Delta J(h)=\int_Q\left(-\sum_{i,j=1}^{N}\lambda a_{ij}(x)\Delta h_{ij}^{(1)}\frac{\partial u(h+\lambda\Delta h)}{\partial x_j}\frac{\partial v}{\partial x_i}-\right.$$

$$-\sum_{i=1}^{N}\lambda b_i(x)\Delta h_i^{(2)}\frac{\partial u(h+\lambda\Delta h)}{\partial x_i}v-\lambda c(x)\Delta h_i^{(3)}u(h+\lambda\Delta h)v\Bigg)dQ+$$

$$+\int\limits_{[0,T]\times\Omega'}\lambda\sum_{k=1}^{m}\Delta h_k^{(4)}\psi_k(t,x_2,\dots,x_n)v(t,\xi_k,x_2,\dots,x_n)dtd\Omega'+$$

$$+\sum_{i=1}^{p}\int\limits_{Q}\alpha_i(t,x)|\Delta u|^2dQ.$$

Dividing both parts by λ and passing to the limit as $\lambda\to+0$, we have

$$\lim_{\lambda\to+0}\frac{\Delta J(h)}{\lambda}=(J'(h),\Delta h)=\int\limits_{Q}\Bigg(-\sum_{i,j=1}^{N}a_{ij}(x)\frac{\partial u(h)}{\partial x_j}\frac{\partial v}{\partial x_i}\Delta h_{ij}^{(1)}-$$

$$-\sum_{i=1}^{N}b_i(x)\frac{\partial u(h)}{\partial x_i}v\Delta h_i^{(2)}-c(x)u(h)v\Delta h_i^{(3)}\Bigg)dQ+ \tag{8}$$

$$+\sum_{k=1}^{m}\int\limits_{[0,T]\times\Omega'}\psi_k v(t,\xi_k,x_2,\dots,x_n)dtd\Omega'\cdot\Delta h_k^{(4)}.$$

Thus, we have proved the following statement.

Theorem 11. *The gradient of the performance criterion (6) is of the following form:*

$$gradJ(h)=$$

$$\left(\left(-\int\limits_{Q}a_{ij}(x)\frac{\partial u(h)}{\partial x_j}\frac{\partial v}{\partial x_i}dQ\right)_{i,j=1}^{N};\left(-\int\limits_{Q}b_i(x)\frac{\partial u(h)}{\partial x_i}vdQ\right)_{i=1}^{N};\right.$$

$$\left.\left(-\int\limits_{Q}c(x)u(h)vdQ\right);\left(\int\limits_{[0,T]\times\Omega'}\psi_k v(t,\xi_k,x_2,\dots,x_n)dtd\Omega'\right)_{k=1}^{m}\right).$$

Remark. *It should been noted that the passing to the limit in (8) was done formally, but its justification can be proved easily.*

Chapter 5

PSEUDO-PARABOLIC SYSTEMS

In order to solve many applied problems of science and engineering it is necessary to study optimal control problems for pseudo-parabolic systems:

$$Lu \equiv (Au)_t + Bu = f(t,x), \tag{1}$$

where A, B are second order elliptic differential operators.

Seemingly one of the first paper on pseudo-parabolic equation is [111], where the equation was obtained from the research of heat transport processes in the heterogeneous environment, as more adequate model of the processes. Pseudo-parabolic equations arise in researches of the fluid and gas filtration in the fissured and porous medium [112-116]; the heat conduction in the heterogeneous environment [111], the ion migration in soil [111-113, 117], the wave propagation in the disperse medium and in the thin elastic glass [118]. Nowadays , there are many papers on pseudo-parabolic equations [51, 52, 90, 119-128].

1. GENERALIZED SOLVABILITY OF PSEUDO-PARABOLIC EQUATIONS (THE DIRICHLET INITIAL BOUNDARY PROBLEM)

Consider equation (1) in a tube $Q \equiv (0,T)\times\Omega$, where Ω is a regular domain in R^n with a piecewise-smooth boundary $\partial\Omega$. The operators A, B are defined as

$$Au \equiv -\sum_{i,j=1}^{n}\left(a_{ij}(x)u_{x_j}\right)_{x_i} + a(x)u, \tag{2à}$$

$$Bu \equiv -\sum_{i,j=1}^{n}\left(b_{ij}(x)u_{x_j}\right)_{x_i} + b(x)u, \tag{2b}$$

where $a_{ij}(x) = a_{ji}(x)$, $b_{ij}(x) = b_{ji}(x)$; $\{a_{ij}\}_{i,j=1}^{n}$, $\{b_{ij}\}_{i,j=1}^{n}$ are continuously differentiable functions in the closed domain $\overline{\Omega}$, and $a(x)$, $b(x)$ are continuous functions in the $\overline{\Omega}$.

We suppose that the differential expression (2a) is positive definite and the differential expression (2b) is nonnegative in the domain Ω i.e.

$$\sum_{i,j=1}^{n} a_{ij}\xi_i\xi_j \geq \alpha \sum_{i=1}^{n} \xi_i^2; \quad \sum_{i,j=1}^{n} b_{ij}\xi_i\xi_j \geq 0, \tag{3}$$

where α is a positive constant, $\xi_i \in R^1$, $i = \overline{1,n}$.

We also suppose that $a(x) \geq 0$, $b(x) \geq 0$.

Introduce into consideration the following spaces. Let W_{bd}^{+} be a completion of the set of the smooth functions in the domain $\overline{Q}$, which satisfy the conditions

$$u\big|_{t=0} = 0; \quad u\big|_{x\in\partial\Omega} = 0, \tag{4}$$

in the norm

$$\|u\|_{W_{bd}^{+}} = \left(\int_Q u_t^2 + \sum_{i,j=1}^{n} a_{ij} u_{x_i} u_{x_j} dQ \right)^{\frac{1}{2}}. \tag{5}$$

Let $W_{bd^+}^{+}$ be a completion of the set of the smooth functions in the domain $\overline{Q}$, which satisfy the adjoint conditions

$$v\big|_{t=T} = 0; \quad v\big|_{x\in\partial\Omega} = 0 \tag{6}$$

in the same norm (5); W_{bd}^{-}, $W_{bd^+}^{-}$ are corresponding negative spaces ($W_{bd}^{+}, \subset L_2(Q) \subset W_{bd}^{-}$, $W_{bd^+}^{+} \subset L_2(Q) \subset W_{bd^+}^{-}$). Let H_{bd}^{+} be a completion of the set of the smooth functions in the domain $\overline{Q}$, which satisfy the conditions (4) by the norm

$$\|u\|_H^2 = \int_Q \left(u^2 + \sum_{i,j=1}^{n} a_{ij} u_{x_i} u_{x_j} \right) dQ,$$

$H^{+}_{bd^{+}}$ be completion of the set of the smooth functions in the domain $\overline{Q}$, which satisfy the conditions (6) in the same norm. Let H^{-}_{bd}, $H^{-}_{bd^{+}}$ be corresponding negative spaces.

Lemma 1. *For all functions* $u(t,x)\in W^{+}_{bd}$ *the following inequality is true*

$$\|Lu\|_{W^{-}_{bd^{+}}} \le C\|u\|_{W^{+}_{bd}}.$$

Proof. First consider a smooth function $u(t,x)$, that satisfies the conditions (4), then applying expansion by continuity of operator $L(\cdot)$ and passing to the limit, we shall obtain the inequality for all functions $u\in W^{+}_{bd}$.

By the definition of the negative norm, we have

$$\|Lu\|_{W^{-}_{bd^{+}}} = \sup_{v\neq 0, v\in W^{+}_{bd^{+}}} \frac{|\langle Lu,v\rangle|}{\|v\|_{W^{+}_{bd^{+}}}} = \sup_{v\neq 0, v\in W^{+}_{bd^{+}}} \frac{|(Lu,v)_{L_2(Q)}|}{\|v\|_{W^{+}_{bd^{+}}}}. \qquad (7)$$

Consider $(Lu,v)_{L_2(Q)}$. Employing the integration by parts, the Schwarz inequality and the conditions (6), we obtain

$$\left|\int_Q a(x)u_t v\,dQ\right| \le \left(\int_Q a(x)v^2 dQ\right)^{1/2}\left(\int_Q a(x)u_t^2 dQ\right)^{1/2} \le C\|v\|_{W^{+}_{bd^{+}}}\|u\|_{W^{+}_{bd}}.$$

Here we apply the inequality

$$\left(\int_Q v^2 dQ\right)^{1/2} \le C\|v\|_{W^{+}_{bd^{+}}}$$

In the same manner, we have

$$\left|\int_Q b(x)uv\,dQ\right| \le \left(\int_Q b(x)v^2 dQ\right)^{1/2}\left(\int_Q b(x)v^2 dQ\right)^{1/2} \le C\|v\|_{W^{+}_{bd^{+}}}\|u\|_{W^{+}_{bd}}$$

Using the integration by parts, we find

$$\left|-\int_Q \sum_{i,j=1}^{n} \left(a_{ij} u_{x_j}\right)_{x_i t} v dQ\right| =$$

$$= \left|-\int_Q \sum_{i,j=1}^{n} \left(a_{ij} u_{x_j t} v\right)_{x_i} dQ + \int_Q \sum_{i,j=1}^{n} a_{ij} u_{x_j t} v_{x_i} dQ\right|.$$

Passing to the integration over surface and taking into account the conditions (6), we conclude that

$$-\int_Q \sum_{i,j=1}^{n} \left(a_{ij} u_{x_j t} v\right)_{x_i} dQ = 0.$$

Using the Schwarz inequality again, we obtain

$$\left|-\int_Q \sum_{i,j=1}^{n} \left(a_{ij} u_{x_i}\right)_{x_j t} v dQ\right| = \left|\int_Q \sum_{i,j=1}^{n} v_{x_i} a_{ij} u_{x_j t} dQ\right| \leq C\|v\|_{W^+_{bd^+}} \|u\|_{W^+_{bd}}.$$

Analogously

$$\left|\int_Q \sum_{i,j=1}^{n} \left(b_{ij} u_{x_j}\right)_{x_i} v dQ\right| \leq C\|v\|_{W^+_{bd^+}} \|u\|_{W^+_{bd}}.$$

Substituting the inequality into the parity (7), we have the desired inequality.

The analogous lemma is valid for the adjoint operator L^*.

Lemma 2. *For all functions* $v(t,x) \in W^+_{bd^+}$ *the following inequality is true*

$$\|L^* v\|_{W^-_{bd^+}} \leq C\|v\|_{W^+_{bd^+}},$$

where $L^* v = -Av_t + Bv$.

It follows from these lemmas that the operator L (and L^*) can be extended to the continuous operator mapping W^+_{bd}, (and $W^+_{bd^+}$, respectively) into $W^-_{bd^+}$ $\left(W^-_{bd}\right)$.

Lemma 3. *For all functions* $u(t,x) \in W^+_{bd}$ *the following inequality is true*

$$\|Lu\|_{W^-_{bd^+}} \geq C\|u\|_{H^+_{bd}}.$$

P r o o f. First consider a smooth function $u(t,x)$, which satisfies conditions (4).

Let $v(t,x)$ be an auxiliary function in the form

$$v(t,x) = -\int_T^t (\tau+1)^{-1} u(\tau,x)d\tau.$$

It is obvious that $v \in W^+_{bd^+}$. Prove the following inequality

$$(Lu,v)_{L_2(Q)} \geq C\|v\|^2_{W^+_{bd^+}}. \tag{8}$$

Using the integration by parts, the relationship between $u(t,x)$ and $v(t,x)$, the boundary conditions, we have

$$\int_Q va(x)u_t dQ = \int_Q (va(x)u)_t dQ + \int_Q (t+1)a(x)v_t^2 dQ \geq$$

$$\geq \int_Q a(x)v_t^2 dQ \geq 0,$$

$$\int_Q vb(x)u dQ = -\int_Q vb(x)(t+1)v_t dQ = -\frac{1}{2}\int_Q \left(b(x)(t+1)v^2\right)_t dQ +$$

$$+\frac{1}{2}\int_Q b(x)v^2 dQ = \frac{1}{2}\int_\Omega b(x)v^2\big|_{t=0} d\Omega + \frac{1}{2}\int_Q b(x)v^2 dQ \geq 0.$$

Next, in much the same manner, we have

$$-\int_Q v\sum_{i,j=1}^n \left(a_{ij}u_{x_j}\right)_{x_i t} dQ \geq C\int_Q \sum_{i,j=1}^n a_{ij}v_{x_i t}v_{x_j t} dQ,$$

$$-\int_Q v\sum_{i,j=1}^n \left(b_{ij}u_{x_j}\right)_{x_i} dQ = -\int_Q (t+1)\sum_{i,j=1}^n v_{x_i} b_{ij} v_{x_j t} dQ. \tag{9}$$

Employing the integration by parts, we have

$$-\int_Q (t+1)\sum_{i,j=1}^{n} b_{ij} v_{x_i} v_{x_j t} dQ =$$

$$= -\frac{1}{2}\int_Q \left((t+1)\sum_{i,j=1}^{n} b_{ij} v_{x_i} v_{x_j}\right)_t dQ + \frac{1}{2}\int_Q \sum_{i,j=1}^{n} b_{ij} v_{x_i} v_{x_j} dQ.$$

Passing to the surface integral, since the symmetry and non-negativity of the matrix $\{b_{ij}\}_{i,j=1}^{n}$, we arrive

$$-\int_Q (t+1)\sum_{i,j=1}^{n} b_{ij} v_{x_i} v_{x_j t} dQ = \tag{10}$$

$$= \frac{1}{2}\int_\Omega \sum_{i,j=1}^{n} b_{ij} v_{x_i} v_{x_j}\bigg|_{t=0} d\Omega + \frac{1}{2}\int_Q \sum_{i,j=1}^{n} b_{ij} v_{x_i} v_{x_j} dQ \geq 0.$$

Substituting (10) into (9) from the obtained inequalities, we have

$$(Lu,v)_{L_2(Q)} \geq C\left(\int_Q \sum_{i,j=1}^{n} a_{ij} v_{x_i t} v_{x_j t} dQ\right),$$

whence, applying the obvious inequality

$$\int_Q \sum_{i,j=1}^{n} a_{ij} v_{x_i t} v_{x_j t} dQ \geq C\|v\|^2_{W^+_{bd^+}},$$

we conclude that the inequality (8) is valid.

By the Schwarz inequality, we have

$$C\|v\|^2_{W^+_{bd^+}} \leq (Lu,v)_{L_2(Q)} \leq \|Lu\|_{W^-_{bd^+}} \|v\|_{W^+_{bd^+}}.$$

Reduce by $\|v\|_{W^+_{bd^+}}$ and take account the relationship between $u(t,x)$ and $v(t,x)$. Whence we have that desired inequality of the lemma is true for a smooth function. Passing to the limit, we obtain the inequality for all functions $u \in W^+_{bd}$.

L e m m a 4. *For any function* $v(t,x) \in W^+_{bd^+}$ *the following inequality holds true*

$$\left\|L^{*}v\right\|_{W_{bd}^{-}} \geq C\left\|v\right\|_{H_{bd^{+}}^{+}}.$$

Proof. Let first $v(t,x) \in W_{bd^{+}}^{+}$ be a smooth function, which satisfies conditions (6). Prove the following inequality

$$(u, L^{*}v)_{L_2(Q)} \geq C\left\|u\right\|_{W_{bd^{+}}^{+}}^{2}, \tag{11}$$

where

$$u(t,x) = \int_{0}^{t}(2T-\tau)^{-1}v(\tau,x)d\tau.$$

Using the integration by parts and the definition of the function $u(t,x)$,we have

$$-\int_{Q} ua(x)v_t dQ = -\int_{Q}(ua(x)v)_t dQ + \int_{Q}(2T-t)a(x)u_t^2 dQ =$$
$$= \int_{Q}(2T-t)a(x)u_t^2 dQ \geq 0.$$

Since the functions $u(t,x)$ and $v(x,t)$ satisfy homogeneous conditions on the border of the set $[0,T]$, we have:

$$\int_{Q} ub(x)vdQ = \int_{Q} ub(x)(2T-t)u_t dQ = \frac{1}{2}\int_{Q}\left(b(x)(2T-t)u^2\right)_t dQ +$$
$$+\frac{1}{2}\int_{Q} b(x)u^2 dQ = \frac{1}{2}\int_{\Omega} b(x)Tu^2\big|_{t=0} d\Omega + \int_{Q} b(x)u^2 dQ \geq 0.$$

Next, we obtain

$$\begin{aligned} &\int_{Q} u\sum_{i,j=1}^{n}\left(a_{ij}v_{tx_j}\right)_{x_i} dQ \geq \int_{Q}(2T-t)\sum_{i,j=1}^{n} a_{ij}u_{x_i t}u_{x_j t}dQ,\\ &-\int_{Q} u\sum_{i,j=1}^{n}\left(b_{ij}v_{x_j}\right)_{x_i} dQ = \int_{Q}(2T-t)\sum_{i,j=1}^{n} b_{ij}u_{x_i}u_{x_i t}dQ. \end{aligned} \tag{12}$$

Since the matrix $\{b_{ij}\}_{i,j=1}^{n}$ is symmetric and nonnegative, we have the inequality

$$\int_Q (2T-t)\sum_{i,j=1}^{n} b_{ij}u_{x_i}u_{x_jt}dQ \geq 0. \qquad (13)$$

Taking into account (12), (13) and the inequality

$$c\left(\int_Q u_t^2 + \sum_{i,j=1}^{n} a_{ij}u_{x_it}u_{x_jt}dQ\right) \leq \int_Q \sum_{i,j=1}^{n} a_{ij}u_{x_it}u_{x_jt}dQ,$$

we obtain (11). Next, in much the same manner as in Lemma 3, applying the Schwarz inequality and passing to the limit, we prove desired inequality for all $v \in W^{+}_{bd^{+}}$.

Based on Lemmas 1-4 and results of Chapter 1, we have the following theorems.

Theorem 1. *For all $f \in H^{-}_{bd^{+}}$ there exists a unique solution of the problem (1), (4) in the sense of Definition 1.1.1.*

Theorem 2. *For all $f \in W^{-}_{bd^{+}}$ there exists a unique solution of the problem (1), (4) in the sense of Definition 1.1.4.*

2. GENERALIZED SOLVABILITY OF PSEUDO-PARABOLIC EQUATIONS (THE NEUMANN INITIAL BOUNDARY PROBLEM)

In this section we shall consider the simpler (comparing to the general equation (1.1)) pseudo-parabolic equations. This simplification enables us to consider the initial boundary value problem with usual von Neumann conditions (compare with the initial boundary conditions of pseudo-hyperbolic systems).

Consider the partial differential equation:

$$Lu = u_t - \Delta(u_t + ku) = f(t,x), \qquad (1)$$

in a tube $Q = (0,T)\times\Omega$, where $u(t,x)$ is a sought function, $x \in \Omega$, $t \in (0,T)$, Ω is a bounded domain in R^n with a smooth border $\partial\Omega$, Δ is a Laplacian, k is a nonnegative constant.

Together with the equation (1), consider the boundary conditions

$$u\big|_{t=0} = 0, \quad \frac{\partial u}{\partial \vec{n}}\Big|_{x\in\partial\Omega} = 0, \tag{2}$$

where $\vec{n}$ is a normal vector to the surface $\partial\Omega$.

Introduce the following denotation. Let W_{bd}^{+} be a completion of the set of the smooth in $\overline{Q}$ functions, which satisfy the condition (2) in the norm

$$\|u\|_{W_{bd}^{+}} = \left(\int_{Q} u_t^2 + \sum_{i=1}^{n} u_{x_i t}^2 \, dQ \right)^{1/2},$$

$W_{bd^+}^{+}$ is an analogous space, but functions satisfy the adjoint conditions

$$v\big|_{t=T} = 0, \quad \frac{\partial v}{\partial \vec{n}}\Big|_{x\in\partial\Omega} = 0, \tag{3}$$

W_{bd}^{-}, $W_{bd^+}^{-}$ are corresponding negative spaces.

Lemma 1. *For all functions* $u(t,x) \in W_{bd}^{+}$ *the following inequality is true*

$$\|Lu\|_{W_{bd^+}^{-}} \le c\|u\|_{W_{bd}^{+}}.$$

Proof. First prove the lemma for smooth functions $u(t,x) \in W_{bd}^{+}$, which satisfy the condition (2), and then passing to the limit, we shall obtain the inequality for all functions $u(t,x) \in W_{bd}^{+}$.

By definition of the negative norm in $W_{bd^+}^{-}$, we have

$$\|Lu\|_{W_{bd^+}^{-}} = \sup_{\substack{v\neq 0,\\ v\in W_{bd^+}^{+}}} \frac{\left|\langle Lu, v\rangle_{W_{bd^+}}\right|}{\|v\|_{W_{bd^+}^{+}}}.$$

Since the bilinear form $\langle \cdot,\cdot \rangle_{W_{bd^+}}$ on the smooth functions coincides with the inner product in $L_2(Q)$, we obtain

$$\|Lu\|_{W^-_{bd^+}} = \sup_{\substack{v\neq 0,\\ v\in W^+_{bd^+}}} \frac{\left|(Lu,v)_{L_2(Q)}\right|}{\|v\|_{W^+_{bd^+}}}. \qquad (4)$$

Consider

$$\left|(Lu,v)_{L_2(Q)}\right| = \left|\int_Q [vu_t + v\Delta u_t + vk\Delta u]dQ\right|.$$

Employing the Schwarz inequality, we have

$$\left|\int_Q vu_t dQ\right| \le \left(\int_Q |u_t|^2 dQ\right)^{1/2} \left(\int_Q |v|^2 dQ\right)^{1/2} \le$$

$$\le \|u\|_{W^+_{bd}} \|v\|_{L_2(Q)} \le c\|u\|_{W^+_{bd}} \|v\|_{W^+_{bd^+}}.$$

Let us show that

$$\left|\int_Q v\Delta u_t dQ\right| \le c\|v\|_{W^+_{bd^+}} \|u\|_{W^+_{bd}},$$

$$\left|\int_Q vk\Delta u dQ\right| \le c\|v\|_{W^+_{bd^+}} \|u\|_{W^+_{bd}}.$$

First consider the integral $\left|\int_Q v\Delta u_t dQ\right|$.

Using partial integration and the conditions (2), we find

$$-\int_Q v\Delta u_t dQ = -\sum_{i=1}^n \int_Q \left(vu_{tx_i}\right)_{x_i} dQ + \sum_{i=1}^n \int_Q v_{x_i} u_{tx_i} dQ. \qquad (5)$$

Passing to the surface integral, we obtain

$$\sum_{i=1}^n \int_Q \left(vu_{tx_i}\right)_{x_i} dQ = \int_{(0,T)\times\partial\Omega} v\frac{\partial u_t}{\partial \vec{n}} d((0,T)\times\partial\Omega) = 0.$$

(since $\frac{\partial u}{\partial \vec{n}}\Big|_{x\in\partial\Omega} = 0$ and, thus, $\frac{\partial u_t}{\partial \vec{n}}\Big|_{x\in\partial\Omega} = 0$).

Applying the Schwarz inequality to (5), we have

$$\left| -\int_Q v\Delta u_t dQ \right| = \left| \sum_{i=1}^{n} \int_Q v_{x_i} u_{tx_i} dQ \right| \le \sum_{i=1}^{n} \left| \int_Q v_{x_i} u_{tx_i} dQ \right| \le$$

$$\le \sum_{i=1}^{n} \left(\int_Q v_{x_i}^2 dQ \right)^{\frac{1}{2}} \left(\int_Q u_{tx_i}^2 dQ \right)^{\frac{1}{2}} \le c\|u\|_{W_{bd}^+} \|v\|_{W_{bd^+}^+}.$$

In the same manner, we prove that

$$\left| \int_Q vk\Delta u dQ \right| \le c\|v\|_{W_{bd^+}^+} \|u\|_{W_{bd}^+}.$$

Finally, we obtain

$$\left|(Lu, v)_{L_2(Q)}\right| \le c\|v\|_{W_{bd^+}^+} \|u\|_{W_{bd}^+}.$$

Returning to (4), we have

$$\|Lu\|_{W_{bd^+}^-} \le c\|u\|_{W_{bd}^+}, \tag{6}$$

as required.

In the same manner, we prove the following lemma.

Lemma 2. *For all functions* $v(t,x) \in W_{bd^+}^+$ *the following inequality holds true*

$$\left\|L^* v\right\|_{W_{bd}^-} \le c\|v\|_{W_{bd^+}^+},$$

where L^* *is a formally adjoint operator*

$$L^* u = -v_t - \Delta(-v_t + kv).$$

Lemma 3. *For all functions* $u(t,x) \in W_{bd}^+$ *the following inequality is true*

$$c\|u\|_{L_2(Q)} \le \|Lu\|_{W_{bd^+}^-}.$$

P r o o f. First prove the lemma for smooth functions $u(t,x)$, which satisfy the condition (2), and then passing to the limit, we obtain the inequality for all functions $u(t,x) \in W^{+}_{bd}$.

Consider an auxiliary operator I_t:

$$v(t,x) = I_t = \int_T^t a^{-1}(s)u(s,x)ds,\ a(s) < 0,\ a_t(s) < 0,\ s \in [0,T].$$

Evaluate

$$(I_t u, Lu)_{L_2(Q)} = (L^* I_t u, u)_{L_2(Q)} = \tag{7}$$
$$= (-v_t + \Delta v_t - \Delta k v, a v_t)_{L_2(Q)} = l_1 + l_2 + l_3 .$$

Applying the integration by parts and the conditions (2) and (3), for functions $u(t,x)$ and $v(t,x)$ we have

$$l_1 = -\int_Q a v_t^2 dQ,$$

$$l_2 = \int_Q a v_t \Delta v_t dQ = -\int_Q a \sum_{i=1}^n v_{tx_i}^2 dQ,$$

$$l_3 = -\int_Q a v_t k \Delta v dQ = \int_Q a k \sum_{i=1}^n v_{x_i} v_{tx_i} dQ = \frac{1}{2}\int_Q \left(ak \sum_{i=1}^n v_{x_i}^2 \right)_t dQ -$$
$$-\frac{1}{2}\int_Q a_t k \sum_{i=1}^n v_{x_i}^2 dQ = -\frac{a(0)}{2}\int_\Omega k \sum_{i=1}^n v_{x_i}^2 |_{t=0} d\Omega - \frac{1}{2}\int_Q a_t k \sum_{i=1}^n v_{x_i}^2 dQ .$$

Substituting the equality into (7) and replacing $a(t)$ by $-(t+1)$, we have

$$(Lu, I_t u)_{L_2(Q)} \geq c\|v\|^2_{W^+_{bd^+}} \geq c\|I_t u\|^2_{W^+_{bd^+}} \geq c\|u\|^2_{L_2(Q)} .$$

From the Schwarz inequality, we obtain

$$(Lu, I_t u)_{L_2(Q)} \leq \|Lu\|_{W^-_{bd^+}} \|I_t u\|_{W^+_{bd^+}} .$$

Reducing each part by $\|I_t u\|_{W^+_{bd^+}}$, we find

$$\|Lu\|_{W^-_{bd^+}} \geq c\|u\|_{L_2(Q)},$$

which is what had to be proved.

Lemma 4. *For all functions* $v(t,x) \in W^+_{bd^+}$ *the following inequality holds true*

$$\|L^* v\|_{W^-_{bd}} \geq c\|v\|_{L_2(Q)}.$$

From Lemmas 1-4 we have the existence and uniqueness theorems.

Theorem 1. *For all functions* $f \in L_2(Q)$ *there exists a unique solution of the problem (1), (2) in the sense of Definition 1.1.1.*

Theorem 2. *For all functions* $f \in W^-_{bd^+}$ *there exists a unique solution of the problem (1), (2) in the sense of Definition 1.1.4.*

3. PULSE CONTROL OF PSEUDO-PARABOLIC SYSTEMS (THE DIRICHLET INITIAL BOUNDARY VALUE PROBLEM)

Apply the obtained results for the optimization problem of pseudo-parabolic systems (the Dirichlet initial boundary value problem). We use the same denotations as in Section 1.

Let the state function satisfies the pseudo-parabolic equation

$$Lu \equiv A\left(\frac{\partial u}{\partial t}\right) + B(u) = f + Ah, \tag{1}$$

$$u|_{t=0} = 0; \quad u|_{x\in\partial\Omega} = 0, \tag{2}$$

Using the template theorems of Section 3.6, fill in the tables for the pseudo-parabolic equation (the Dirichlet initial boundary value problem).

Table 1.

^{1}.	Operator	Space N	Space $W^{-}(Q)$
1.	$L(\cdot)$	H^{+}_{bd}	$W^{-}_{bd^{+}}$
2.	$L(\cdot)$	W^{+}_{bd}	$H^{-}_{bd^{+}}$

Table 2.

^{1}.	Operator $A_i(\cdot)$	Space $W^{-}(Q)$
1.	$A_1(\cdot)$	$W^{-}_{bd^{+}}$
2.	$A_3(\cdot)$	$W^{-}_{bd^{+}}, H^{-}_{bd^{+}}$

Table 2 (continuation).

^{1}.	Operator $A_i(\cdot)$	Space $W^{-}(Q)$
3.	$A_4(\cdot)$	$W^{-}_{bd^{+}}$
4.	$A_5(\cdot)$	$W^{-}_{bd^{+}}, H^{-}_{bd^{+}}$

Table 3.

^{1}.	Operator $L(\cdot)$	Space $W^{-}(Q)$	Space $W^{+}(Q)$
1.	$L(\cdot)$	$W^{-}_{bd^{+}}$	$W^{+}_{bd^{+}}$

Table 4 is empty.

Table 5.

^{1}.	Operator $L(\cdot)$.	Space $W^{-}(Q)$
1.	$L(\cdot)$	$W^{-}_{bd^{+}}$

Table 6.

i.	Operator $A_i(\cdot)$	Space $W^-(Q)$
1.	$A_1(\cdot)$	$W^-_{bd^+}$
2.	$A_3(\cdot)$	$W^-_{bd^+}, H^-_{bd^+}$
3.	$A_4(\cdot)$	$W^-_{bd^+}$
4.	$A_5(\cdot)$	$W^-_{bd^+}, H^-_{bd^+}$

Table 7.

i.	Exponent α	Space $W^-(Q)$	Map $A_{i,\varepsilon}(\cdot)$
1.	1/2	$W^-_{bd^+}$	$A_{1,\varepsilon}(\cdot)$
2.	1/2	$W^-_{bd^+}$	$A_{2,\varepsilon}(\cdot)$
3.	1/2	$W^-_{bd^+}, H^-_{bd^+}$	$A_{3,\varepsilon}(\cdot)$
4.	1/2	$W^-_{bd^+}$	$A_{4,\varepsilon}(\cdot)$
5.	1/2	$W^-_{bd^+}, H^-_{bd^+}$	$A_{5,\varepsilon}(\cdot)$

4. PULSE CONTROL OF PSEUDO-PARABOLIC SYSTEMS (THE NEUMANN INITIAL BOUNDARY VALUE PROBLEM)

Apply the obtained results for the optimization problem of pseudo-parabolic systems (the Neumann initial boundary value problem). We use the same denotations as in Section 2.

Let the state function satisfies the pseudo-parabolic equation

$$Lu \equiv \frac{\partial u}{\partial t} - \Delta\left(\frac{\partial u}{\partial t} + ku\right) = f + Ah, \tag{1}$$

$$u|_{t=0} 0, \quad \frac{\partial u}{\partial \vec{n}}\bigg|_{x\in\partial\Omega} = 0, \tag{2}$$

Using the template theorems of Section 3.6, fill in the tables for the pseudo-parabolic equation (the Neumann initial boundary value problem).

Table 1.

¹.	Operator	Space N	Space $W^-(Q)$
1.	$L(\cdot)$	$L_2(Q)$	$W^-_{bd^+}$
2.	$L(\cdot)$	W^+_{bd}	$L_2(Q)$

Table 2.

¹.	Operator $A_i(\cdot)$	Space $W^-(Q)$
1.	$A_1(\cdot)$	$W^-_{bd^+}$
2.	$A_3(\cdot)$	$W^-_{bd^+}$
3.	$A_4(\cdot)$	$W^-_{bd^+}$
4.	$A_5(\cdot)$	$W^-_{bd^+}$

Table 3.

¹.	Operator $L(\cdot)$	Space $W^-(Q)$	Space $W^+(Q)$
1.	$L(\cdot)$	$W^-_{bd^+}$	$W^+_{bd^+}$

Table 4 is empty.

Table 5.

^{1}.	Operator $L(\cdot)$	Space $W^-(Q)$
1.	$L(\cdot)$	$W^-_{bd^+}$

Table 6.

^{1}.	Operator $A_i(\cdot)$	Space $W^-(Q)$
1.	$A_1(\cdot)$	$W^-_{bd^+}$
2.	$A_3(\cdot)$	$W^-_{bd^+}$
3.	$A_4(\cdot)$	$W^-_{bd^+}$
4.	$A_5(\cdot)$	$W^-_{bd^+}$

Table 7.

^{1}.	Exponent α	Space $W^-(Q)$	Map $A_{i,\varepsilon}(\cdot)$
1.	1/2	$W^-_{bd^+}$	$A_{1,\varepsilon}(\cdot)$
2.	1/2	$W^-_{bd^+}$	$A_{2,\varepsilon}(\cdot)$
3.	1/2	$W^-_{bd^+}$	$A_{3,\varepsilon}(\cdot)$
4.	1/2	$W^-_{bd^+}$	$A_{4,\varepsilon}(\cdot)$

Chapter 6

HYPERBOLIC SYSTEMS

1. GENERALIZED SOLVABILITY OF HYPERBOLIC SYSTEMS (THE DIRICHLET INITIAL BOUNDARY VALUE PROBLEM)

This Section is devoted to the research of hyperbolic partial differential equations. A lot of the mechanic and physics problems such as a chord, bar and membrane oscillation, an electromagnetic oscillation etc. are described by the hyperbolic equations [98, 129-133].

Consider a linear partial differential equation

$$Lu = u_{tt} + Bu = f(t,x), \tag{1}$$

in a tube domain $Q = (0,T)\times\Omega$, Ω is a bounded domain of $x_1, x_2, \ldots, x_n$ variation with a smooth domain boundary $\partial\Omega$. The elliptic operator B does not depend on the temporary variable t and is of the following form

$$Bu = -\sum_{i,j=1}^{n}\left(a_{ij}(x)u_{x_j}\right)_{x_i} + \sum_{i=1}^{n} b_i(x)u_{x_i} + c(x)u$$

Suppose that the matrix $A = \{a_{ij}(x)\}_{i,j=1}^{n}$ is symmetric $a_{ij}(x) = a_{ji}(x)$, the matrix cells $a_{ij}(x)$ and the functions $b_i(x)$ are continuously differentiable in the domain $\overline{\Omega}$ and for all vectors $\xi = (\xi_1, \xi_2, \ldots, \xi_n) \in R^n$ the following inequalities hold true

$$\sum_{i,j=1}^{n} a_{ij}(x)\xi_i\xi_j \geq \alpha\sum_{i=1}^{n}\xi_i^2, \qquad \alpha = const > 0,$$

$$c(x) \geq \sum_{i=1}^{n}\frac{\partial b_i(x)}{\partial x_i}, \quad c(x) \geq 0, \ \forall x \in \overline{\Omega}.$$

Since the functions $b_i(x)$ are continuous on the compact set $\overline{\Omega}$, there exists a constant $c_b > 0$ such that $|b_i(x)| < c_b$ for all $x \in \overline{\Omega}$, $i = \overline{1,n}$.

We denote by W^+_{bd} a completion of the set of smooth in $\overline{Q}$ functions satisfying the following conditions

$$u|_{t=0} = u_t|_{t=0} = 0,\ u|_{x\in\partial\Omega} = 0. \tag{2}$$

in the norm

$$\|u\|_W = \left(\int_Q u_t^2 + \sum_{i=1}^n u_{x_i}^2\, dQ\right)^{\frac{1}{2}}.$$

Analogously let $W^+_{bd^+}$ be a completion of the set of sn. functions satisfying the following conditions

$$v|_{t=T} = v_t|_{t=T} = 0,\ v|_{x\in\partial\Omega} = 0 \tag{3}$$

in the same norm

$$\|v\|_W = \left(\int_Q v_t^2 + \sum_{i=1}^n v_{x_i}^2\, dQ\right)^{\frac{1}{2}}.$$

Let W^-_{bd}, $W^-_{bd^+}$ be corresponding negative spaces.

Lemma 1. *For all $u(t,x) \in W^+_{bd}$ and $v(t,x) \in W^+_{bd^+}$ the following inequalities*

$$\|Lu\|_{W^-_{bd^+}} \le C\|u\|_{W^+_{bd}},$$

$$\|L^*v\|_{W^-_{bd}} \le C\|v\|_{W^+_{bd^+}},$$

hold true, where an operator L^* is formally adjoint to the operator L:

$$L^*v = v_{tt} - \sum_{i,j=1}^n \left(a_{ij}(x)v_{x_j}\right)_{x_i} - \sum_{i=1}^n \left(b_i(x)v\right)_{x_i} + c(x)v.$$

Proof. Consider first a simpler case, when $u(t,x)$ is a continuous function satisfying the conditions (2). By the negative norm definition, we have

$$\|Lu\|_{W^-_{bd^+}} = \sup_{\substack{v\neq 0,\\ v\in W^+_{bd^+}}} \frac{\left|\langle Lu, v\rangle_{W_{bd^+}}\right|}{\|v\|_{W^+_{bd^+}}} = \sup_{\substack{v\neq 0,\\ v\in W^+_{bd^+}}} \frac{\left|(Lu,v)_{L_2(Q)}\right|}{\|v\|_{W^+_{bd^+}}}, \tag{4}$$

since for smooth functions $u(t,x)$ satisfying the condition (2), the bilinear form coincides with the inner product in $L_2(Q)$.

Consider the term of the right-hand side of the fraction (4):

$$\left|(Lu,v)_{L_2(Q)}\right| = \left|\left(u_{tt} - \sum_{i,j=1}^{n}\left(a_{ij}u_{x_j}\right)_{x_i} + \sum_{i=1}^{n} b_i u_{x_i} + cu, v\right)_{L_2(Q)}\right|.$$

Applying the partial integration, the Schwarz inequality and the initial conditions (2), (3), we obtain

$$\left|\int_Q u_{tt} v\,dQ\right| = \left|\int_Q (u_t v)_t\,dQ - \int_Q u_t v_t\,dQ\right| = \left|\int_Q u_t v_t\,dQ\right| \leq$$

$$\leq \left(\int_Q u_t^2\,dQ\right)^{1/2}\left(\int_Q v_t^2\,dQ\right)^{1/2} \leq \|u\|_{W^+_{bd}}\|v\|_{W^+_{bd^+}}.$$

Next,

$$\left|\int_Q v\sum_{i,j=1}^{n}\left(a_{ij}u_{x_j}\right)_{x_i} dQ\right| = \left|\int_Q \sum_{i,j=1}^{n}\left(v a_{ij}u_{x_j}\right)_{x_i} dQ - \int_Q \sum_{i,j=1}^{n} v_{x_i} a_{ij} u_{x_j}\,dQ\right| \leq$$

$$\leq C\left(\int_Q \sum_{i=1}^{n} v_{x_i}^2\,dQ\right)^{1/2}\left(\int_Q u_{x_i}^2\,dQ\right)^{1/2} \leq C\|v\|_{W^+_{bd^+}}\|u\|_{W^+_{bd}}.$$

In the same manner, we have

$$\left|\left(\sum_{i=1}^{n} b_i u_{x_i}, v\right)_{L_2(Q)}\right| \leq \sum_{i=1}^{n}\left\|b_i u_{x_i}\right\|_{L_2(Q)}\|v\|_{L_2(Q)} \leq C\|u\|_{W^+_{bd}}\|v\|_{W^+_{bd^+}},$$

$$\left|(cu,v)_{L_2(Q)}\right| \le \|cu\|_{L_2(Q)}\|v\|_{L_2(Q)} \le C\|u\|_{W^+_{bd}}\|v\|_{W^+_{bd^+}}.$$

Returning to the equality (4), we obtain the lemma assertion for smooth functions $u(t,x)$, which satisfy the condition (2). Passing to the limit, we obtain the lemma assertion for all functions $u(t,x)\in W^+_{bd}$. The second inequality is proved in the same manner.

Lemma 2. *For all functions* $u(t,x)\in W^+_{bd}$ *and* $v(t,x)\in W^+_{bd^+}$ *the following inequalities hold true*

$$\|u\|_{L_2(Q)} \le C\|Lu\|_{W^-_{bd^+}},$$

$$\|v\|_{L_2(Q)} \le C\|L^*v\|_{W^-_{bd}}.$$

Proof. Consider the first inequality. Suppose first that a function $u(t,x)$ is smooth and satisfies the condition (2). Consider the following integral operator

$$v(t,x) = -\int_T^t e^{-\omega\tau}u(\tau,x)d\tau,$$

where ω is a positive constant.

It is obvious that $u = -e^{\omega t}v_t$. Consider

$$(Lu,v)_{L_2(Q)} = (u_{tt},v)_{L_2(Q)} - \left(\sum_{i,j=1}^n \left(a_{ij}u_{x_j}\right)_{x_i}, v\right)_{L_2(Q)} +$$

$$+\left(\sum_{i=1}^n b_i u_{x_i}, v\right)_{L_2(Q)} + (cu,v)_{L_2(Q)}. \qquad (5)$$

Estimate the every summand in the right-hand side.

Applying the integration by parts, we have

$$(u_{tt},v)_{L_2(Q)} = \int_Q (u_t v)_t dQ - \int_Q u_t v_t dQ = \int_\Omega u_t v\Big|_{t=0}^{t=T} d\Omega - \int_Q u_t v_t dQ.$$

Since $u_t(0,x) = v(T,x) = 0$, we obtain

$$(u_{tt}, v)_{L_2(Q)} = -\int_Q u_t v_t dQ.$$

Taking into account that $u = -e^{\omega t} v_t$ and so $u_t = -e^{\omega t} v_{tt} - \omega e^{\omega t} v_t$, $v_t(0, x) = 0$, we arrive

$$(u_{tt}, v)_{L_2(Q)} = -\int_Q u_t v_t dQ = \int_Q e^{\omega t} v_t v_{tt} dQ + \int_Q e^{\omega t} \omega v_t^2 dQ =$$

$$= \frac{1}{2}\int_Q \left(e^{\omega t} v_t^2\right)_t dQ - \frac{1}{2}\int_Q e^{\omega t} \omega v_t^2 dQ + \int_Q e^{\omega t} \omega v_t^2 dQ =$$

$$= \frac{1}{2}\int_\Omega e^{\omega T} v_t^2(T, x) d\Omega + \frac{1}{2}\int_Q e^{\omega t} \omega v_t^2 dQ \geq \frac{1}{2}\int_Q e^{\omega t} \omega v_t^2 dQ.$$

The other summands of the equality (5) are estimated in the same manner as the previous one. Consider the second summand:

$$-\left(\sum_{i,j=1}^{n} \left(a_{ij} u_{x_j}\right)_{x_i}, v\right)_{L_2(Q)} = -\int_Q \sum_{i,j=1}^{n} \left(v a_{ij} u_{x_j}\right)_{x_i} dQ +$$

$$+ \int_Q \sum_{i,j=1}^{n} a_{ij} u_{x_j} v_{x_i} dQ = \int_Q \sum_{i,j=1}^{n} a_{ij} u_{x_j} v_{x_i} dQ =$$

$$= -\int_Q \sum_{i,j=1}^{n} a_{ij} e^{\omega t} v_{x_j t} v_{x_i} dQ = -\frac{1}{2}\int_Q \left(\sum_{i,j=1}^{n} a_{ij} e^{\omega t} v_{x_i} v_{x_j}\right)_t dQ +$$

$$+ \frac{1}{2}\int_Q \sum_{i,j=1}^{n} a_{ij} \omega e^{\omega t} v_{x_j} v_{x_i} dQ = \frac{1}{2}\int_\Omega \sum_{i,j=1}^{n} a_{ij} v_{x_j} v_{x_i}\Big|_{t=0} d\Omega +$$

$$+ \frac{1}{2}\int_Q \sum_{i,j=1}^{n} a_{ij} \omega e^{\omega t} v_{x_j} v_{x_i} dQ \geq \frac{\omega\alpha}{2}\int_Q \sum_{i=1}^{n} e^{\omega t} v_{x_i}^2 dQ.$$

Next, estimate the third summand of (5):

$$\left(\sum_{i=1}^{n} b_i u_{x_i}, v\right)_{L_2(Q)} = \sum_{i=1}^{n}\int_Q \left(b_i u v\right)_{x_i} dQ - \sum_{i=1}^{n}\int_Q b_i u v_{x_i} dQ -$$

$$-\sum_{i=1}^{n}\int_Q (b_i)_{x_i} uv dQ = \sum_{i=1}^{n}\int_Q b_i e^{\omega t} v_t v_{x_i} dQ + \sum_{i=1}^{n}\int_Q (b_i)_{x_i} e^{\omega t} v_t v dQ \geq$$

$$\geq -\sum_{i=1}^{n}\int_Q c_b e^{\omega t} |v_t| \cdot |v_{x_i}| dQ + \frac{1}{2}\sum_{i=1}^{n}\int_Q \left((b_i)_{x_i} e^{\omega t} v^2\right)_t dQ -$$

$$-\frac{1}{2}\sum_{i=1}^{n}\int_Q (b_i)_{x_i} \omega e^{\omega t} v^2 dQ = -\sum_{i=1}^{n}\int_Q c_b e^{\omega t} |v_t| \cdot |v_{x_i}| dQ -$$

$$-\frac{1}{2}\sum_{i=1}^{n}\int_\Omega (b_i)_{x_i} v^2\big|_{t=0} d\Omega - \frac{1}{2}\sum_{i=1}^{n}\int_Q b_{x_i} \omega e^{\omega t} v^2 dQ .$$

Examine the last summand of (5):

$$(cu, v)_{L_2(Q)} = -\int_Q e^{\omega t} c v_t v dQ = -\frac{1}{2}\int_Q \left(e^{\omega t} c v^2\right)_t dQ +$$

$$+\frac{1}{2}\int_Q \omega e^{\omega t} c v^2 dQ = \frac{1}{2}\int_\Omega c v^2\big|_{t=0} d\Omega + \frac{1}{2}\int_Q \omega e^{\omega t} c v^2 dQ .$$

Applying the all acquired estimation, we have

$$(Lu, v)_{L_2(Q)} \geq \frac{1}{2}\int_Q \omega e^{\omega t} v_t^2 dQ + \frac{\omega\alpha}{2}\sum_{i=1}^{n}\int_Q e^{\omega t} v_{x_i}^2 dQ -$$

$$-\sum_{i=1}^{n}\int_Q c_b e^{\omega t} |v_t| \cdot |v_{x_i}| dQ + \frac{1}{2}\int_\Omega \left(c - \sum_{i=1}^{n}(b_i)_{x_i}\right) v^2\big|_{t=0} d\Omega +$$

$$+\frac{1}{2}\int_Q \left(c - \sum_{i=1}^{n}(b_i)_{x_i}\right) \omega e^{\omega t} v^2 dQ .$$

Since $c(x) \geq \sum_{i=1}^{n}\dfrac{\partial b_i(x)}{\partial x_i}$, we arrive

$$(Lu, v)_{L_2(Q)} \geq \frac{1}{2}\int_Q \omega e^{\omega t} v_t^2 dQ + \frac{\omega\alpha}{2}\sum_{i=1}^{n}\int_Q e^{\omega t} v_{x_i}^2 dQ -$$

$$-\sum_{i=1}^{n}\int_Q c_b e^{\omega t}|v_t|\cdot|v_{x_i}|dQ=\left(\frac{1}{4}\int_Q \omega e^{\omega t}v_t^2dQ+\frac{\omega\alpha}{4}\sum_{i=1}^{n}\int_Q e^{\omega t}v_{x_i}^2dQ\right)+$$

$$+\left(\frac{1}{4}\int_Q \omega e^{\omega t}v_t^2dQ-\sum_{i=1}^{n}\int_Q c_b e^{\omega t}|v_t|\cdot|v_{x_i}|dQ+\frac{\omega\alpha}{4}\sum_{i=1}^{n}\int_Q e^{\omega t}v_{x_i}^2dQ\right).$$

The term

$$\left(\frac{1}{4}\int_Q \omega e^{\omega t}v_t^2dQ-\sum_{i=1}^{n}\int_Q c_b e^{\omega t}|v_t|\cdot|v_{x_i}|dQ+\frac{\omega\alpha}{4}\sum_{i=1}^{n}\int_Q e^{\omega t}v_{x_i}^2dQ\right)$$

can be made positive by choosing the constant ω. It suffices to put $\omega=c_b\sqrt{\frac{4n}{\alpha}}$. We have

$$\left(\frac{1}{4}\int_Q \omega e^{\omega t}v_t^2dQ-\sum_{i=1}^{n}\int_Q c_b e^{\omega t}|v_t|\cdot|v_{x_i}|dQ+\frac{\omega\alpha}{4}\sum_{i=1}^{n}\int_Q e^{\omega t}v_{x_i}^2dQ\right)=$$

$$=c_b e^{\omega t}\sum_{i=1}^{n}\int_Q\left(\frac{1}{2\sqrt{n\alpha}}v_t^2-|v_t|\cdot|v_{x_i}|+\frac{\sqrt{\alpha n}}{2}v_{x_i}^2\right)dQ=$$

$$=\frac{c_b e^{\omega t}}{2\sqrt{n\alpha}}\sum_{i=1}^{n}\int_Q\left(|v_t|-\sqrt{\alpha n}|v_{x_i}|\right)^2dQ\geq 0.$$

Thus,

$$(Lu,v)_{L_2(Q)}\geq\left(\frac{1}{4}\int_Q \omega e^{\omega t}v_t^2dQ+\frac{\omega\alpha}{4}\sum_{i=1}^{n}\int_Q e^{\omega t}v_{x_i}^2dQ\right)\geq C\|v\|^2_{W^+_{bd^+}}.$$

Employing to the left-hand side the Schwarz inequality, we obtain

$$\|Lu\|_{W^-_{bd^+}}\|v\|_{W^+_{bd^+}}\geq(Lu,v)_{L_2(Q)}\geq C\|v\|^2_{W^+_{bd^+}}.$$

Dividing by $\|v\|_{W^+_{bd^+}}$, we arrive the following inequality:

$$\|Lu\|_{W^-_{bd^+}}\geq C\|v\|_{W^+_{bd^+}}.$$

Taking into account the definition of the function $v(t,x)$, we have

$$\|v\|_{W^+_{bd^+}} \geq C\|u\|_{L_2(Q)},$$

whence we finally obtain

$$\|Lu\|_{W^-_{bd^+}} \geq C\|u\|_{L_2(Q)}.$$

This inequality can be extended to any function $u(t,x) \in W^+_{bd}$ by passing to limit.

The second inequality is proved in the same manner.

Theorem 1. *For all functions $f \in L_2(Q)$ there exists a unique solution of the problem (1), (2) in the sense of Definition 1.1.1.*

Theorem 2. *For all functions $f \in W^-_{bd^+}$, there exists a unique solution of the problem (1), (2) in the sense of Definition 1.1.4.*

2. GENERALIZED SOLVABILITY OF HYPERBOLIC SYSTEMS (THE NEUMANN INITIAL BOUNDARY VALUE PROBLEM)

Suppose that a state function is described by the hyperbolic equation

$$Lu = u_{tt} + Bu = f(t,x), \tag{1}$$

where B is an elliptic operator in the following form

$$Bu = -\sum_{i,j=1}^{n}\left(a_{ij}(x)u_{x_j}\right)_{x_i}, \tag{2}$$

the functions $a_{ij} = a_{ji}(x)$, $\{a_{ij}\}_{i,j=1}^{n}$ are continuously differentiable in the domain $\overline{\Omega}$. The operator B is uniformly elliptic in $\overline{\Omega}$:

$$\sum_{i,j=1}^{n} a_{ij}\xi_i\xi_j \geq \alpha\sum_{i=1}^{n}\xi_i^2, \ \forall \xi_i \in R, \ i = \overline{1,n}. \tag{3}$$

The equation (1) is considered in the tube domain $Q=(0,T)\times\Omega$, $\Omega\subset R^n$.

Introduce the following denotations. Let W_{bd}^{+} be a completion of the set of smooth in $\overline{Q}$ functions, which satisfy the following conditions

$$u\big|_{t=\ddot{0}} = u_t\big|_{t=0} = 0\,, \quad \sum_{i,j=1}^{n} \dot{a}_{ij} u_{x_i} n_{x_j}\Bigg|_{x\in\partial\Omega} = 0\,, \tag{4}$$

in the norm

$$\|u\|_{W_{bd}^{+}}^{2} = \int_{Q} u_t^2 + \sum_{i,j=1}^{n} a_{ij} u_{x_i} u_{x_j}\, dQ\,, \tag{5}$$

where $\vec{n}=\left(n_{x_1},\ldots,n_{x_n}\right)$ is a normal vector to the surface of the domain Ω at a point $\left(x_1,x_2,\ldots,x_n\right)$.

The boundary condition

$$\sum_{i,j=1}^{n} a_{ij} u_{x_i} n_{x_j}\Bigg|_{x\in\partial\Omega} = 0\,,$$

can be written in the form

$$\sum_{i,j=1}^{n} a_{ij} u_{x_i} n_{x_j} = (\mathrm{grad}\,u,\vec{\mu}) = \frac{\partial u}{\partial\vec{\mu}}\,,$$

where $(\cdot,\cdot)$ is the inner product in R^n, μ is the following co-normal vector

$$\vec{\mu} = \frac{A\cdot\vec{n}}{|A\cdot\vec{n}|}\,,$$

wherå A is a matrix of elements $a_{ij}(x)$.

Let $W_{bd^{+}}^{+}$ be a completion of the set of smooth in $\overline{Q}$ functions, which satisfy the following conditions

$$v\big|_{t=T} = v_t\big|_{t=T} = 0\,, \quad \sum_{i,j=1}^{n} a_{ij} v_{x_i} n_{x_j}\Bigg|_{x\in\partial\Omega} = 0\,, \tag{6}$$

in the norm (5). According to the pairs of sets $L_2(Q), W^+_{bd}$ and $L_2(Q), W^+_{bd^+}$ consider the negative spaces $W^-_{bd}, W^-_{bd^+}$.

Prove a priori inequalities in the negative norms for a hyperbolic system.

Lemma 1. *For all functions $u(t,x) \in W^+_{bd}$ the following inequality holds true*

$$\|Lu\|_{W^-_{bd^+}} \le c\|u\|_{W^+_{bd}}\,.$$

Proof. Prove first the inequality for smooth functions $u(t,x)$, which satisfy the conditions (4).

By the definition of the negative norm, we have

$$\|Lu\|_{W^-_{bd^+}} = \sup_{\substack{v\neq 0\\ v\in W^+_{bd^+}}} \frac{\left|\langle Lu, v\rangle_{W_{bd^+}}\right|}{\|v\|_{W^+_{bd^+}}} = \sup_{\substack{v\neq 0\\ v\in W^+_{bd^+}}} \frac{\left|(Lu, v)_{L_2(Q)}\right|}{\|v\|_{W^+_{bd^+}}}\,, \tag{7}$$

since for a smooth function $u(t,x)$ the bilinear form $\langle\cdot,\cdot\rangle_{W_{bd^+}}$ equals to the inner product $(\cdot,\cdot)_{L_2(Q)}$ in the space $L_2(Q)$.

Consider the equality (7). Applying partial integration and the conditions (4), (6), we have

$$\int_Q v u_{tt}\, dQ = -\int_Q v_t u_t\, dQ\,,$$

$$-\int_Q v \sum_{i,j=1}^{n} \left(a_{ij} u_{x_j}\right)_{x_i} dQ = \int_Q \sum_{i,j=1}^{n} a_{ij} v_{x_i} u_{x_j}\, dQ\,.$$

Adding the equalities and using the Schwarz inequality, we find

$$(Lu, v)_{L_2(Q)} \le c\|u\|_{W^+_{bd}} \|v\|_{W^+_{bd^+}}\,.$$

Substituting the inequality into (7) and dividing by $\|v\|_{W^+_{bd^+}}$, we obtain the desired inequality for all smooth functions $u(t,x)$, which satisfy the condition (4).

Considering the continuous extension of operator L on the space W^+_{bd}, we have the desired inequality for all functions from W^+_{bd}.

Lemma 2. *For all functions $v(t,x) \in W^+_{bd^+}$ the following inequality holds true*

$$\left\|L^* v\right\|_{W^-_{bd}} \leq c\|v\|_{W^+_{bd^+}},$$

where L^* is the adjoint operator.

The proof of Lemma 2 is completely analogous to the previous lemma.

The inequalities from Lemma 1 and 2 show that the operators L and L^* map continuously the spaces W^+_{bd} and $W^+_{bd^+}$ into $W^-_{bd^+}$ and W^-_{bd}, respectively.

Lemma 3. *For all functions $u(t,x) \in W^+_{bd}$ the following inequality holds true*

$$\|Lu\|_{W^-_{bd^+}} \geq c\|u\|_{L_2(Q)}.$$

Proof. We first prove the lemma for a smooth function, which satisfies the conditions (4).

Consider an auxiliary function $v(t,x)$

$$v(t,x) = -\int_T^t (\tau+1)^{-1} u(\tau,x)\,d\tau.$$

For all functions $v(t,x)$ we shall prove the following inequality

$$(Lu,v)_{L_2(Q)} \geq c\|v\|^2_{W^+_{bd^+}}. \tag{8}$$

Consider the inner product

$$(Lu,v)_{L_2(Q)} = (u_{tt} + Bu, v)_{L_2(Q)}.$$

Employing the integration by parts and relation between the function $u(t,x)$ and $v(t,x)$, we have

$$\int_Q v u_{tt} dQ = \int_Q (v u_t)_t dQ + \int_Q v_t (t+1) v_{tt} dQ + \int_Q v_t^2 dQ = \tag{9}$$

$$= \int_\Omega v u_t \Big|_{t=0}^{t=T} d\Omega + \frac{1}{2}\int_\Omega (t+1) v_t^2 \Big|_{t=0}^{t=T} d\Omega + \frac{1}{2}\int_Q v_t^2 dQ \geq \frac{1}{2}\int_Q v_t^2 dQ.$$

Next, in the same manner, we have

$$(Bu,v)_{L_2(Q)} = -\int_Q v \sum_{i,j=1}^{n} \left(a_{ij} u_{x_j}\right)_{x_i} dQ = \tag{10}$$

$$= -\int_Q \sum_{i,j=1}^{n} \left(v a_{ij} u_{x_j}\right)_{x_i} dQ + \int_Q \sum_{i,j=1}^{n} v_{x_i} a_{ij} u_{x_j} dQ =$$

$$= -\int_\Gamma v \sum_{i,j=1}^{n} a_{ij} u_{x_j} n_{x_i} d\Gamma - \int_Q \sum_{i,j=1}^{n} (t+1) a_{ij} v_{x_i} v_{x_j t} dQ =$$

$$= -\int_Q \sum_{i,j=1}^{n} (t+1) a_{ij} v_{x_i} v_{x_j t} dQ,$$

where $\Gamma = (0,T) \times \partial\Omega$.

Taking into account symmetry of the matrix $\{a_{ij}\}_{i,j=1}^{n}$ and applying integration by parts, we find

$$-\int_Q \sum_{i,j=1}^{n} (t+1) a_{ij} v_{x_i} v_{x_j t} dQ \geq \frac{1}{2}\int_Q \sum_{i,j=1}^{n} a_{ij} v_{x_i} v_{x_j} dQ.$$

Substituting the inequality into (10), we arrive

$$(Bu,v)_{L_2(Q)} \geq \frac{1}{2}\int_Q \sum_{i,j=1}^{n} a_{ij} v_{x_i} v_{x_j} dQ. \tag{11}$$

From (9) and (11) we have inequality (8).

Employing the Schwarz inequality to the left-hand side of (8), we obtain

$$(Lu,v)_{L_2(Q)} \le \|v\|_{W^+_{bd^+}} \|Lu\|_{W^-_{bd^+}}. \tag{12}$$

Comparing formulae (8) and (12), dividing by $\|v\|_{W^+_{bd^+}}$ and taking into account the relation between $\dot{u}(t,x)$ and $v(t,x)$, we have the assertion of Lemma 3 for a smooth function $u(t,x)$, which satisfies the condition (4). Passing to the limit, we obtain the assertion for all functions $u(t,x) \in W^+_{bd}$.

Lemma 4. *For all functions* $v(t,x) \in W^+_{bd^+}$ *the following inequality holds true*

$$\|L^* v\|_{W^-_{bd}} \ge c\|v\|_{L_2(Q)}.$$

P r o o f. Let first $v(t,x)$ be a smooth function that satisfies the condition (6). Consider the following auxiliary function $u(t,x)$:

$$u(t,x) = \int_0^t (2T-\tau)^{-1} v(\tau,x)d\tau.$$

Show the inequality

$$\left(u, L^* v\right)_{L_2(Q)} \ge c\|u\|^2_{W^+_{bd}}. \tag{13}$$

Applying the integration by parts, relation between $u(t,x)$ and $v(t,x)$, we have

$$\begin{aligned}\int_Q uv_{tt}dQ = \int_\Omega uv_t\Big|_{t=0}^{t=T} d\Omega + \int_Q u_t^2 dQ - \int_Q (2T-t)u_t u_{tt} dQ \ge \\ \ge \frac{1}{2}\int_Q u_t^2 dQ.\end{aligned} \tag{14}$$

In the same manner, consider the other summands of $\left(u, L^* v\right)_{L_2(Q)}$:

$$-\left(u, \sum_{i,j=1}^{n}\left(a_{ij}v_{x_j}\right)_{x_i}\right)_{L_2(Q)} = -\int_{\Gamma} u \sum_{i,j=1}^{n} A_{ij} v_{x_j} n_{x_i} d\Gamma + \int_{Q} \sum_{i,j=1}^{n} a_{ij} u_{x_i} v_{x_j} dQ =$$

$$= \int_{Q} (2T-t) \sum_{i,j=1}^{n} a_{ij} u_{x_i} u_{x_j} dQ \geq \frac{1}{2} \int_{Q} \sum_{i,j=1}^{n} a_{ij} u_{x_i} u_{x_j} dQ. \quad (15)$$

Substituting (14) and (15) into (13), we obtain the assertion of Lemma 4 for a smooth function $v \in W^{+}_{bd^{+}}$. Passing to the limit, we obtain the assertion for all functions $v \in W^{+}_{bd^{+}}$.

Applying the theorems of Section 1.1, we have the following theorems for the problem (1), (4).

Theorem 1. *For all functions $f \in L_2(Q)$ there exists a unique solution of the problem (1), (4) by Definition 1.1.1.*

Theorem 2. *For all functions $f \in W^{-l}_{bd^{+}}(Q)$ there exists an unique solution of the problem (1), (4) by Definition 1.1.4.*

3. GALERKIN METHOD FOR HYPERBOLIC SYSTEMS

In this section we consider the application of the Galerkin method for approximate solving the hyperbolic equation (the Dirichlet initial boundary value problem).

We shall find the approximate solution of the equation

$$\frac{\partial^2 u}{\partial t^2} - \sum_{i,j=1}^{n} \frac{\partial}{\partial x_i}\left(a_{ij}(x)\frac{\partial u}{\partial x_j}\right) + \sum_{i=1}^{n} b_i(x)\frac{\partial u}{\partial x_i} + c(x)u = f, \quad (1)$$

$$u\big|_{t=0} = u_t\big|_{t=0} = 0, \; u\big|_{x\in\partial\Omega} = 0 \quad (2)$$

in the form

$$u_k(t,x) = \sum_{i=1}^{k} g_i(t)\omega_i(x),$$

where $\{\omega_i(x)\}$ is an orthonormalized basis of the space $L_2(\Omega)$. The functions ω_i satisfy the following conditions $\omega_i(x)|_{x\in\partial\Omega}=0$, $f\in L_2(Q)$; $g_i(t)$ are chosen as a solution of the Cauchy problem for the following set of the linear ordinary differential equations:

$$\frac{d^2g_j(t)}{dt^2}+\sum_{i=1}^{k}g_i(t)(B\omega_i,\omega_j)_{L_2(\Omega)}=(f,\omega_j)_{L_2(\Omega)},$$
$$g_i(0)=\frac{dg_i(0)}{dt}=0, i=\overline{1,k},\ j=\overline{1,k}. \tag{3}$$

From the set of equations (3) it follows that

$$(Lu_k,\omega_j)_{L_2(\Omega)}=(f,\omega_j)_{L_2(\Omega)},$$
$$u_k(0,x)=\frac{\partial u_k(0,x)}{\partial t}=0, i=\overline{1,k},\ j=\overline{1,k}. \tag{4}$$

Lemma 1. *The following inequality holds true*

$$\|u_k\|_{W_{bd}^{+}}\le C\|f\|_{L_2(Q)}.$$

Proof. Multiplying both the left and right hand sides of the equation (4) by $e^{-\omega t}\frac{dg_j}{dt}$, where the constant ω was defined in Section 1, and summing up over j from 1 to k, we obtain:

$$\left(Lu_k,e^{-\omega t}\frac{\partial u_k}{\partial t}\right)_{L_2(Q)}\equiv\left(\frac{\partial^2 u_k}{\partial t^2},e^{-\omega t}\frac{\partial u_k}{\partial t}\right)_{L_2(Q)}-$$
$$-\left(\sum_{i,j=1}^{n}\frac{\partial}{\partial x_i}\left(a_{ij}\frac{\partial u_k}{\partial x_j}\right),e^{-\omega t}\frac{\partial u_k}{\partial t}\right)_{L_2(Q)}+\left(\sum_{i=1}^{n}b_i\frac{\partial u_k}{\partial x_i},e^{-\omega t}\frac{\partial u_k}{\partial t}\right)_{L_2(Q)}+$$
$$+\left(cu_k,e^{-\omega t}\frac{\partial u_k}{\partial t}\right)_{L_2(Q)}=\left(f,e^{-\omega t}\frac{\partial u_k}{\partial t}\right)_{L_2(Q)}. \tag{5}$$

Applying the integration by parts, passing to the surface integration and taking account the boundary conditions

$$u_k\big|_{t=0} = \frac{\partial u_k}{\partial t}\big|_{t=0} = 0,\ u_k\big|_{x\in\partial\Omega} = 0,$$

we find

$$\left(\frac{\partial^2 u_k}{\partial t^2}, e^{-\omega t}\frac{\partial u_k}{\partial t}\right)_{L_2(Q)} = \frac{1}{2}\int_Q \left(e^{-\omega t}\left(\frac{\partial u_k}{\partial t}\right)^2\right)_t dQ +$$

$$+\frac{1}{2}\int_Q \omega e^{-\omega t}\left(\frac{\partial u_k}{\partial t}\right)^2 dQ = \frac{1}{2}\int_\Omega e^{-\omega T}\left(\frac{\partial u_k}{\partial t}\right)^2\Big|_{t=T} d\Omega +$$

$$+\frac{1}{2}\int_Q \omega e^{-\omega t}\left(\frac{\partial u_k}{\partial t}\right)^2 dQ \geq \frac{1}{2}\int_Q \omega e^{-\omega t}\left(\frac{\partial u_k}{\partial t}\right)^2 dQ.$$

Next, in the same manner, we have

$$-\left(\sum_{i,j=1}^{n}\frac{\partial}{\partial x_i}\left(a_{ij}\frac{\partial u_k}{\partial x_j}\right), e^{-\omega t}\frac{\partial u_k}{\partial t}\right)_{L_2(Q)} =$$

$$= -\int_Q \sum_{i,j=1}^{n}\frac{\partial}{\partial x_i}\left(a_{ij}\frac{\partial u_k}{\partial x_j}e^{-\omega t}\frac{\partial u_k}{\partial t}\right)dQ + \int_Q \sum_{i,j=1}^{n} e^{-\omega t}a_{ij}\frac{\partial u_k}{\partial x_j}\frac{\partial^2 u_k}{\partial x_i\partial t}dQ =$$

$$= -\int_{[0,T]\times\partial\Omega}\sum_{i,j=1}^{n} a_{ij}e^{-\omega t}\frac{\partial u_k}{\partial x_j}\frac{\partial u_k}{\partial t}n_{x_i}\,d([0,T]\times\partial\Omega) +$$

$$+\int_Q \sum_{i,j=1}^{n} e^{-\omega t}a_{ij}\frac{\partial u_k}{\partial x_j}\frac{\partial^2 u_k}{\partial x_i\partial t}dQ.$$

Since $u_k\big|_{x\in\partial\Omega} = 0$ and thus $\frac{\partial u_k}{\partial t}\big|_{x\in\partial\Omega} = 0$, we conclude that first integral of the right-hand side is equal to zero. Consider the second integral of the right-hand side:

$$-\left(\sum_{i,j=1}^{n}\frac{\partial}{\partial x_i}\left(a_{ij}\frac{\partial u_k}{\partial x_j}\right),e^{-\omega t}\frac{\partial u_k}{\partial t}\right)_{L_2(Q)}=\int_Q\sum_{i,j=1}^{n}e^{-\omega t}a_{ij}\frac{\partial u_k}{\partial x_j}\frac{\partial^2 u_k}{\partial x_i\partial t}dQ=$$

$$=\frac{1}{2}\int_Q\left(\sum_{i,j=1}^{n}e^{-\omega t}a_{ij}\frac{\partial u_k}{\partial x_j}\frac{\partial u_k}{\partial x_i}\right)_t dQ+\frac{1}{2}\int_Q\omega\sum_{i,j=1}^{n}e^{-\omega t}a_{ij}\frac{\partial u_k}{\partial x_j}\frac{\partial u_k}{\partial x_i}dQ=$$

$$=\frac{1}{2}\int_\Omega\sum_{i,j=1}^{n}e^{-\omega T}a_{ij}\frac{\partial u_k}{\partial x_j}\frac{\partial u_k}{\partial x_i}\Big|_{t=T}d\Omega+\frac{1}{2}\int_Q\omega\sum_{i,j=1}^{n}e^{-\omega t}a_{ij}\frac{\partial u_k}{\partial x_j}\frac{\partial u_k}{\partial x_i}dQ\geq$$

$$\geq\frac{1}{2}\int_Q\omega\sum_{i,j=1}^{n}e^{-\omega t}a_{ij}\frac{\partial u_k}{\partial x_j}\frac{\partial u_k}{\partial x_i}dQ\geq\frac{1}{2}\int_Q\omega\alpha\sum_{i=1}^{n}e^{-\omega t}\left(\frac{\partial u_k}{\partial x_i}\right)^2 dQ.$$

Consider the third and fourth summands of (5):

$$\left(\sum_{i=1}^{n}b_i\frac{\partial u_k}{\partial x_i},e^{-\omega t}\frac{\partial u_k}{\partial t}\right)_{L_2(Q)}\geq-\int_Q c_b e^{-\omega t}\sum_{i=1}^{n}\left|\frac{\partial u_k}{\partial x_i}\right|\cdot\left|\frac{\partial u_k}{\partial t}\right|dQ,$$

$$\left(cu_k,e^{-\omega t}\frac{\partial u_k}{\partial t}\right)_{L_2(Q)}=\frac{1}{2}\int_Q\left(ce^{-\omega t}(u_k)^2\right)_t dQ+$$

$$+\frac{1}{2}\int_Q c\omega e^{-\omega t}(u_k)^2 dQ=\frac{1}{2}\int_\Omega ce^{-\omega t}(u_k)^2\Big|_{t=T}d\Omega+$$

$$+\frac{1}{2}\int_Q c\omega e^{-\omega t}(u_k)^2 dQ\geq 0.$$

Summing up the obtained inequalities, we have

$$\left(Lu_k,e^{-\omega t}\frac{\partial u_k}{\partial t}\right)_{L_2(Q)}\geq\frac{1}{2}\int_Q\omega e^{-\omega t}\left(\frac{\partial u_k}{\partial t}\right)^2 dQ+$$

$$+\frac{1}{2}\int_Q\omega\alpha\sum_{i=1}^{n}e^{-\omega t}\left(\frac{\partial u_k}{\partial x_i}\right)^2 dQ-\int_Q c_b e^{-\omega t}\sum_{i=1}^{n}\left|\frac{\partial u_k}{\partial x_i}\right|\cdot\left|\frac{\partial u_k}{\partial t}\right|dQ=$$

$$=\frac{1}{4}\int_Q \omega e^{-\omega t}\left(\frac{\partial u_k}{\partial t}\right)^2 dQ+\frac{1}{4}\int_Q \omega\alpha\sum_{i=1}^n e^{-\omega t}\left(\frac{\partial u_k}{\partial x_i}\right)^2 dQ+$$

$$+\frac{1}{4}\int_Q e^{-\omega t}\left(\omega\left(\frac{\partial u_k}{\partial t}\right)^2-4c_b\sum_{i=1}^n\left|\frac{\partial u_k}{\partial x_i}\right|\cdot\left|\frac{\partial u_k}{\partial t}\right|+\omega\alpha\sum_{i=1}^n\left(\frac{\partial u_k}{\partial x_i}\right)^2\right)dQ.$$

Taking into account the value of the constant $\omega = c_b\sqrt{\frac{4n}{\alpha}}$, we obtain that

$$\left(Lu_k, e^{-\omega t}\frac{\partial u_k}{\partial t}\right)_{L_2(Q)}\geq\frac{1}{4}\int_Q \omega e^{-\omega t}\left(\frac{\partial u_k}{\partial t}\right)^2 dQ+$$

$$+\frac{1}{4}\int_Q \omega\alpha\sum_{i=1}^n e^{-\omega t}\left(\frac{\partial u_k}{\partial x_i}\right)^2 dQ+$$

$$\frac{c_b}{\sqrt{4n\alpha}}\sum_{i=1}^n\int_Q e^{-\omega t}\left(\left(\frac{\partial u_k}{\partial t}\right)^2-2\sqrt{n\alpha}\left|\frac{\partial u_k}{\partial x_i}\right|\cdot\left|\frac{\partial u_k}{\partial t}\right|+n\alpha\left(\frac{\partial u_k}{\partial x_i}\right)^2\right)dQ\geq$$

$$\geq c\int_Q\left(\frac{\partial u_k}{\partial t}\right)^2+\sum_{i=1}^n\left(\frac{\partial u_k}{\partial x_i}\right)^2 dQ+$$

$$+\frac{c_b}{\sqrt{4n\alpha}}\sum_{i=1}^n\int_Q e^{-\omega t}\left(\left|\frac{\partial u_k}{\partial t}\right|-\sqrt{n\alpha}\left|\frac{\partial u_k}{\partial x_i}\right|\right)^2 dQ\geq c\|u_k\|^2_{W^+_{bd}}.$$

Thus,

$$\left(f, e^{-\omega t}\frac{\partial u_k}{\partial t}\right)_{L_2(Q)}=\left(Lu_k, e^{-\omega t}\frac{\partial u_k}{\partial t}\right)_{L_2(Q)}\geq c\|u_k\|^2_{W^+_{bd}}.$$

By the Schwarz inequality, we have

$$\|u_k\|_{W^+_{bd}} \cdot \|f\|_{L_2(Q)} \geq \left\|\frac{\partial u_k}{\partial t}\right\|_{L_2(Q)} \cdot \|f\|_{L_2(Q)} \geq \left(f, e^{-\omega t}\frac{\partial u_k}{\partial t}\right)_{L_2(Q)} =$$

$$= \left(Lu_k, e^{-\omega t}\frac{\partial u_k}{\partial t}\right)_{L_2(Q)} \geq c\|u_k\|^2_{W^+_{bd}}.$$

Dividing by $\|u_k\|_{W^+_{bd}}$, we obtain the desired inequality.

From the lemma it follows that the sequence $u_k(t,x)$ is bounded in the space W^+_{bd}, so there exists a weakly convergent to $u_0 \in W^+_{bd}$ subsequence $\{u_{k_l}(t,x)\}$. To prove the strong convergence, consider the following spaces.

Let H^+ be a completion of the set of smooth in $\overline{\Omega}$ functions, which satisfy boundary conditions $u|_{x\in\partial\Omega} = 0$, in the norm

$$\|u\|^2_{H^+} = \sum_{i,j=1}^{n}\int_{\Omega} a_{ij}\frac{\partial u}{\partial x_i}\frac{\partial u}{\partial x_j}d\Omega.$$

Denote by H_1 and H_2 completions of the set of smooth in $\overline{Q}$ functions, which satisfy conditions (2), in the following inner products:

$$(u,v)_{H_1} = \int_Q \frac{\partial^2 u}{\partial t^2}\cdot\frac{\partial^2 v}{\partial t^2} + \sum_{i,j=1}^{n} a_{ij}\frac{\partial^2 u}{\partial t\partial x_i}\frac{\partial^2 v}{\partial t\partial x_j}dQ,$$

$$(u,v)_{H_2} = \frac{1}{2}\left(\left(Lu, e^{-\omega t}v_t\right)_{L_2(Q)} + \left(Lv, e^{-\omega t}u_t\right)_{L_2(Q)}\right),$$

respectively.

Integrating by parts and passing to the surface integration, it is easy to prove the following lemma.

Lemma 2. *For all smooth in $\overline{Q}$ functions $u(t,x)$, which satisfy the conditions (2), the following inequality holds true*

$$\|u\|_{H_1} \geq c_1\|u\|_{H_2} \geq c_2\|u\|_{W^+_{bd}}.$$

Assume that function $f(t,x)$ satisfies conditions $f|_{t=0} = 0$ and $f_t \in L_2(Q)$.

Differentiating the both right-hand and left-hand sides of (3) with respect to t, we obtain

$$\left(\frac{\partial^3 u_k}{\partial t^3}, \omega_j\right)_{L_2(\Omega)} + \left(B\frac{\partial u_k}{\partial t}, \omega_j\right)_{L_2(\Omega)} = \left(f_t, \omega_j\right)_{L_2(\Omega)}.$$

Multiplying the both right-hand and left-hand sides by $e^{-\omega t}\dfrac{d^2 g_j}{dt^2}$, summing up over j from 1 to k and integrating over t from 0 to T, we have

$$\left(\frac{\partial^3 u_k}{\partial t^3}, e^{-\omega t}\frac{\partial^2 u_k}{\partial t^2}\right)_{L_2(\Omega)} + \left(B\frac{\partial u_k}{\partial t}, e^{-\omega t}\frac{\partial^2 u_k}{\partial t^2}\right)_{L_2(\Omega)} = \tag{6}$$

$$= \left(f_t, e^{-\omega t}\frac{\partial^2 u_k}{\partial t^2}\right)_{L_2(\Omega)}.$$

Consider the integrands of the left-hand side of (6).

$$\left(\frac{\partial^3 u_k}{\partial t^3}, e^{-\omega t}\frac{\partial^2 u_k}{\partial t^2}\right)_{L_2(Q)} = \frac{1}{2}\int_Q \left(e^{-\omega t}\left(\frac{\partial^2 u_k}{\partial t^2}\right)^2\right)_t dQ +$$

$$+\frac{1}{2}\int_Q \omega e^{-\omega t}\left(\frac{\partial^2 u_k}{\partial t^2}\right)^2 dQ =$$

$$= \frac{1}{2}\int_\Omega e^{-\omega T}\left(\frac{\partial^2 u_k}{\partial t^2}\right)^2\Big|_{t=T} d\Omega + \frac{1}{2}\int_Q \omega e^{-\omega t}\left(\frac{\partial^2 u_k}{\partial t^2}\right)^2 dQ,$$

We have $\dfrac{\partial^2 u_k}{\partial t^2}\Big|_{t=0} = 0$ by choosing the function $f(t,x)$.

Next, we have

$$-\left(\sum_{i,j=1}^{n}\frac{\partial}{\partial x_i}\left(a_{ij}\frac{\partial^2 u_k}{\partial t\partial x_j}\right), e^{-\omega t}\frac{\partial^2 u_k}{\partial t^2}\right)_{L_2(Q)} =$$

$$\int_Q\sum_{i,j=1}^{n}e^{-\omega t}a_{ij}\frac{\partial^2 u_k}{\partial t\partial x_j}\frac{\partial^3 u_k}{\partial x_i\partial t^2}dQ=\frac{1}{2}\int_Q\left(\sum_{i,j=1}^{n}e^{-\omega t}a_{ij}\frac{\partial^2 u_k}{\partial t\partial x_j}\frac{\partial^2 u_k}{\partial x_i\partial t}\right)_t dQ+$$

$$+\frac{1}{2}\int_Q\omega\sum_{i,j=1}^{n}e^{-\omega t}a_{ij}\frac{\partial^2 u_k}{\partial t\partial x_j}\frac{\partial^2 u_k}{\partial t\partial x_i}dQ=$$

$$=\frac{1}{2}\int_\Omega\sum_{i,j=1}^{n}e^{-\omega T}a_{ij}\frac{\partial^2 u_k}{\partial t\partial x_j}\frac{\partial^2 u_k}{\partial t\partial x_i}\Big|_{t=T}d\Omega+$$

$$+\frac{1}{2}\int_Q\omega\sum_{i,j=1}^{n}e^{-\omega t}a_{ij}\frac{\partial^2 u_k}{\partial t\partial x_j}\frac{\partial^2 u_k}{\partial t\partial x_i}dQ\geq$$

$$\geq\frac{1}{2}\int_Q\omega\sum_{i,j=1}^{n}e^{-\omega t}a_{ij}\frac{\partial^2 u_k}{\partial t\partial x_j}\frac{\partial^2 u_k}{\partial t\partial x_i}dQ\geq\frac{1}{2}\int_Q\omega\alpha\sum_{i=1}^{n}e^{-\omega t}\left(\frac{\partial^2 u_k}{\partial t\partial x_i}\right)^2 dQ.$$

In the same manner as in Lemma 1, consider the other integrands.

$$\left(\sum_{i=1}^{n}b_i\frac{\partial^2 u_k}{\partial t\partial x_i}, e^{-\omega t}\frac{\partial^2 u_k}{\partial t^2}\right)_{L_2(Q)}\geq-\int_Q c_b e^{-\omega t}\sum_{i=1}^{n}\left|\frac{\partial^2 u_k}{\partial t\partial x_i}\right|\cdot\left|\frac{\partial^2 u_k}{\partial t^2}\right|dQ.$$

$$\left(c\frac{\partial u_k}{\partial t}, e^{-\omega t}\frac{\partial^2 u_k}{\partial t^2}\right)_{L_2(Q)}=\frac{1}{2}\int_Q\left(ce^{-\omega t}\left(\frac{\partial u_k}{\partial t}\right)^2\right)_t dQ+$$

$$+\frac{1}{2}\int_Q c\omega e^{-\omega t}\left(\frac{\partial u_k}{\partial t}\right)^2 dQ=$$

$$=\frac{1}{2}\int_\Omega ce^{-\omega t}\left(\frac{\partial u_k}{\partial t}\right)^2\Big|_{t=T}d\Omega+\frac{1}{2}\int_Q c\omega e^{-\omega t}\left(\frac{\partial u_k}{\partial t}\right)^2 dQ\geq 0.$$

Summing up the obtained inequalities, we have

$$\left(L\left(\frac{\partial u_k}{\partial t}\right), e^{-\omega t}\frac{\partial^2 u_k}{\partial t^2}\right)_{L_2(Q)} \geq \frac{1}{2}\int_Q \omega e^{-\omega t}\left(\frac{\partial^2 u_k}{\partial t^2}\right)^2 dQ +$$

$$+\frac{1}{2}\int_Q \omega\alpha\sum_{i=1}^{n} e^{-\omega t}\left(\frac{\partial^2 u_k}{\partial t\partial x_i}\right)^2 dQ - \int_Q c_b e^{-\omega t}\sum_{i=1}^{n}\left|\frac{\partial^2 u_k}{\partial t\partial x_i}\right|\cdot\left|\frac{\partial^2 u_k}{\partial t^2}\right| dQ =$$

$$= \frac{1}{4}\int_Q \omega e^{-\omega t}\left(\frac{\partial^2 u_k}{\partial t^2}\right)^2 dQ + \frac{1}{4}\int_Q \omega\alpha\sum_{i=1}^{n} e^{-\omega t}\left(\frac{\partial^2 u_k}{\partial t\partial x_i}\right)^2 dQ +$$

$$+\frac{1}{4}\sum_{i=1}^{n}\int_Q e^{-\omega t}\left(n\omega\left(\frac{\partial^2 u_k}{\partial t^2}\right)^2 - 4c_b\left|\frac{\partial^2 u_k}{\partial t\partial x_i}\right|\left|\frac{\partial^2 u_k}{\partial t^2}\right| + \omega\alpha\left(\frac{\partial^2 u_k}{\partial t\partial x_i}\right)^2\right) dQ.$$

Taking into account the value of the constant $\omega = c_b\sqrt{\frac{4n}{\alpha}}$, we obtain

$$\left(L\left(\frac{\partial u_k}{\partial t}\right), e^{-\omega t}\frac{\partial^2 u_k}{\partial t^2}\right)_{L_2(Q)} \geq c\int_Q \left(\frac{\partial^2 u_k}{\partial t^2}\right)^2 + \sum_{i=1}^{n}\left(\frac{\partial^2 u_k}{\partial t\partial x_i}\right)^2 dQ +$$

$$+\frac{c_b}{\sqrt{4n\alpha}}\sum_{i=1}^{n}\int_Q e^{-\omega t}\left(\left|\frac{\partial^2 u_k}{\partial t^2}\right| - \sqrt{n\alpha}\left|\frac{\partial^2 u_k}{\partial t\partial x_i}\right|\right)^2 dQ \geq c\|u_k\|_{H_1}^2.$$

Since (6), we conclude that

$$\left(f_t, e^{-\omega t}\frac{\partial^2 u_k}{\partial t^2}\right)_{L_2(Q)} = \left(L\left(\frac{\partial u_k}{\partial t}\right), e^{-\omega t}\frac{\partial^2 u_k}{\partial t^2}\right)_{L_2(Q)} \geq c\|u_k\|_{H_1}^2.$$

Applying to the right-hand side the Schwartz inequality, we find

$$\|f_t\|_{L_2(Q)} \geq c\|u_k\|_{H_1}. \tag{7}$$

Thus, the sequence $\{u_k\}$ is a weakly compact set in the spaces $H_1, H_2, W_{bd}^{+}, L_2(Q)$, so there exists weakly convergent to $u_0 \in H_1$ subsequence. From (7), we have

$$\left\| \frac{\partial u_k}{\partial t} \right\|_{L_2(Q)} \le c_1 \left\| \frac{\partial^2 u_k}{\partial t^2} \right\|_{L_2(Q)} \le c_1 \|u_k\|_{H_1} \le c_2 .$$

Whence, there exists a weakly convergent to $\frac{\partial u_0}{\partial t} \in L_2(Q)$ subsequence $\frac{\partial u_{k_l}}{\partial t}$. In the same manner the weak convergence of the other partial derivative sequences may be proved.

Thus, we obtain

$$\lim_{l \to \infty} \left\| u_{k_l} - u_0 \right\|_{H_2}^2 = \lim_{l \to \infty} \left(\left\| u_{k_l} \right\|_{H_2}^2 - 2\left(u_{k_l}, u_0\right)_{H_2} + \left\| u_0 \right\|_{H_2}^2 \right) = \\ = \lim_{l \to \infty} \left(\left\| u_{k_l} \right\|_{H_2}^2 - \left\| u_0 \right\|_{H_2}^2 \right) \tag{8}$$

Observe that for all functions $u \in H_1$

$$\|u\|_{H_2}^2 = \int_0^T \left\langle Lu, e^{-\omega t} u_t \right\rangle dt ,$$

where $\langle \cdot, \cdot \rangle$ is the bilinear form being generated in the triple of Hilbert spaces $H^+ \subset L_2(\Omega) \subset H^-$.

Since the functions u_k satisfy equality

$$\|u_k\|_{H_2}^2 = \left(Lu_k, e^{-\omega t} \frac{\partial u_k}{\partial t} \right)_{L_2(Q)} = \left(f, e^{-\omega t} \frac{\partial u_k}{\partial t} \right)_{L_2(Q)},$$

rewrite the relation (8) in the following form:

$$\lim_{l \to \infty} \left\| u_{k_l} - u_0 \right\|_{H_2}^2 = \left(f, e^{-\omega t} \frac{\partial u_0}{\partial t} \right)_{L_2(Q)} - \int_0^T \left\langle Lu, e^{-\omega t} \frac{\partial u_0}{\partial t} \right\rangle dt . \tag{9}$$

On the other hand, from the equality (3) we have that right-hand side of (9) equals to zero.

Thus, the sequence $\left\{ u_{k_l} \right\}$ is strong convergent to u_0 in the space H_2 and, hence, in space W_{bd}^{+}. From Lemma 1.1 it follows that the sequence $\left\{Lu_{k_l}\right\}$ is fundamental in the complete space $W^{-}_{bd^+}$, so $\lim\limits_{l\to\infty} Lu_{k_l} = Lu_0$.

Multiplying the both right-hand and left-hand sides of (4) by a smooth function $g(t)$ ($g(T)=0$) and integrating over t from 0 to T, we obtain:

$$\left(Lu_{k_l}, g\omega_j\right)_{L_2(Q)} = \left(f, g\omega_j\right)_{L_2(Q)}.$$

Approaching the limit as $l\to\infty$, we have

$$\left\langle Lu_0, g\omega_j\right\rangle_{W_{bd^+}} = \left(f, g\omega_j\right)_{L_2(Q)}.$$

Since the set $\left\{g\omega_j\right\}$ is total in the space $W^{+}_{bd^+}$, we conclude that $Lu_0 = f$ in the space $W^{-}_{bd^+}$, thus the function $u_0(t,x)$ is a solution of the problem $Lu=f$ in the sense of Definition 1.1.2 and, hence, in the sense of Definition 1.1.1. We note that there was no necessity to choose a subsequence $u_{k_l}(t,x)$ because of uniqueness of the solution. The obtained result we formulate in the following theorem.

Theorem 1. *For a function $f(t,x)$: $f\in L_2(Q)$, $f|_{t=0}=0$ the approximate sequence (3) converges to the solution of the problem (1), (2) in the sense of Definition 1.1.1 in the norm of the space W^{+}_{bd}.*

Taking into account the density of the considered functions $f(t,x)$ in the space $L_2(Q)$, it is easy to prove the following theorem.

Theorem 2. *For all functions $f\in L_2(Q)$ the approximate sequence (3) converges to the solution of problem (1), (2) in the sense of Definition 1.1.1 in the norm of space W^{+}_{bd}.*

Consider theapplication of the Galerkin method when the right-hand side of the state equation belongs to the negative space $W^{-}_{bd^+}$.

Let f_0 be an arbitrary function from $W^-_{bd^+}$. By virtue of the density of the space $L_2(Q)$ in the space $W^-_{bd^+}$, there exists a sequence of functions $f_l \in L_2(Q)$ such that $\|f_l - f_0\|_{W^-_{bd^+}} \xrightarrow[l\to\infty]{} 0$. In this case, we consider the approximate sequence in the following form:

$$\langle Lu_k^l, \omega_j \rangle = \langle f_l, \omega_j \rangle, \tag{10}$$

where

$$u_k^l = \sum_{i=1}^{k} g_{k,i}^l(t)\omega_i(x), \tag{11}$$

the function $g_{k,i}^l(t)$ is a solution of the following set of ordinary differential equations

$$\frac{d^2 g_{k,i}^l(t)}{dt^2} + \sum_{i=1}^{k} g_{k,i}^l(t)(B\omega_i, \omega_j)_{L_2(\Omega)} = (f_l, \omega_j)_{L_2(\Omega)},$$
$$g_{k,i}^l(0) = \frac{dg_{k,i}^l(0)}{dt} = 0, i = \overline{1,k}, j = \overline{1,k}. \tag{12}$$

By Theorem 2 the sequence $\{u_k^l\}_{k=1}^{\infty}$ converges to the solution u_0^l of the problem $Lu = f_l$ in the norm of space W^+_{bd}. Consider the sequence $\{u_0^l\}$. By the Lemma 1.1.3, we have

$$\|u_0^l - u_0\|_{L_2(Q)} \le c\|f_l - f_0\|_{W^-_{bd^+}} \xrightarrow[l\to\infty]{} 0.$$

Thus the following theorem holds true.

Theorem 3. *For all functions $f(t,x) \in W^-_{bd^+}$ the approximate sequence (11) converges to the solution of the problem (1), (2) in the sense of Definition 1.1.4 as $k \to \infty, l \to \infty$ in the norm of space $L_2(Q)$.*

4. PULSE OPTIMAL CONTROL OF HYPERBOLIC SYSTEMS (THE DIRICHLET INITIAL BOUNDARY VALUE PROBLEM)

Using the template theorems from Section 3.6, we shall fill in the following tables.

Table 1.

N	Operator	Space N	Space $W^-(Q)$
1.	L	$L_2(Q)$	$W^-_{bd^+}$
2.	L	W^+_{bd}	$L_2(Q)$

Table 2.

N	Operator $A_i(\cdot)$	Space $W^-(Q)$
1.	$A_1(\cdot)$	W^+_{bd}
2.	$A_3(\cdot)$	W^+_{bd}
3.	$A_5(\cdot)$	W^+_{bd}

Table 3.

N	Operator L	Space $W^-(Q)$	Space $W^+(Q)$
1.	L	$W^-_{bd^+}$	$W^+_{bd^+}$

Table 4 is empty.

Table 5.

N	Operator L	Space $W^-(Q)$
1.	L	$W^-_{bd^+}$

Table 6.

N	Operator L	Map $A_i(\cdot)$	Space $W^-(Q)$
1.	L	$A_1(\cdot)$	$W^-_{bd^+}$
2.	L	$A_3(\cdot)$	$W^-_{bd^+}$
3.	L	$A_5(\cdot)$	$W^-_{bd^+}$

Table 7.

N	Exponent α	Space $W^-(Q)$	Map $A_{i,\varepsilon}(\cdot)$
1.	1/2	$W^-_{bd^+}$	$A_{1,\varepsilon}(\cdot)$
2.	1/2	$W^-_{bd^+}$	$A_{3,\varepsilon}(\cdot)$
3.	1/2	$W^-_{bd^+}$	$A_{5,\varepsilon}(\cdot)$

Chapter 7

PSEUDO-HYPERBOLIC SYSTEMS

1. GENERALIZED SOLVABILITY OF PSEUDO-HYPERBOLIC SYSTEMS (THE DIRICHLE INITIAL BOUNDARY VALUE PROBLEM)

1.1 Formulation of the problem. Main notations.

In this chapter we shall consider the problems of optimization of the systems described by the pseudo-hyperbolic equations. Such equations arise, for example, in the investigations of mass transport in heterogeneous porous media [134]. In addition, many processes are described by non-stationary equations with small viscosity. For instance, torsion oscillations of metallic cylinder with inner friction, propagation of perturbations in viscous and elastic rod, one-dimensional flow of isotropic viscous liquid, sound propagation in viscous gas and similar processes are described by the model equations in the following form:

$$\frac{\partial^2 u}{\partial t^2} = \gamma \frac{\partial^3 u}{\partial t \partial x^2} + \frac{\partial^2 u}{\partial x^2}, \tag{1}$$

where $\gamma = \text{const} > 0$ is a parameter, $\gamma \dfrac{\partial^3 u}{\partial t \partial x^2}$ is a small viscosity.

There are well-known the equations of viscous and elastic medium [135]

$$\rho \frac{\partial^2 u}{\partial t^2} = (\Lambda + 2M)\text{div}(\text{grad}u) - M\text{rotrot}u, \tag{2}$$

where $\Lambda = \lambda + \lambda' \dfrac{\partial}{\partial t}$, $M = \gamma + \gamma' \dfrac{\partial}{\partial t}$, λ, γ are the Lame constants λ', γ' are the viscosity constants, ρ is density.

For example, the equations of longitudinal oscillations of viscous and elastic rod has the following form [136]:

$$\frac{\partial^2 u}{\partial t^2} - b^2 \frac{\partial^3 u}{\partial t \partial x^2} - a^2 \frac{\partial^2 u}{\partial x^2} = 0,$$

where $a = \sqrt{\frac{E}{\rho}}, b = \sqrt{\frac{\gamma}{\rho}}$, ρ is density of the medium, E is the elasticity coefficient, γ is the viscosity coefficient.

The equation of the propagation of initial perturbations in viscous gas has the following form:

$$\frac{\partial^2 u}{\partial t^2} - c^2 \frac{\partial^3 u}{\partial t \partial x^2} - \frac{4}{3}\gamma \frac{\partial^2 u}{\partial x^2} = 0,$$

where c is the sound velocity in the absence of viscosity, γ is the cinematic coefficient of viscosity.

Pseudo-hyperbolic systems were investigated in papers [137-140,172,176].

Let us consider the system with distributed parameters and pseudo-hyperbolic state equation.

Problem 1. *Find a function* $u(t,x)$ *satisfying the following equation*

$$L_1 u = \frac{\partial^2 u}{\partial t^2} - \sum_{i,j=1}^{N} \frac{\partial^2}{\partial t \partial x_i}\left(a_{ij}(t,x)\frac{\partial u}{\partial x_j}\right) + \frac{\partial}{\partial t}\left(\sum_{i=1}^{N} a_i(t,x)\frac{\partial u}{\partial x_i}\right) +$$
$$+ \frac{\partial}{\partial t}\left(a(t,x)u\right) - \sum_{i,j=1}^{N} \frac{\partial}{\partial x_i}\left(b_{ij}(x)\frac{\partial u}{\partial x_j}\right) + \tag{3}$$
$$+ \sum_{i=1}^{N} c_i(t,x)\frac{\partial u}{\partial x_i} + b(t,x)u = f(t,x),$$

where $u(t,x)$ *is defined in the domain* $Q = (0,T)\times\Omega$, $\Omega \subset R^N$ (Ω *is a bounded domain with a sufficiently smooth bound* $\partial\Omega$).

Suppose that

$$a_{ij}(t,x)=a_{ji},\ a_{ij}(t,x)\in C^{2,1}_{t,x}(\overline{Q});\ b_{ij}(x)=b_{ji},\ \left\{b_{ij}(x)\right\}^{N}_{i,j=1}\quad —$$

are continuously differentiable in the closed domain $\overline{\Omega}$ functions, $\{a_i(t,x)\}^N_{i=1}\in C^{2,1}_{t,x}(\overline{Q})$; $a(t,x)\in C^{2,0}_{t,x}(\overline{Q})$; $b(t,x)\in C^{1,0}_{t,x}(\overline{Q})$; $\left\{c_i(t,x)\right\}^N_{i=1}$ are continuously differentiable in the closed domain $\overline{Q}$ functions;

$$\sum_{i,j=1}^{N}a_{ij}(t,x)\xi_i\xi_j\geq\lambda_a\sum_{i=1}^{N}\xi_i^2,\ \sum_{i,j=1}^{N}b_{ij}(x)\xi_i\xi_j\geq 0,\forall\xi_i\in R^1,$$

$$i=\overline{1,N};\ \lambda_a=\text{const}>0;$$

$$\sum_{i,j=1}^{N}\frac{\partial a_{ij}(t,x)}{\partial t}\xi_i\xi_j\geq 0,\ \sum_{i,j=1}^{N}\frac{\partial^2 a_{ij}(t,x)}{\partial t^2}\xi_i\xi_j\leq 0,\forall\xi_i\in R^1, i=\overline{1,N};$$

$$2a(t,x)\geq\sum_{i=1}^{N}\frac{\partial a_i}{\partial x_i},\frac{\partial a(t,x)}{\partial t}\geq\sum_{i=1}^{N}\frac{\partial^2 a_i}{\partial t\partial x_i},\frac{\partial^2 a(t,x)}{\partial t^2}\leq\sum_{i=1}^{N}\frac{\partial^3 a_i}{\partial t^2\partial x_i};$$

$$b(t,x)\geq\lambda_b,\lambda_b=\text{const}>0;$$

$$|c_i(t,x)|+|a_i(t,x)|\leq\lambda_c=\text{const}>0, i=\overline{1,N};$$

$$b(t,x)\geq 2\sum_{i=1}^{N}\frac{\partial c_i(t,x)}{\partial x_i},0\leq\frac{\partial b(t,x)}{\partial t}\leq\sum_{i=1}^{V}\frac{\partial^2 c_i(t,x)}{\partial t\partial x_i}.$$

The adjoint equation has the following form

$$L_1^{*}v=\frac{\partial^2 v}{\partial t^2}+\sum_{i,j=1}^{N}\frac{\partial}{\partial x_i}\left(a_{ij}(t,x)\frac{\partial^2 v}{\partial x_j\partial t}\right)+$$

$$+\sum_{i=1}^{N}\frac{\partial}{\partial x_i}\left(a_i(t,x)\frac{\partial v}{\partial t}\right)-a(t,x)\frac{\partial v}{\partial t}-\sum_{i,j=1}^{N}\frac{\partial}{\partial x_i}\left(b_{ij}(x)\frac{\partial v}{\partial x_j}\right)-\tag{4}$$

$$-\sum_{i=1}^{N}\frac{\partial}{\partial x_i}(c_i(t,x)v)+b(t,x)v=g(t,x).$$

Let the system state satisfies the following conditions:

$$u\big|_{t=0} = \frac{\partial u}{\partial t}\Big|_{t=0} = 0 \ , \quad u\big|_{x\in\partial\Omega} = 0. \tag{5}$$

Let us introduce the following notations.

W^{+}_{bd} is a completion of the space of smooth functions, which satisfy the conditions (5), in the norm

$$\|u\|^2_{W^{+}_{bd}} = \int_Q \left(u_t^2 + \sum_{i=1}^{N} u_{tx_i}^2 \right) dQ; \tag{6}$$

$W^{+}_{bd^{+}}$ is the same space, but the functions satisfy the following conditions

$$v\big|_{t=T} = \frac{\partial v}{\partial t}\Big|_{t=T} = 0 , \quad v\big|_{x\in\partial\Omega} = 0 ; \tag{7}$$

$H^{+}_{bd}, H^{+}_{bd^{+}}$ are the analogous spaces obtained by completing of smooth functions satisfying the conditions (5) and (7), respectively, in the norm

$$\|u\|^2_{H^{+}_{bd}} = \int_Q (u^2 + \sum_{i=1}^{N} u_{x_i}^2) dQ ;$$

$W^{-}_{bd}, W^{-}_{bd^{+}}, H^{-}_{bd}, H^{-}_{bd^{+}}$ are the correspondent negative spaces.

The following imbeddings are valid:

$$W^{+} \subset H^{+} \subset L_2(Q) \subset H^{-} \subset W^{-},$$

and, moreover, imbeddings are dense and the imbedding operators are completely continuous.

Problem 2. *Consider the same equation when its coefficients do not depend on time*

$$L_2 u = \frac{\partial^2 u}{\partial t^2} - \sum_{i,j=1}^{N} \frac{\partial^2}{\partial t \partial x_i}\left(a_{ij}(x)\frac{\partial u}{\partial x_j}\right) + \frac{\partial}{\partial t}\left(\sum_{i=1}^{N} a_i(x)\frac{\partial u}{\partial x_i}\right) +$$
$$+ \frac{\partial u}{\partial t}(a(x)u) - \sum_{i,j=1}^{N} \frac{\partial}{\partial x_i}\left(b_{ij}(x)\frac{\partial u}{\partial x_j}\right) + \tag{8}$$
$$+ \sum_{i=1}^{N} c_i(x)\frac{\partial u}{\partial x_i} + b(x)u = f(t,x),$$

where $u(t,x)$ *is defined in the domain* $Q = (0,T)\times\Omega$, $\Omega \subset R^N$ (Ω *is a bounded domain with a sufficiently smooth bound* $\partial\Omega$).

Suppose that

$$a_{ij}(x) = a_{ji}; b_{ij}(x) = b_{ji}, \ \{a_{ij}(x)\}_{i,j=1}^{N}; \{b_{ij}(x)\}_{i,j=1}^{N};$$

$\{a_i(x)\}_{i=1}^{N}$; $\{c_i(x)\}_{i=1}^{N}$ are continuously differentiable in the closed domain $\overline{\Omega}$ functions; $a(x)$, $b(x)$ are continuous in $\overline{\Omega}$ functions;

$$\sum_{i,j=1}^{N} \dot{a}_{ij}(x)\xi_i\xi_j \geq \lambda_a \sum_{i=1}^{N} \xi_i^2, \ \sum_{i,j=1}^{N} b_{ij}(x)\xi_i\xi_j \geq 0, \forall \xi_i \in R^1,$$
$$i = \overline{1,N}; \lambda_a = \text{const} > 0;$$
$$b(x) \geq \lambda_b, \lambda_b = \text{const} > 0;$$
$$|c_i(x)| + |a_i(x)| \leq \lambda_c = \text{const} > 0, i = \overline{1,N};$$
$$b(x) \geq 2\sum_{i=1}^{N} \frac{\partial c_i(x)}{\partial x_i}; \ 2a(x) \geq \sum_{i=1}^{N} \frac{\partial a_i(x)}{\partial x_i}.$$

In Problem 2 we shall consider the same boundary conditions and chains of Hilbert spaces as in Problem 1.

Problem 3. *We shall investigate the equation*

$$L_3 u = \frac{\partial^2 u}{\partial t^2} + M\frac{\partial u}{\partial t} + A\left(\frac{\partial u}{\partial t}\right) + B(u) = f(t,x), \tag{9}$$

where $$A(\cdot) = -\sum_{i,j=1}^{N} \frac{\partial}{\partial x_i}\left(A_{ij} \frac{\partial}{\partial x_j} \right); \quad B(\cdot) = -\sum_{i,j=1}^{N} \frac{\partial}{\partial x_i}\left(B_{ij} \frac{\partial}{\partial x_j} \right);$$

$u(t, x)$ *is defined in the domain* $Q = (0,T)\times\Omega$, $\Omega \subset R^N$ (Ω *is a bounded domain with sufficiently smooth bound* $\partial\Omega$).

Suppose that $A_{ij}(x) = A_{ji}$, $B_{ij}(x) = B_{ji}$, $\{A_{ij}\}_{i,j=1}^{N}$, $\{B_{ij}\}_{i,j=1}^{N}$ are continuously differentiable in the closed domain $\overline{\Omega}$ functions, $M = M(x)$ is continuous in the closed domain $\overline{\Omega}$ function;

$$\sum_{i,j=1}^{N} A_{ij}\xi_i\xi_j \geq \lambda_A \sum_{i=1}^{N} \xi_i^2, \ \lambda_B^{-1} \sum_{i,j=1}^{N} \xi_i^2 \geq \sum_{i,j=1}^{N} B_{ij}\xi_i\xi_j \geq 0,$$

$$\forall \xi_i \in R^1, i = \overline{1, N}; M \geq 0,$$

where λ_A, λ_B are positive constants.

Let us introduce the following denotations: $\overline{W}_{bd}^{+}$ is the completion of the space of smooth functions, which satisfy the condition (5), in the norm

$$\|u\|_{\overline{W}_{bd}^{+}} = \left(\int_Q (u_{tt}^2 + \sum_{i=1}^{N} u_{tx_i}^2) dQ \right)^{1/2};$$

$\overline{W}_{bd^+}^{+}$ is the same space, but the functions satisfy the conditions of the adjoint problem (7); $\overline{W}_{bd}^{-}, \overline{W}_{bd^+}^{-}$ are the corresponding negative spaces.

1.2 Properties of pseudo-hyperbolic operator in Hilbert space

Let us investigate the properties of the operators L_1 and L_1^*. At first, we shall show that the operator L_1 is extendable to the operator, which continuously maps the whole space W_{bd}^{+} into space $W_{bd^+}^{-}$.

Lemma 1. *For the functions $u(t,x)\in W^{+}_{bd}$ the following a priori inequalities holds true:*

$$\|L_1 u\|_{W^{-}_{bd^+}} \le C\|u\|_{W^{+}_{bd}}. \tag{10}$$

where C hereinafter is some positive constant.

Proof. At first, we shall prove the lemma for smooth functions $u(t,x)$, which satisfy the condition (5́), and later we shall obtain the validity of the lemma for any $u(t,x)\in W^{+}_{bd}$ extending in the correspondent way the operator L_1 and passing to the limit.

By definition of the negative norm, we have

$$\|L_1 u\|_{W^{-}_{bd^+}} = \sup_{v\neq 0}\frac{\left|\langle L_1 u, v\rangle_{W_{bd^+}}\right|}{\|v\|_{W^{+}_{bd^+}}} = \sup_{v\neq 0}\frac{\left|(L_1 u, v)_{L_2(Q)}\right|}{\|v\|_{W^{+}_{bd^+}}},$$

$$v(t,x)\in W^{+}_{bd^+}, \tag{11}$$

as far as for smooth $u(t,x)$, which satisfy the condition (5), the bilinear form coincide with the inner product in $L_{2}(Q)$. Let us consider the numerator in the right-hand side of (11):

$$\left|(L_1 u, v)_{L_2(Q)}\right| = \left|\left(\frac{\partial^2 u}{\partial t^2} - \sum_{i,j=1}^{N}\frac{\partial^2}{\partial t\partial x_i}\left(a_{ij}\frac{\partial u}{\partial x_j}\right) + \frac{\partial}{\partial t}(au) - \right.\right.$$

$$\left.\left. - \sum_{i,j=1}^{N}\frac{\partial}{\partial x_i}\left(b_{ij}\frac{\partial u}{\partial x_j}\right) + \sum_{i=1}^{N} c_i\frac{\partial u}{\partial x_i} + bu, v\right)_{L_2(Q)}\right|.$$

Using the integration by parts, the integral Cauchy inequality and granting the initial conditions, we can write

$$\left|\int_Q u_{tt} v\, dQ\right| = \left|\int_Q (u_t v)_t\, dQ - \int_Q u_t v_t\, dQ\right| = \left|\int_Q u_t v_t\, dQ\right| \le$$

$$\le \left(\int_Q u_t^2 dQ\right)^{1/2} \left(\int_Q v_t^2 dQ\right)^{1/2} \le \|u\|_{W_{bd}^+} \|v\|_{W_{bd^+}^+} .$$

Next,

$$\left|\int_Q v \sum_{i,j=1}^N \frac{\partial^2}{\partial t \partial x_i}\left(a_{ij}\frac{\partial u}{\partial x_j}\right) dQ\right| =$$

$$= \left|\int_Q \sum_{i,j=1}^N \left(v\frac{\partial}{\partial x_i}\left(a_{ij}\frac{\partial u}{\partial x_j}\right)\right)_t' dQ - \int_Q \sum_{i,j=1}^N v_t \frac{\partial}{\partial x_i}\left(a_{ij}\frac{\partial u}{\partial x_j}\right) dQ\right| =$$

$$= \left|-\int_Q \sum_{i,j=1}^N \left(v_t a_{ij} u_{x_j}\right)_{x_i} dQ + \int_Q \sum_{i,j=1}^N v_{tx_i} a_{ij} u_{x_j} dQ\right| \le$$

$$\le C_1 \left(\int_Q \sum_{i=1}^N v_{tx_i}^2 dQ\right)^{1/2} \left(\int_Q u_{x_i}^2 dQ\right)^{1/2} \le$$

$$\le C_1 \|v\|_{W_{bd^+}^+} \|u\|_{H_{bd}^+} \le C_2 \|v\|_{W_{bd^+}^+} \|u\|_{W_{bd}^+} .$$

In a similar way,

$$\left|\int_Q v \frac{\partial}{\partial t}(au) dQ\right| = \left|\int_Q (vau)_t dQ - \int_Q v_t au dQ\right| \le$$

$$\le C_3 \left(\int_Q v_t^2 dQ\right)^{1/2} \left(\int_Q u^2 dQ\right)^{1/2} \le$$

$$\le C_3 \|v\|_{W_{bd^+}^+} \|u\|_{H_{bd}^+} \le C_3' \|v\|_{W_{bd^+}^+} \|u\|_{W_{bd}^+} .$$

$$\left|\int_Q v \sum_{i,j=1}^N \frac{\partial}{\partial x_i}\left(b_{ij}\frac{\partial u}{\partial x_j}\right) dQ\right| =$$

$$\left|\int_Q \sum_{i,j=1}^N \left(v b_{ij} u_{x_j}\right)_{x_i} dQ - \int_Q \sum_{i,j=1}^N v_{x_i} b_{ij} u_{x_j} dQ\right| \le$$

$$\leq C_4\left(\int_Q v_{x_i}^2\,dQ\right)^{1/2}\left(\int_Q u_{x_j}^2\,dQ\right)^{1/2} \leq C_4\,\|v\|_{H^+_{bd^+}}\,\|u\|_{H^+_{bd}} \leq$$

$$\leq C_4'\,\|v\|_{W^+_{bd^+}}\,\|u\|_{W^+_{bd}}\,.$$

$$\left|\int_Q v\sum_{i=1}^N c_i u^-_{x_i}\,dQ\right| \leq C_5\left(\int_Q v^2 dQ\right)^{1/2}\left(\int_Q \sum_{i=1}^N u_{x_i}^2\,dQ\right)^{1/2} \leq$$

$$\leq C_5\|v\|_{H^+_{bd^+}}\,\|u\|_{H^+_{bd}} \leq C_5'\|v\|_{W^+_{bd^+}}\,\|u\|_{W^+_{bd}}\,.$$

$$\left|\int_Q vbud Q\right| \leq C_6\left(\int_Q v^2 dQ\right)^{1/2}\left(\int_Q u^2 dQ\right)^{1/2} \leq$$

$$\leq C_6\|v\|_{H^+_{bd^+}}\,\|u\|_{H^+_{bd}} \leq C_6'\|v\|_{W^+_{bd^+}}\,\|u\|_{W^+_{bd}}\,.$$

From the written above we obtain

$$\left|(v, L_1 u)_{L_2(Q)}\right| \leq C\|v\|_{W^+_{bd^+}}\,\|u\|_{W^+_{bd}}\,.$$

Using obtained inequality, we have

$$\|L_1 u\|_{W^-_{bd^+}} = \sup_{v\neq 0}\frac{\left|(L_1 u, v)_{L_2(Q)}\right|}{\|v\|_{W^+_{bd^+}}} \leq \sup_{v\neq 0}\frac{C\|v\|_{W^+_{bd^+}}\|u\|_{W^+_{bd}}}{\|v\|_{W^+_{bd^+}}} = C\|u\|_{W^+_{bd}}$$

for smooth functions $u(t, x)$, which satisfy the condition (5).

Using the inequality (10), we can extend the operator L_1, which is defined on smooth functions satisfying the condition (5) to the whole space W^+_{bd} (continuity extension).

For extended operator we shall save the previous notation and hereinafter we shall consider only extended operator.

Passing to the limit, we obtain the validity of the lemma for any functions $u(t,x)\in W^+_{bd}$.

Remark 1. *The inequality (10) imply that the linear operator* L_1 *continuously maps the whole space* W_{bd}^{+} *into* $W_{bd^+}^{-}$.

In a similar way we can prove that extended operator L_1^* continuously maps the whole space $W_{bd^+}^{+}$ into the space W_{bd}^{-}.

Lemma 2. *For all functions* $v(t,x) \in W_{bd^+}^{+}$ *the following inequality holds true*

$$\left\| L_1^* v \right\|_{W_{bd}^-} \leq C \left\| v \right\|_{W_{bd^+}^+}.$$

Statement 1. *Let* $u \in W_{bd}^{+}, v \in W_{bd^+}^{+}$, *then the following equality holds true*

$$\langle v, L_1 u \rangle_{W_{bd^+}} = \left\langle L_1^* v, u \right\rangle_{W_{bd}}.$$

For Problem 2 the following lemmas hold true.

Lemma 3. *For all functions* $u(t,x) \in W_{bd}^{+}$ *the following a priori inequality holds true:*

$$\left\| L_2 u \right\|_{W_{bd^+}^-} \leq C \left\| u \right\|_{W_{bd}^+}.$$

The proof of Lemma 3 is analogous to the proof of Lemma 1.

Lemma 4. *For all functions* $v(t,x) \in W_{bd^+}^{+}$ *the following inequality holds true*

$$\left\| L_2^* v \right\|_{W_{bd}^-} \leq C \left\| v \right\|_{W_{bd}^+}.$$

The proof of Lemma 4 is similar to the proof of Lemma 2.

For Problem 3 the following lemmas hold true.

Lemma 5. *For all functions* $u(t,x) \in \overline{W}_{bd}^{+}$ *the following a priori inequality holds true*

$$\left\| L_3^* v \right\|_{W_{bd}^-} \leq C \left\| v \right\|_{W_{bd}^+}.$$

The proof of Lemma 5 is similar to the proof of Lemma 1.

Lemma 6. *For all functions* $v(t,x) \in \overline{W}^{+}_{bd^{+}}$ *the following inequality holds true*

$$\|L_3^{*}v\|_{\overline{W}^{-}_{bd}} \le C\|v\|_{\overline{W}^{+}_{bd^{+}}} .$$

The proof of Lemma 6 is similar to the proof of Lemma 2.

1.3. The existence and uniqueness of solution of initial boundary value problem

To investigate the solvability of an initial boundary value problem we shall use equipped Hilbert spaces, a priori estimations in negative norms, and modified method of deriving of energetic inequalities in negative spaces.

Problem 4. We shall investigate the solvability of the following problems

$$L_1 u = f, f \in H^{-}_{bd^{+}}, u \in W^{+}_{bd}; \tag{12}$$

$$L_1^{*} v = g, g \in H^{-}_{bd}, v \in W^{+}_{bd^{+}}, \tag{13}$$

where u, v are unknown elements, and f, g are given elements.

The solutions of the problems (12), (13) we shall consider as generalized solutions in the following sense.

Definition 1. *The generalized solution of the problem (12) is a function* $u(t,x) \in W^{+}_{bd}$ *such that there exists a sequence of smooth functions* $u_i(t,x)$ *which satisfy the condition (5) and*

$$\|u_i - u\|_{W^{+}_{bd}} \xrightarrow[i\to\infty]{} 0, \|L_1 u_i - f\|_{W^{-}_{bd^{+}}} \xrightarrow[i\to\infty]{} 0 .$$

Definition 2. *The generalized solution of the problem (12) is a function* $u(t,x) \in W^{+}_{bd}$ *such that the identity*

$$\left(L_1^{*}v, u\right)_{L_2(Q)} = \langle f, v\rangle_{H^{-}_{bd^{+}}}$$

holds true for any smooth functions v(t,x) satisfying the condition (7).

In a similar way we can introduce the definition of a generalized solution of the adjoint problem.

Lemma 7. *For any function* $u(t,x) \in W^{+}_{bd}$ *the following inequality holds true*

$$\|L_1 u\|_{W^{-}_{bd^{+}}} \geq C\|u\|_{H^{+}_{bd}}. \tag{14}$$

Proof. At first, let us prove the inequality for smooth functions $u(t,x)$ satisfying the condition (5). Introduce an auxiliary function $v(t,x)$ in the following way:

$$v(t,x) = \int_t^T \frac{u(\hat{o},x)}{p(\hat{o})} d\hat{o},$$

where

$$\frac{1}{p(t)} = \frac{\sin\frac{\pi}{2}\sqrt{\frac{T-t}{T}}}{e^{\frac{N\lambda_c^2}{\lambda_a\lambda_b}t}}.$$

The definition of the function $v(t,x)$ implies that $u = -pv_t$ $\forall t \in [0,T)$.

Note that

$$p(t) \geq 1, \ \frac{\partial p(t)}{\partial t} > \frac{N\lambda_c^2}{\lambda_a\lambda_b} p(t), \forall t \in [0,T).$$

Let us prove the validity of the following inequality:

$$(L_1 u, v)_{L_2(Q)} \geq C_1\|v\|^2_{W^{+}_{bd^{+}}}. \tag{15}$$

Consider

$$(L_1 u, v)_{L_2(Q)} = I_1 + I_2 + I_3 + I_4 + I_5 + I_6 + I_7.$$

Using the integration by parts, relation between $u(t,x)$ and $v(t,x)$, and initial conditions, we obtain that

$$I_1 = (v,u_{tt})_{L_2(Q)} = \int_Q (vu_t)_t dQ - \int_Q v_t u_t dQ =$$

$$= \int_Q p v_t v_{tt} dQ + \int_Q p_t v_t^2 dQ =$$

$$= \frac{1}{2}\int_Q \left(p v_t^2\right)_t dQ - \frac{1}{2}\int_{Q'} p_t v_t^2 dQ + \int_Q p_t v_t^2 dQ \geq$$

$$\geq \frac{1}{2}\int_Q p_t v_t^2 dQ \geq C_2 \int_Q v_t^2 dQ.$$

Granting initial and boundary conditions, we have

$$I_2 = -\left(v, \sum_{i,j=1}^{N} \frac{\partial^2}{\partial t \partial x_i}\left(a_{ij}\frac{\partial u}{\partial x_j}\right)\right)_{L_2(Q)} =$$

$$= -\int_Q \left(v \sum_{i,j=1}^{N} \frac{\partial}{\partial x_i}\left(a_{ij}\frac{\partial u}{\partial x_j}\right)\right)_t dQ + \int_Q v_t \sum_{i,j=1}^{N} \frac{\partial}{\partial x_i}\left(a_{ij}\frac{\partial u}{\partial x_j}\right) dQ =$$

$$= \int_Q \sum_{i,j=1}^{N} \left(v_t a_{ij} u_{x_j}\right)_{x_i} dQ + \int_Q p \sum_{i,j=1}^{N} v_{tx_i} a_{ij} v_{tx_j} dQ \geq$$

$$\geq \frac{1}{2}\lambda_a \int_Q \sum_{i=1}^{N} v_{tx_i}^2 dQ + \frac{1}{2}\lambda_a \int_Q p \sum_{i=1}^{N} v_{tx_i}^2 dQ.$$

In a similar way,

$$I_3 = \left(v, \frac{\partial}{\partial t}\left(\sum_{i=1}^{N} a_i \frac{\partial u}{\partial x_i}\right)\right)_{L_2(Q)} =$$

$$= \int_Q \left(v \sum_{i=1}^{N} a_i u_{x_i}\right)_t dQ + \int_Q p v_t \sum_{i=1}^{N} a_i v_{tx_i} dQ =$$

$$=\frac{1}{2}\int_Q p\sum_{i=1}^N \left(a_i v_t^2\right)_{x_i} dQ-\frac{1}{2}\int_Q p\sum_{i=1}^N \frac{\partial a_i}{\partial x_i} v_t^2\, dQ=-\frac{1}{2}\int_Q p\sum_{i=1}^N \frac{\partial a_i}{\partial x_i} v_t^2\, dQ.$$

$$I_4=\left(v,\frac{\partial}{\partial t}(au)\right)_{L_2(Q)}=\int_Q (vau)_t dQ-\int_Q av_t u dQ=\int_Q pav_t^2 dQ.$$

Granting the fact that the matrix $\left\{b_{ij}\right\}_{i,j=1}^N$ is symmetric and nonnegative we conclude that the following relation is valid:

$$I_5=-\left(v,\sum_{i,j=1}^N \frac{\partial}{\partial x_i}\left(b_{ij}u_{x_j}\right)\right)_{L_2(Q)}=$$

$$=-\int_Q \sum_{i,j=1}^N \left(vb_{ij}u_{x_j}\right)_{x_i} dQ-\int_Q p\sum_{i,j=1}^N v_{x_i} b_{ij} v_{tx_j} dQ=$$

$$=-\frac{1}{2}\int_Q \left(p\sum_{i,j=1}^N v_{x_i} b_{ij} v_{x_j}\right)_t dQ+\frac{1}{2}\int_Q p_t \sum_{i,j=1}^N v_{x_i} b_{ij} v_{x_j} dQ=$$

$$=\frac{1}{2}\int_\Omega p(0)\sum_{i,j=1}^N v_{x_i}(0,x) b_{ij} v_{x_j}(0,x) d\Omega+\frac{1}{2}\int_Q p_t \sum_{i,j=1}^N v_{x_i} b_{ij} v_{x_j} dQ\geq 0.$$

Next,

$$I_6=\left(v,\sum_{i=1}^N c_i \frac{\partial u}{\partial x_i}\right)_{L_2(Q)}=$$

$$=-\int_Q p\sum_{i=1}^N vc_i v_{tx_i} dQ\geq -\lambda_c \int_Q p\sum_{i=1}^N |v|\cdot\left|v_{tx_i}\right| dQ.$$

$$I_7=(v,bu)_{L_2(Q)}=-\int_Q pbvv_t dQ=$$

$$=-\frac{1}{2}\int_Q \left(pbv^2\right)_t dQ+\frac{1}{2}\int_Q p_t b v^2 dQ+\frac{1}{2}\int_Q p b_t v^2 dQ=$$

$$=\frac{1}{2}\int_{\Omega}p(0)b(0,x)v^2(0,x)d\Omega+\frac{1}{2}\int_{Q}p_tbv^2dQ+\frac{1}{2}\int_{Q}pb_tv^2dQ\geq$$

$$\geq\frac{1}{2}\int_{Q}p_tbv^2dQ\geq\frac{\lambda_c^2}{2\lambda_a}\int_{Q}p\sum_{i=1}^{N}v^2dQ.$$

Note that it is possible to prove the convergence of the improper integrals, which are the terms of the sums.

Thus,

$$(v,L_1u)_{L_2(Q)}\geq C_1\int_{Q}(v_t^2+\sum_{i,j=1}^{N}v_{tx_i}^2)dQ+$$

$$+\frac{1}{2}\int_{Q}p\sum_{i=1}^{N}\left(\frac{\lambda_c^2}{\lambda_a}v^2-2\lambda_c|v|\cdot|v_{tx_i}|+\lambda_a v_{tx_i}^2\right)dQ+$$

$$+\int_{Q}p\left(a-\frac{1}{2}\sum_{i=1}^{N}\frac{\partial a_i}{\partial x_i}\right)v_t^2dQ\geq$$

$$\geq C_1\|v\|_{W_{bd^+}^+}^2+\frac{1}{2}\int_{Q}p\sum_{i=1}^{N}\left(\frac{\lambda_c}{\sqrt{\lambda_a}}|v|-\sqrt{\lambda_a}|v_{tx_i}|\right)^2dQ\geq C_1\|v\|_{W_{bd^+}^+}^2.$$

The proved relations imply that (15) holds true. Applying the Schwarz inequality to (15), we obtain:

$$C_1\|v\|_{W_{bd^+}^+}^2\leq(v,L_1u)_{L_2(Q)}\leq\|v\|_{W_{bd^+}^+}\|Lu\|_{W_{bd^+}^-}.$$

Reducing by $\|v\|_{W_{bd^+}^+}$,we have the following inequality:

$$C_1\|v\|_{W_{bd^+}^+}\leq\|L_1u\|_{W_{bd^+}^-}.$$

Next, let us consider

$$\|v\|_{W_{bd^+}^+}=\left(\int_{Q}\left(v_t^2+\sum_{i=1}^{N}v_{tx_i}^2\right)dQ\right)^{1/2}=\left(\int_{Q}\frac{1}{p^2(t)}\left(u^2+\sum_{i=1}^{N}u_{x_i}^2\right)dQ\right)^{1/2}$$

Applying to the right-hand side the theorem of the mean, we have

$$\left(\int_Q \frac{1}{p^2(t)}\left(u^2+\sum_{i=1}^N u_{x_i}^2\right)dQ\right)^{1/2} =$$

$$= \frac{1}{p(\theta)}\left(\int_Q\left(u^2+\sum_{i=1}^N u_{x_i}^2\right)dQ\right)^{1/2} = \frac{1}{p(\theta)}\|u\|_{H_{bd}^+},$$

where θ is some mean value from the interval $(0,T)$.

Granting the relation between $u(t,x)$ and $v(t,x)$, we prove that the inequality holds true for smooth functions $u(t,x)$ satisfying the condition (5):

$$\|L_1 u\|_{W_{bd^+}^-} \geq C_1\|v\|_{W_{bd^+}^+} \geq C\|u\|_{i_{bd}^+}.$$

The validity of the inequality (14) for any $u(t,x) \in W_{bd}^+$ can be proved by passing to the limit.

Lemma 8. *For any function* $v(t,x) \in W_{bd^+}^+$ *the following inequality holds true*

$$\|L_1^* v\|_{W_{bd}^-} \geq C\|v\|_{H_{bd^+}^+}. \tag{16}$$

P r o o f. At first, let us prove the inequality for smooth functions $v(t,x)$ satisfying the conditions (16). Introduce an auxiliary function $u(t,x)$ in the following way:

$$u(t,x) = \int_0^t \frac{v(\hat{o},x)}{d(\hat{o})}d\hat{o}, \text{ where } \frac{1}{d(t)} = \frac{\sin\frac{\eth}{2}\sqrt{\frac{t}{T}}}{e^{-\frac{2N\ddot{e}_c^2}{\ddot{e}_a\ddot{e}_b}t}}.$$

The definition of the function $u(t,x)$ implies that $v = du_t$ $\forall t \in [0,T)$.

Note that

$$d(t) \geq e^{-\frac{2N\lambda_c^2}{\lambda_a\lambda_b}T}, \; -\frac{\partial d(t)}{\partial t} > \frac{2N\lambda_c^2}{\lambda_a\lambda_b} d(t), \forall t \in [0, T).$$

Let us prove that the following inequality holds true

$$(u, L_1^* v)_{L_2(Q)} \geq C_1 \|u\|^2_{W_{bd}^+}. \tag{17}$$

Consider

$$\left(u, L_1^* v\right)_{L_2(Q)} = I_1 + I_2 + I_3 + I_4 + I_5 + I_6 + I_7.$$

Using the operation of differentiation by parts, relation between $u(t, x)$ and $v(t, x)$, and initial conditions, we obtain

$$I_1 = (u, v_{tt})_{L_2(Q)} = \int_Q (uv_t)_t dQ - \int_Q v_t u_t dQ =$$

$$= -\int_Q du_t u_{tt} dQ + \int_Q d_t u_t^2 dQ =$$

$$= -\frac{1}{2}\int_Q \left(du_t^2\right)_t dQ + \frac{1}{2}\int_Q d_t u_t^2 dQ - \int_Q d_t u_t^2 dQ =$$

$$= -\frac{1}{2}\int_Q d_t u_t^2 dQ \geq C_2 \int_Q v_t^2 dQ.$$

Granting initial and boundary conditions, we have

$$I_2 = \left(u, \sum_{i,j=1}^N \frac{\partial}{\partial x_i}\left(a_{ij}\frac{\partial^2 v}{\partial x_j \partial t}\right)\right)_{L_2(Q)} = \int_Q \sum_{i,j=1}^N \left(u\left(a_{ij}\frac{\partial^2 v}{\partial x_j \partial t}\right)\right)_{x_i} dQ -$$

$$-\int_Q \sum_{i,j=1}^N u_{x_i} a_{ij} v_{x_j t} dQ = -\int_Q \sum_{i,j=1}^N \left(u_{x_i} a_{ij} v_{x_j}\right)_t dQ + \int_Q \sum_{i,j=1}^N u_{tx_i} a_{ij} v_{x_j} dQ +$$

$$+\int_Q \sum_{i,j=1}^N u_{x_i} \frac{\partial a_{ij}}{\partial t} v_{x_j} dQ = \int_Q d \sum_{i,j=1}^N u_{tx_i} a_{ij} u_{tx_j} dQ + \int_Q d \sum_{i,j=1}^N u_{x_i} \frac{\partial a_{ij}}{\partial t} u_{tx_j} dQ =$$

$$= \int_Q d\sum_{i,j=1}^N u_{tx_i} a_{ij} u_{tx_j} + \frac{1}{2}\int_Q \sum_{i,j=1}^N \left(du_{x_i} \frac{\partial a_{ij}}{\partial t} u_{x_j} \right)_t dQ -$$

$$-\frac{1}{2}\int_Q d_t \sum_{i,j=1}^N u_{x_i} \frac{\partial a_{ij}}{\partial t} u_{x_j} dQ - \frac{1}{2}\int_Q d\sum_{i,j=1}^N u_{x_i} \frac{\partial^2 a_{ij}}{\partial t^2} u_{x_j} dQ \geq$$

$$\geq \tilde{C}_3 \int_Q \sum_{i=1}^N u_{tx_i}^2 dQ + \frac{1}{2}\lambda_a \int_Q d\sum_{i=1}^N u_{tx_i}^2 dQ .$$

In a similar way,

$$I_3 = \left(u, \sum_{i=1}^N \frac{\partial}{\partial x_i}\left(a_i \frac{\partial v}{\partial t} \right) \right)_{L_2(Q)} =$$

$$= \int_Q \sum_{i=1}^N (u a_i v_t)_{x_i} dQ - \int_Q \sum_{i=1}^N u_{x_i} a_i v_t dQ =$$

$$= -\int_Q \left(\sum_{i=1}^N u_{x_i} a_i v \right)_t dQ + \int_Q d\sum_{i=1}^N u_{tx_i} a_i u_t d Q + \int_Q d\sum_{i=1}^N u_{x_i} \frac{\partial a_i}{\partial t} u_t d Q =$$

$$= \frac{1}{2}\int_Q d\sum_{i=1}^N \left(a_i u_t^2 \right)_{x_i} dQ - \frac{1}{2}\int_Q d\sum_{i=1}^N \frac{\partial a_i}{\partial x_i} u_t^2 dQ + \int_Q \left(d\sum_{i=1}^N u_{x_i} \frac{\partial a_i}{\partial t} u \right)_t dQ -$$

$$-\int_Q d_t \sum_{i=1}^N u_{x_i} \frac{\partial a_i}{\partial t} u dQ - \int_Q d\sum_{i=1}^N u_{tx_i} \frac{\partial a_i}{\partial t} u dQ - \int_Q d\sum_{i=1}^N u_{x_i} \frac{\partial^2 a_i}{\partial t^2} u dQ =$$

$$= -\frac{1}{2}\int_Q d\sum_{i=1}^N \frac{\partial a_i}{\partial x_i} u_t^2 dQ + \int_\Omega d(T) \sum_{i=1}^N u_{x_i}(T,x) \frac{\partial a_i(T,x)}{\partial t} u(T,x) d\Omega -$$

$$-\frac{1}{2}\int_Q d_t \sum_{i=1}^N \left(\frac{\partial a_i}{\partial t} u^2 \right)_{x_i} dQ + \frac{1}{2}\int_Q d_t \sum_{i=1}^N \frac{\partial^2 a_i}{\partial t \partial x_i} u^2 dQ$$

$$-\int_Q d\sum_{i=1}^N u_{tx_i} \frac{\partial a_i}{\partial t} u dQ - \frac{1}{2}\int_Q d\sum_{i=1}^N \left(\frac{\partial^2 a_i}{\partial t^2} u^2 \right)_{x_i} dQ +$$

$$+\frac{1}{2}\int_Q d\sum_{i=1}^N \frac{\partial^3 a_i}{\partial^2 t\partial x_i}u^2 dQ = -\frac{1}{2}\int_Q d\sum_{i=1}^N \frac{\partial a_i}{\partial x_i}u_t^2 dQ -$$

$$-\int_Q d\sum_{i=1}^N u_{tx_i}\frac{\partial a_i}{\partial t}udQ + \frac{1}{2}\int_Q d_t\sum_{i=1}^N \frac{\partial^2 a_i}{\partial t\partial x_i}u^2 dQ$$

$$+\frac{1}{2}\int_Q d\sum_{i=1}^N \frac{\partial^3 a_i}{\partial t^2\partial x_i}u^2 dQ +$$

$$+\frac{1}{2}\int_\Omega d(T)\sum_{i=1}^N \left(\frac{\partial a_i(T,x)}{\partial t}u^2(T,x)\right)_{x_i} d\Omega -$$

$$-\frac{1}{2}\int_Q d(T)\sum_{i=1}^N \frac{\partial^2 a_i(T,x)}{\partial t\partial x_i}u^2(T,x)dQ =$$

$$= \int_Q d\sum_{i=1}^N u_{tx_i}\frac{\partial a_i}{\partial t}udQ + \frac{1}{2}\int_Q d_t\sum_{i=1}^N \frac{\partial^2 a_i}{\partial t\partial x_i}u^2 dQ +$$

$$+\frac{1}{2}\int_Q d\sum_{i=1}^N \frac{\partial^3 a_i}{\partial t^2\partial x_i}u^2 dQ -$$

$$-\frac{1}{2}\int_Q d(T)\sum_{i=1}^N \frac{\partial^2 a_i(T,x)}{\partial t\partial x_i}u^2(T,x)dQ - \frac{1}{2}\int_Q d\sum_{i=1}^N \frac{\partial a_i}{\partial x_i}u_t^2 dQ.$$

$$I_4 = -\int_Q uav_t dQ = -\int_Q (uav)_t dQ + \int_Q u_t avdQ + \int_Q ua_t vdQ =$$

$$= \int_Q dau_t^2 dQ + \frac{1}{2}\int_Q \left(da_t u^2\right)_t dQ - \frac{1}{2}\int_Q d_t a_t u^2 dQ -$$

$$-\frac{1}{2}\int_Q da_{tt}u^2 dQ \geq \frac{1}{2}\int_\Omega d(T)a_t(T)u^2(T,x)d\Omega + \int_Q dau_t^2 dQ -$$

$$-\frac{1}{2}\int_Q d_t a_t u^2 dQ - \frac{1}{2}\int_Q da_{tt}u^2 dQ.$$

It is follows from the fact that the matrix $\left\{b_{ij}\right\}_{i,j=1}^{N}$ is symmetric and nonnegative that

$$I_5 = -\left(u, \sum_{i,j=1}^{N}\frac{\partial}{\partial x_i}\left(b_{ij}v_{x_j}\right)\right)_{L_2(Q)} =$$

$$= -\int_Q \sum_{i,j=1}^{N}\left(ub_{ij}v_{x_j}\right)_{x_i} dQ + \int_Q d\sum_{i,j=1}^{N} u_{x_i} b_{ij} u_{tx_j} dQ =$$

$$= \frac{1}{2}\int_Q \left(d\sum_{i,j=1}^{N} u_{x_i} b_{ij} u_{x_j}\right)_t dQ - \frac{1}{2}\int_Q d_t \sum_{i,j=1}^{N} u_{x_i} b_{ij} u_{x_j} dQ =$$

$$= \frac{1}{2}\int_\Omega d(T)\sum_{i,j=1}^{N} u_{x_i}(T,x) b_{ij} u_{x_j}(T,x) d\Omega -$$

$$-\frac{1}{2}\int_Q d_t \sum_{i,j=1}^{N} u_{x_i} b_{ij} u_{x_j} dQ \ge 0.$$

Next,

$$I_6 = -\left(u, \sum_{i=1}^{N}\frac{\partial}{\partial x_i}(c_i v)\right)_{L_2(Q)} = -\int_Q \sum_{i=1}^{N}(uc_i v)_{x_i} dQ +$$

$$+\int_Q d\sum_{i=1}^{N} u_{x_i} c_i u_t dQ = \int_Q \left(d\sum_{i=1}^{N} u_{x_i} c_i u\right)_t dQ - \int_Q d_t \sum_{i=1}^{N} u_{x_i} c_i u dQ -$$

$$-\int_Q d\sum_{i=1}^{N} u_{tx_i} c_i u dQ - \int_Q d\sum_{i=1}^{N} u_{x_i}\frac{\partial c_i}{\partial t} u dQ =$$

$$= \int_\Omega d(T)\sum_{i=1}^{N} u_{x_i}(T,x) c_i(T,x) u(T,x) d\Omega -$$

$$-\frac{1}{2}\int_Q d_t \sum_{i=1}^{N}\left(c_i u^2\right)_{x_i} dQ + \frac{1}{2}\int_Q d_t \sum_{i=1}^{N}\frac{\partial c_i}{\partial t} u^2 dQ -$$

$$-\int_Q d\sum_{i=1}^N c_i u_{tx_i} u dQ - \frac{1}{2}\int_Q d\sum_{i=1}^N \left(\frac{\partial c_i}{\partial t}u^2\right)_{x_i} dQ +$$

$$+\frac{1}{2}\int_Q d\sum_{i=1}^N \frac{\partial^2 c_i}{\partial t\partial x_i}u^2 dQ = \frac{1}{2}\int_\Omega d(T)\sum_{i=1}^N \left(u^2(T,x)c_i(T,x)\right)_{x_i} d\Omega -$$

$$-\frac{1}{2}\int_\Omega d(T)\sum_{i=1}^N \frac{\partial c_i(T,x)}{\partial x_i}u^2(T,x)d\Omega +$$

$$+\frac{1}{2}\int_Q d_t\sum_{i=1}^N \frac{\partial c_i}{\partial x_i}u^2 dQ + \frac{1}{2}\int_Q d\sum_{i=1}^N \frac{\partial^2 c_i}{\partial t\partial x_i}u^2 dQ - \int_Q d\sum_{i=1}^N c_i u_{tx_i} u dQ \geq$$

$$\geq \frac{1}{2}\int_Q d_t\sum_{i=1}^N \frac{\partial c_i}{\partial x_i}u^2 dQ + \frac{1}{2}\int_Q d\sum_{i=1}^N \frac{\partial^2 c_i}{\partial t\partial x_i}u^2 dQ - \int_Q d\sum_{i=1}^N c_i u_{tx_i} u dQ -$$

$$-\frac{1}{2}\int_\Omega d(T)\sum_{i=1}^N \frac{\partial c_i(T,x)}{\partial x_i}u^2(T,x)d\Omega .$$

$$I_7 = (u, bv)_{L_2(Q)} = \int_Q dbuu_t dQ =$$

$$= \frac{1}{2}\int_Q \left(dbu^2\right)_t dQ - \frac{1}{2}\int_Q d_t bu^2 dQ - \frac{1}{2}\int_Q db_t u^2 dQ =$$

$$= \frac{1}{2}\int_\Omega d(T)b(T,x)u^2(T,x)d\Omega - \frac{1}{2}\int_Q d_t bu^2 dQ - \frac{1}{2}\int_Q db_t u^2 dQ \geq$$

$$\geq \frac{1}{4}\int_\Omega d(T)b(T,x)u^2(T,x)d\Omega - \frac{1}{4}\int_Q d_t bu^2 dQ +$$

$$+\frac{\lambda_c^2}{2\lambda_a}\int_Q d\sum_{i=1}^N u^2 dQ - \frac{1}{2}\int_Q db_t u^2 dQ \geq$$

$$\geq \frac{1}{4}\int_\Omega d(T)b(T,x)u^2(T,x)d\Omega + \frac{\lambda_c^2}{2\lambda_a}\int_Q d\sum_{i=1}^N u^2 dQ - \frac{1}{2}\int_Q db_t u^2 dQ .$$

Note that it is possible to prove the convergence of the improper integrals, which are the terms of the sums.

Thus,

$$\left(u, L_1^* v\right)_{L_2(Q)} \geq C_1 \int_Q (u_t^2 + \sum_{i=1}^{N} u_{tx_i}^2) dQ +$$

$$+\frac{\lambda_-^2}{2\lambda_a} \int_Q d \sum_{i=1}^{N} u^2 dQ - \int_Q d \sum_{i=1}^{N} u_{tx_i} \left(c_i + \frac{\partial a_i}{\partial t} \right) u dQ +$$

$$+\frac{1}{2} \lambda_a \int_Q d \sum_{i=1}^{N} u_{tx_i}^2 dQ +$$

$$+\frac{1}{2} \int_\Omega d(T) \left(a_t(T,x) - \sum_{i=1}^{N} \frac{\partial^2 a_i(T,x)}{\partial t \partial x_i} \right) u^2(T,x) d\Omega -$$

$$-\frac{1}{2} \int_Q d_t \left(a_t - \sum_{i=1}^{N} \frac{\partial^2 a_i}{\partial t \partial x_i} \right) u^2 dQ + \frac{1}{2} \int_Q d \left(\sum_{i=1}^{N} \frac{\partial^3 a_i}{\partial t^2 \partial x_i} - a_{tt} \right) u^2 dQ +$$

$$+\frac{1}{4} \int_\Omega d(T) \left(b(T,x) - 2\sum_{i=1}^{N} \frac{\partial c_i(T,x)}{\partial x_i} \right) u^2(T,x) d\Omega -$$

$$-\frac{1}{2} \int_Q d_t \left(b - \sum_{i=1}^{N} \frac{\partial c_i}{\partial x_i} \right) u^2 dQ + \frac{1}{2} \int_Q d \left(\sum_{i=1}^{N} \frac{\partial^2 c_i}{\partial t \partial x_i} - b_t \right) u^2 dQ \geq$$

$$\geq C_1 \|u\|_{W_{bd}^+}^2 + \frac{1}{2} \int_Q d \sum_{i=1}^{N} \left(\frac{\lambda_c^2}{\lambda_a} u^2 - 2\lambda_c |u| \cdot |u_{tx_i}| + \lambda_a u_{tx_i}^2 \right) dQ \geq$$

$$\geq C_1 \|u\|_{W_{bd}^+}^2 + \frac{1}{2} \int_Q d \sum_{i=1}^{N} \left(\frac{\lambda_c}{\sqrt{\lambda}} |u| - \sqrt{\lambda_a} |u_{tx_i}| \right)^2 dQ \geq C_1 \|u\|_{W_{bd}^+}^2 .$$

The relations mentioned above implies the fact that (17) holds true. Applying the Schwartz inequality to (17), we obtain:

$$C_1 \|u\|_{W_{bd}^+}^2 \leq \left(u, L_1^* v\right) \leq \|u\|_{W_{bd}^+} \left\|L_1^* v\right\|_{W_{bd}^-} .$$

Reducing by $\|u\|_{W^{+}_{bd^{+}}}$ and accounting the relation between $v(t,x)$ and $u(t,x)$, we obtain that the inequality holds true for smooth functions $v(t,x)$ satisfying the conditions (7):

$$\|L_1^* v\|_{W^{-}_{bd}} \geq C_1 \|u\|_{W^{+}_{bd}} \geq C \|v\|_{H^{+}_{bd^{+}}} .$$

The validity of the inequality (16) for any function $v(t,x) \in W^{+}_{bd^{+}}$ is proved by passing to the limit.

Theorem 1. *For any function $f(t,x) \in H^{-}_{bd^{+}}$ there exists a unique generalized solution of the problem (12) in the sense of Definition 1. The similar statement holds true for the adjoint problem.*

Proof. Theorem 1 follows from general Theorem 1.1.1

Next, we shall investigate the solvability of the problem

$$L_1 u = f, f \in W^{-}_{bd^{+}}, u \in H^{+}_{bd}; \tag{18}$$

$$L_1^* v = g, g \in W^{-}_{bd}, v \in H^{+}_{bd^{+}}, \tag{19}$$

where $u(t,x)$, $v(t,x)$ are unknown elements, and f, g are given elements.

By the solutions of the problems (18), (19) we shall mean the generalized solutions in the following sense.

Definition 3. *The generalized solution of the problem (18) is a function $u(t,x) \in H^{+}_{bd}$ such that there exists a sequence of smooth functions $u_i(t,x)$ which satisfy the condition (5) and*

$$\|u_i - u\|_{H^{+}_{bd}} \xrightarrow[i \to \infty]{} 0, \ \|L_1 u_i - f\|_{W^{-}_{bd^{+}}} \xrightarrow[i \to \infty]{} 0 .$$

It should been stressed that the difference between generalized solution in the sense of Definition 1 and the generalized solution in the sense of Definition 3 is that the corresponding sequences of smooth functions must converge in the different metrics.

Definition 4. *The generalized solution of the problem (18) is a function* $u(t,x) \in H^{+}_{bd}$ *such that the identity*

$$\left(L_1^{*}v, u\right)_{L_2(Q)} = \langle f, v\rangle_{W^{-}_{bd^{+}}}$$

holds true for any smooth function $v(t,x)$ *satisfying the conditions (7).*

In a similar way, we can define a generalized solutiion of the adjoint problem.

Theorem 2. *For any function* $f(t,x) \in W^{-}_{bd^{+}}$ *there exists a unique generalized solution of the problem (18) in the sense of Definition 3. Analogous statement holds true for the adjoint problem.*

Remark. *The inequalities (14), (16) are not only sufficient but also necessary conditions of the existence of a unique generalized solution.*

Corollary. *The inequalities (14), (16) imply more positive inequalities:*

$$\|u\|_{W^{+}_{bd}} \le C\|L_1 u\|_{H^{-}_{bd^{+}}} \text{ for any } u(t,x) \in W^{+}_{bd}, L_1 u \in H^{-}_{bd^{+}}, \tag{20}$$

$$\|v\|_{W^{+}_{bd^{+}}} \le C\|L_1^{*} v\|_{H^{-}_{bd}} \text{ for any } v(t,x) \in W^{+}_{bd^{+}}, L_1^{*} v \in H^{-}_{bd}. \tag{21}$$

Problem 5. *We shall investigate the solvability of the following problems*

$$L_2 u = f, f \in H^{-}_{bd^+}, u \in W^{+}_{bd}; \tag{22}$$

$$L^*_2 v = g, g \in H^{-}_{bd}, v \in W^{+}_{bd^+}, \tag{23}$$

where $u(t,x)$, $v(t,x)$ *are unknown elements, and* f, g *are given elements.*

By solution of problem (22), (23) we shall mean a generalized solution in the following sense.

Definition 5. *The generalized solution of the problem (21) is a function* $u(t,x) \in W^{+}_{bd}$ *such that there exists a sequence of smooth functions* $u_i(t,x)$ *satisfying the conditions (5) and*

$$\|u_i - u\|_{W^{+}_{bd}} \xrightarrow[i\to\infty]{} 0, \; \|L_2 u_i - f\|_{W^{-}_{bd^+}} \xrightarrow[i\to\infty]{} 0.$$

The generalized solution of the adjoint problem is defined in a similar way.

Lemma 9. *For any functions* $u(t,x) \in W^{+}_{bd}$ *and* $v(t,x) \in W^{+}_{bd^+}$ *the following inequalities hold true*

$$\|L_2 u\|_{W^{-}_{bd^+}} \geq C\|u\|_{i^{+}_{bd}}, \tag{24}$$

$$\|L^*_2 v\|_{W^{-}_{bd}} \geq C\|v\|_{H^{+}_{bd^+}}. \tag{25}$$

The proof of Lemma 9 is similar to the proofs of Lemmas 7 and 8.

Theorem 3. *For any function* $f(t,x) \in H^{-}_{bd^+}$ *there exists a unique generalized solution of the problem (22) in the sense of Definition 5. Analogous statement holds true for the adjoint problem.*

Proof of Theorem 3 is analogous to the proof of Theorem 1.

Next, we shall investigate the solvability of the following problems

$$L_2 u = f, f \in W^-_{bd^+}, u \in H^+_{bd};\tag{26}$$

$$L_2^* v = g, g \in W^-_{bd}, v \in H^+_{bd^+},\tag{27}$$

where $u(t, x)$, $v(t, x)$ are unknown elements, and f, g are given elements.

By solutions of the problems (26), (27) we shall mean generalized solutions in the following sense.

Definition 6. *The generalized solution of the problem (26) is a function* $u(t,x) \in H^+_{bd}$ *such that there exists a sequence of smooth fucntions* $u_i(t, x)$ *satisfying the conditions (5) and*

$$\|u_i - u\|_{H^+_{bd}} \xrightarrow[i\to\infty]{} 0, \quad \|L_2 u_i - f\|_{W^-_{bd^+}} \xrightarrow[i\to\infty]{} 0.$$

In a similar way, the generalized solution of the adjoint problem is defined. It should been stressed that the difference between the generalized solution in the sense of Definition 5 and the gerneralized solution in the sense of Definition 6 is that the corresponding sequences of smooth functions must converge in the different metrics.

Theorem 4. *For any function* $f(t,x) \in W^-_{bd^+}$ *there exists a unique generalized solution of the problem (26) in the sense of Definition 6. The analogous statement holds true for the adjoint problem.*

Remark. *The inequalities (24), (25) are not only sufficient, but also necessary conditions of the existence of a unique generalized solution.*

Corollary. *From the inequalities (24), (25) follows more positive inequalities*

$$\|u\|_{W^{+}_{bd}} \le C\|L_2 u\|_{H^{-}_{bd^+}} \text{ for any } u(t,x)\in W^{+}_{bd}, L_2 u \in H^{-}_{bd^+}, \quad (28)$$

$$\|v\|_{W^{+}_{bd^+}} \le C\|L_2^{*} v\|_{H^{-}_{bd}} \text{ for any } v(t,x)\in W^{+}_{bd^+}, L_2^{*} v \in H^{-}_{bd}. \quad (29)$$

Problem 6. *We shall investigate the solvability of the following problems*

$$L_3 u = f, f\in L_2(Q), u\in \overline{W}^{+}_{bd}; \quad (30)$$

$$L_3^{*} v = g, g\in L_2(Q), v\in \overline{W}^{+}_{bd^+}, \quad (31)$$

where $u(t, x)$, $v(t, x)$ *are unknown elements, and* f, g *are given elements.*

By solution of the problems (30), (31) we shall mean a generalized solution in the following sense.

Definition 7. *The generalized solution of the problem (30) is a function* $u(t,x)\in \overline{W}^{+}_{bd}$ *such that there exists a sequence of smooth functions satisfying the conditions (5) and*

$$\|u_i - u\|_{\overline{W}^{+}_{bd}} \xrightarrow[i\to\infty]{} 0, \ \|L_3 u_i - f\|_{\overline{W}^{-}_{bd^+}} \xrightarrow[i\to\infty]{} 0.$$

The generalized solution for the adjoint problem is defined in a similar way.

Lemma 10. *For any functions* $u(t,x)\in \overline{W}^{+}_{bd}$ *the following inequality holds true*

$$\|L_3 u\|_{\overline{W}^{-}_{bd^+}} \ge C\|u\|_{L_2(Q)}. \quad (32)$$

Proof. Let us prove the inequality for smooth functions $u(t, x)$ satisfying the conditions (5). Introduce an auxiliary operator in the following way:

$$v(t,x)=\int_T^t \varphi(\xi)\int_T^\xi \psi(\eta)u(\eta,x)d\eta d\xi,$$

where

$$\varphi(t)=\exp(-\exp(\lambda_A^{-1}\lambda_B^{-1}(T-t))),$$
$$\psi(t)=\exp(\exp(\lambda_A^{-1}\lambda_B^{-1}(T-t))+\lambda_A^{-1}\lambda_B^{-1}(T-t)).$$

If the function $u(t,x)$ satisfies the boundary conditions (5), then the function $v(t,x)$ defined above satisfies (7), furthermore, the following relation holds true:

$$u(t,x)=\exp(\lambda_A^{-1}\lambda_B^{-1}(t-T))v_{tt}(t,x)-\lambda_A^{-1}\lambda_B^{-1}v_t(t,x). \quad (33)$$

Denote the expression

$$(L_3u,v)_{L_2(Q)}=(u,L_3^*v)_{L_2(Q)}=$$
$$=(\exp(\lambda_A^{-1}\lambda_B^{-1}(t-T))v_{tt}-\lambda_A^{-1}\lambda_B^{-1}v_t,L_3^*v)_{L_2(Q)}.$$

Applying the operation of integration by parts, the Ostrogradsky-Gauss formula, and granting the boundary conditions in the passing to the integration on the boundary, we obtain

$$(L_3u,v)_{L_2(Q)}=(\exp(\lambda_A^{-1}\lambda_B^{-1}(t-T))v_{tt},v_{tt})_{L_2(Q)}+$$
$$+\frac{1}{2}\lambda_A^{-1}\lambda_B^{-1}(\exp(\lambda_A^{-1}\lambda_B^{-1}(t-T))v_t,A(v_t))_{L_2(Q)}+$$
$$+I_1+I_2+I_3+I_4+I_5+I_6+I_7+I_8,$$

where

$$I_1=-(\exp(\lambda_A^{-1}\lambda_B^{-1}(t-T))v_t,B(v_t))_{L_2(Q)}+$$
$$+\lambda_A^{-1}\lambda_B^{-1}(v_t,A(v_t))_{L_2(Q)}\geq$$
$$\geq-(v_t,B(v_t))_{L_2(Q)}+\lambda_A^{-1}\lambda_B^{-1}(v_t,A(v_t))_{L_2(Q)}\geq$$

$$\geq -\lambda_B^{-1}\int_Q \sum_{i=1}^{n} v_{tx_i}^2 \, dQ + \lambda_B^{-1}\int_Q \sum_{i=1}^{n} v_{tx_i}^2 \, dQ = 0;$$

$$I_2 = \frac{1}{2}\lambda_A^{-2}\lambda_B^{-2}(\exp(\lambda_A^{-1}\lambda_B^{-1}(t-T))v, B(v))_{L_2} \geq 0;$$

$$I_3 = \frac{1}{2}\lambda_A^{-1}\lambda_B^{-1}(v, B(v))\Big|_{\substack{L_2(\Omega)\\ t=0}} \geq 0;$$

$$I_4 = \frac{1}{2}\exp(-\lambda_A^{-1}\lambda_B^{-1}T)(v_t, A(v_t))_{\substack{L_2(\Omega)\\ t=0}} -$$

$$-\frac{1}{2}\lambda_A^{-1}\lambda_B^{-1}\exp(-\lambda_A^{-1}\lambda_B^{-1}T)(v_t, B(v_t))_{\substack{L_2(\Omega)\\ t=0}} \geq$$

$$\geq \frac{1}{2}\lambda_A \exp(-\lambda_A^{-1}\lambda_B^{-1}T)\int_Q \left(\sum_{i=1}^{N} v_{tx_i}^2\right)\Bigg|_{t=0} d\Omega -$$

$$-\frac{1}{2}\lambda_A \exp(-\lambda_A^{-1}\lambda_B^{-1}T)\int_Q \left(\sum_{i=1}^{N} v_{tx_i}^2\right)\Bigg|_{t=0} d\Omega = 0;$$

$$I_5 = -\exp(-\lambda_A^{-1}\lambda_B^{-1}T)(v_t, B(v))\Big|_{\substack{L_2(\Omega)\\ t=0}} +$$

$$+\frac{1}{2}\lambda_A^{-1}\lambda_B^{-1}\exp(-\lambda_A^{-1}\lambda_B^{-1}T)(v, B(v))\Big|_{\substack{L_2(\Omega)\\ t=0}} +$$

$$+\frac{1}{2}\lambda_A\lambda_B \exp(-\lambda_A^{-1}\lambda_B^{-1}T)(v_t, B(v_t))\Big|_{\substack{L_2(\Omega)\\ t=0}} =$$

$$= \frac{1}{2}\lambda_A^{-1}\lambda_B^{-1}\exp(-\lambda_A^{-1}\lambda_B^{-1}T)((\lambda_A\lambda_B v_t - v), B(\lambda_A\lambda_B v_t - v))\Big|_{\substack{L_2(\Omega)\\ t=0}} \geq 0;$$

$$I_6 = -(\exp(\lambda_A^{-1}\lambda_B^{-1}(t-T))v_{tt}, Mv_t)_{L_2(Q)} =$$

$$= \exp(-\lambda_A^{-1}\lambda_B^{-1}T)v_t, Mv_t)\Big|_{\substack{L_2(\Omega)\\ t=0}} +$$

$$+\lambda_A^{-1}\lambda_B^{-1}(\exp(\lambda_A^{-1}\lambda_B^{-1}(t-T))v_t, Mv_t)_{L_2(Q)} +$$

$$+(\exp(\lambda_A^{-1}\lambda_B^{-1}(t-T))v_t, Mv_{tt})_{L_2(Q)} =$$

$$=\frac{1}{2}\exp(-\lambda_A^{-1}\lambda_B^{-1}T)(v_t,Mv_t)\Big|_{\substack{L_2(\Omega)\\ t=0}}+$$

$$+\frac{1}{2}\lambda_A^{-1}\lambda_B^{-1}(\exp\lambda_A^{-1}\lambda_B^{-1}(t-T)v_t,Mv_t)_{L_2}\geq 0;$$

$$I_7=\lambda_A^{-1}\lambda_B^{-1}(v_t,Mv_t)_{L_2(Q)}\geq 0;$$

$$I_8=-\lambda_A^{-1}\lambda_B^{-1}(v_t,v_{tt})_{L_2(Q)}=\lambda_A^{-1}\lambda_B^{-1}(v_t,v_t)\Big|_{\substack{L_2(\Omega)\\ t=0}}+$$

$$+\lambda_A^{-1}\lambda_B^{-1}(v_t,v_t)_{L_2(Q)}+\lambda_A^{-1}\lambda_B^{-1}(v_{tt},v_t)_{L_2(Q)}=$$

$$=\frac{1}{2}\lambda_A^{-1}\lambda_B^{-1}(v_t,v_t)\Big|_{\substack{L_2(\Omega)\\ t=0}}+\frac{1}{2}\lambda_A^{-1}\lambda_B^{-1}(v_t,v_t)_{L_2(Q)}\geq 0.$$

Therefore,

$$(L_3u,v)_{L_2(Q)}\geq C_1\|v\|^2_{\overline{W}^+_{bd^+}}. \tag{34}$$

Applying the Schwartz inequality to the left-hand side of (34), we have:

$$\left|(L_3u,v)_{L_2(Q)}\right|\leq\|L_3u\|_{\overline{W}^-_{bd^+}}\|v\|_{\overline{W}^+_{bd}}.$$

Hence,

$$C_1\|v\|_{\overline{W}^+_{bd^+}}\leq\|L_3u\|_{\overline{W}^-_{bd^+}}.$$

From the relation (33) we obtain

$$\|u(t,x)\|_{L_2(Q)}=\left\|\exp\left(\lambda_A^{-1}\lambda_B^{-1}(t-T)\right)v_{tt}(t,x)-\lambda_A^{-1}\lambda_B^{-1}v_t(t,x)\right\|_{L_2(Q)}\leq$$

$$\leq C_2\|v_{tt}\|_{L_2(Q)}\leq C_3\|v\|_{\overline{W}^+_{bd^+}}.$$

Thus, the inequality (32) holds true for smooth functions $u(t,x)$ satisfying the conditions (5). The validity of (32) for any $u(t,x)\in\overline{W}^+_{bd}$ is proved by passing to the limit.

Lemma 11. *For any functions $v(t,x) \in \overline{W}^{+}_{bd^+}$ the following inequality holds true*

$$\left\| L_3^* v \right\|_{\overline{W}_{bd}^-} \geq C \left\| v \right\|_{L_2(Q)} \tag{35}$$

Proof. The proof of Lemma 11 is similar to the proof of Lemma 10. The integral operator in this case has the following form

$$u(t,x) = \int_0^t \varphi_1(\xi) \int_0^\xi \psi_1(\eta) v(\eta, x) d\eta d\xi,$$

where

$\varphi_1(t) = \exp(-\exp(\lambda_A^{-1}\lambda_B^{-1}t)), \psi(t) = \exp(\exp(\lambda_A^{-1}\lambda_B^{-1}t) + \lambda_A^{-1}\lambda_B^{-1}t).$

Theorem 5. *For any function $f(t,x) \in L_2'(Q)$ there exists a unique generalized solution of the problem (30) in the sense of Definition 7. The analogous statement holds true for the adjoint problem.*

The proof of Theorem 5 is similar to the proof of Theorem 1.

Next, we shall investigate the solvability of the following problems

$$L_3 u = f, f \in \overline{W}^-_{bd^+}, u \in L_2(Q), \tag{36}$$

$$L_3^* v = g, g \in \overline{W}^-_{bd}, v \in L_2(Q), \tag{37}$$

where $u(t,x)$, $v(t,x)$ are unknown elements, and f, g are given elements. By solution of problems (36), (37) we shall mean a generalized solution in the following sense.

Definition 8. *The generalized solution of the problem (36) is a function* $u(t,x) \in L_2(Q)$ *such that there exists a sequence of smooth functions* $u_i(t,x)$ *satisfying the conditions (5), and*

$$\|u_i - u\|_{L_2(Q)} \xrightarrow[i\to\infty]{} 0, \ \|L_3 u_i - f\|_{\overline{W}_{bd^+}^{-}} \xrightarrow[i\to\infty]{} 0.$$

The definition for the adjoint problem is similar.

It should been stressed that the difference between the generalized solution in the sense of Definition 7 and the generalized solution in the sense of Definition 8 is that the corresponding sequences of smooth functions must converge in the different metrics.

Theorem 6. *For any function* $u(t,x) \in \overline{W}_{bd^+}^{-}$ *there exists a unique generalized solution of the problem (36) in the sense of Definition 8. The analogous statement holds true for the adjoint problem.*

The proof of Theorem 6 is similar to the proof of Theorem 2.

Remark. *The inequalities (32), (35) are not only sufficient, but also necessary conditions of the existence of a unique generalized solution.*

Corollary. *The inequalities (32), (35) implies the positive inequalities (obtained by many authors):*

$$\|u\|_{\overline{W}_{bd}^{+}} \le C\|L_3 u\|_{L_2(Q)} \text{ for any } u(t,x) \in \overline{W}_{bd}^{+}, L_3 u \in L_2(Q), \tag{38}$$

$$\|v\|_{\overline{W}_{bd^+}^{+}} \le C\|L_3^* v\|_{L_2(Q)} \text{ for any } v(t,x) \in \overline{W}_{bd^+}^{+}, L_3^* v \in L_2(Q). \tag{39}$$

2. GENERALIZED SOLVABILITY OF PSEUDO-HYPERBOLIC SYSTEMS (THE NEUMANN INITIAL BOUNDARY VALUE PROBLEM)

2.1 Main notations and auxiliary statements

Consider the linear partial differential equation:

$$Lu = u_{tt} + A(u_t) + B(u) + Cu_t + Du = f(t,x), \tag{1}$$

in a tube domain $Q = (0,T) \times \Omega$, where $u(t,x)$ is an unknown function depending on spatial variable $x \in \Omega$ and time $t \in (0,T)$, Ω is a bounded simply connected domain in R^n with smooth boundary $\partial\Omega$.

The operators $A(\cdot)$ and $B(\cdot)$ do not depends on t and they are defined by the following differential expressions:

$$A(\cdot) = -\sum_{i,j=1}^{n} \frac{\partial}{\partial x_i}\left(A_{ij} \frac{\partial}{\partial x_j} \right) \tag{2}$$

and

$$B(\cdot) = -\sum_{i,j=1}^{n} \frac{\partial}{\partial x_i}\left(B_{ij} \frac{\partial}{\partial x_j} \right). \tag{3}$$

Functions $A_{ij}(x)$, $B_{ij}(x)$, $C(x)$, $D(x)$ $(i,j = \overline{1,n})$ are defined in a closed domain $\overline{\Omega}$. Let $A_{ij}(x) = A_{ji}(x)$, $B_{ij}(x) = B_{ji}(x)$ be continuously differentiable, and $C(x)$, $D(x)$ be continuous in the closed domain $\overline{\Omega}$ functions.

We shall suppose that the differential expression (2) is uniformly elliptic, and the expression (3) is non-negative in $\overline{\Omega}$, i.e. for any $\xi = (\xi_1, \xi_2, \ldots, \xi_n) \in R^n$ and any $x \in \overline{\Omega}$ the following relations hold true

$$\alpha_A^{-1} \sum_{i=1}^{n} \xi_i^2 \geq \sum_{i,j=1}^{n} A_{ij}(x) \xi_i \xi_j \geq \alpha_A \sum_{i=1}^{n} \xi_i^2 ,$$

$$\alpha_B^{-1} \sum_{i=1}^{n} \xi_i^2 \geq \sum_{i,j=1}^{n} B_{ij}(x) \xi_i \xi_j \geq 0 ,$$

where $\alpha_A, \alpha_B > 0$ are constants which do not depend on x and ξ.

Suppose also that

$$C'(x) \geq 0 , \; D(x) \geq 0 .$$

Definition 1. *$L(\cdot)$ - the gradient of the function $u(t, x)$ (denoted as $\mathrm{grad}_L u(t, x)$) is a vector-column*

$$\mathrm{grad}_L u(t, x) = \left(\frac{\partial_L u}{\partial x_1}, \frac{\partial_L u}{\partial x_2}, \ldots, \frac{\partial_L u}{\partial x_n} \right)^{\mathrm{T}} ,$$

where

$$\frac{\partial_L u}{\partial x_1} = \sum_{i=1}^{n} A_{1i}(x) \frac{\partial^2 u}{\partial x_i \partial t} + B_{1i}(x) \frac{\partial u}{\partial x_i} ,$$

$$\frac{\partial_L u}{\partial x_n} = \sum_{i=1}^{n} A_{ni}(x) \frac{\partial^2 u}{\partial x_i \partial t} + B_{ni}(x) \frac{\partial u}{\partial x_i} ,$$

In the matrix form

$$\mathrm{grad}_L u(t, x) = A \mathrm{grad} u_t + B \mathrm{grad} u ,$$

where A and B are matrices $n \times n$ consisting of the elements $\{A_{ij}(x)\}$ and $\{B_{ij}(x)\}$, $\mathrm{grad} u = \left(\frac{\partial u}{\partial x_1}, \frac{\partial u}{\partial x_2}, \ldots, \frac{\partial u}{\partial x_n} \right)^T .$

Definition 2. *$L(\cdot)$-derivative of the function $u(t,x)$ with respect to the normal $\vec{n}$ to the surface $\partial\Omega$ (denoted as $\frac{\partial_L u}{\partial\vec{n}}$) is the following inner product in R^n:*

$$\frac{\partial_L u}{\partial\vec{n}} = (\text{grad}_L u, \vec{n}).$$

In the expanded form $L(\cdot)$ -derivative can be written as

$$\frac{\partial_L u}{\partial\vec{n}} = \sum_{i,j=1}^{n} A_{ij}(x)\frac{\partial^2 u}{\partial x_j \partial t}\cdot n_{x_i} + B_{ij}(x)\frac{\partial u}{\partial x_j}\cdot n_{x_i},$$

where n_{x_i} is i-th component of the vector $\vec{n}$ at the point $(x_1, x_2, \ldots, x_n)$ of the surface $\partial\Omega$.

Let us introduce the following notations: $L_2(Q)$ is the space of measurable square integrable on the set Q functions, $D(L)$ is the set of smooth (infinitely differentiable) in $\overline{Q}$ functions, which satisfy the following conditions:

$$u\big|_{t=0} = \frac{\partial u}{\partial t}\Big|_{t=0} = 0, \frac{\partial_L u}{\partial\vec{n}}\Big|_{x\in\partial\Omega} = 0. \tag{4}$$

We shall assume that the operator $L(\cdot)$ maps $L_2(Q)$ into $L_2(Q)$ and has the domain of definition $D(L) \subset L_2(Q)$. Note that the set $D(L)$ is dense in the space $L_2(Q)$, and therefore it is possible to define correctly the adjoint operator $L^*(\cdot): L_2(Q) \to L_2(Q)$, whose domain of definition is the set of functions $v(t,x)$ for which there exists an element f (such element is equal to $L^*(v)$) that the following identity holds true

$$(Lu, v)_{L_2(Q)} = (u, f)_{L_2(Q)}$$

for any $u \in D(L)$.

Thus, the formally adjoint operator has the following form:

$$L^{*}v = \frac{\partial^2 v}{\partial t^2} - A(\frac{\partial v}{\partial t}) + B(v) - C\frac{\partial v}{\partial t} + Dv, \tag{5}$$

with boundary conditions

$$v\big|_{t=T} = \frac{\partial v}{\partial t}\Big|_{t=T} = 0, \quad \frac{\partial_{L^{*}} v}{\partial \vec{n}}\Big|_{x\in\partial\Omega} = 0. \tag{6}$$

The notations introduced above give the possibility to consider the second initial boundary value problem.

Problem 1. *To find a function* $u(t,x)$, *which satisfies the equation (1) in the domain* Q *and the conditions (4) on the boundary* ∂Q.

Problem 2. *To find a function* $v(t,x)$ *which satisfies the equation (5) in the domain* Q *and the conditions (6) on the boundary* ∂Q.

Note that simultaneously with the common equation of pseudo-hyperbolic type (1) we shall investigate also some simpler version of this equation, for which it is possible to obtain more interesting results.

Let

$$L_1 u \equiv \frac{\partial^2 u}{\partial t^2} + A\left(\frac{\partial u}{\partial t} + k_1 u\right) + C\frac{\partial u}{\partial t} + Du, \tag{7}$$

where the operator $A(\cdot)$ and the functions $C(x)$, $D(x)$ satisfy the same conditions as in the case of the operator $L(\cdot)$, and k_1 be a non-negative constant. Let the domain of definition of the operator $L_1(\cdot)$ is the set $D(L_1)$ of smooth in $\overline{Q}$ functions $u(t,x)$, which satisfy conditions

$$u\big|_{t=0} = \frac{\partial u}{\partial t}\Big|_{t=0} = 0, \frac{\partial u}{\partial \mu_A}\Big|_{x\in\partial\Omega} = 0, \tag{8}$$

where μ_A is a co-normal to the surface $\partial\Omega$, which is defined as:

$$\vec{\mu}_A = \frac{A\vec{n}}{|A\vec{n}|},$$

where $\vec{n}$ is an outward normal to the surface $\partial\Omega$, A is a matrix $n \times n$ consisting of the elements $\{A_{ij}(x)\}$.

Note that despite of the difference between the boundary conditions (4) and (8) all results, which shall been obtained for the operator $L(\cdot)$, can be easily adopted for the operator $L_1(\cdot): L_2(Q) \to L_2(Q)$.

We shall consider the operator

$$L_2 u \equiv \frac{\partial^2 u}{\partial t^2} + A\left(\frac{\partial u}{\partial t} + k_1 u\right) + k_2 \frac{\partial u}{\partial t} + k_3 u, \tag{9}$$

where $k_1, k_2, k_3 \geq 0$.

In this case the domain of definition of the operator $L_2(\cdot)$ is the set $D(L)$. Thus, the operator $L_1(\cdot)$ converts to the operator $L_2(\cdot)$, if in $L_1(\cdot)$ we put $C(x) = k_2$, $D(x) = k_3$. Since, all results concerning the operator $L(\cdot)$ are valid for the equation (9). But in contrast to the operators $L(\cdot)$ and $L_1(\cdot)$, which map $L_2(Q) \to L_2(Q)$, the operator $L_2(\cdot)$ can be considered in other pairs of spaces also. Denote by $W_2^{0,1}$ the completion of the set of smooth in $\overline{Q}$ functions with respect to the norm generated by the following inner product:

$$(u,v)_{W_2^{0,1}} = \left(\int_Q uv + \sum_{i,j=1}^{n} A_{ij}(x) u_{x_i} v_{x_j}\, dQ\right)^{1/2}. \tag{10}$$

Since the boundary conditions (8) are not held in this norm (10), it is clear that the set $D(L_2)$ is dense in the space $W_2^{0,1}$ as far as by completing the set $D(L_2)$ in the norm (10) we shall obtain the same space $W_2^{0,1}$ (it is the same situation as in the case of density of finite functions of the class C_0^∞ in the space $L_2(Q)$). Taking into account the density of $D(L_2)$ in $W_2^{0,1}$, we shall consider that the operator $L_2(\cdot)$ maps $W_2^{0,1} \to W_2^{0,1}$ and also we shall define the formally adjoint operator

$$L_2^* v \equiv \frac{\partial^2 v}{\partial t^2} + A\left(-\frac{\partial v}{\partial t} + k_1 v \right) - k_2 \frac{\partial v}{\partial t} + k_3 v \tag{11}$$

with the conditions

$$v\big|_{t=T} = \frac{\partial v}{\partial t}\Big|_{t=T} = 0, \frac{\partial v}{\partial \bar{\mu}_A}\Big|'_{x \in \partial\Omega} = 0 . \tag{12}$$

Now it is possible to formulate the following problems.

Problem 3. *To find a function* $u(t,x)$ *satisfying the equation (9) in the domain* Q *and the conditions (8) on the boundary* ∂Q.

Problem 4. *To find a function* $v(t,x)$ *satisfying the equation (11) in the domain* Q *and the conditions (12) on the boundary* ∂Q.

Since the right-hand side of Problems 1-4 can be a discontinuous function or even a Schwarz distribution then it is possible that there are no any classic solutions of this equations, therefore we must consider some generalizations of the solutions and the extensions of the operators $L(\cdot)$, $L_1(\cdot)$ and $L_2(\cdot)$, respectively.

Let us introduce the following notations. Let W_1^+, W_2^+, H_2^+ be completions of the set $D(L)$ in the norms

$$\|u\|_{W_1^+} = \left(\int_Q u_{tt}^2 + \sum_{i=1}^n u_{x_i t}^2 dQ \right)^{1/2}, \tag{13}$$

$$\|u\|_{W_2^+} = \left(\int_Q u_t^2 + \sum_{i=1}^n u_{x_i t}^2 dQ \right)^{1/2}, \tag{14}$$

$$\|u\|_{H_2^+} = \left(\int_Q u^2 + \sum_{i=1}^{n} u_{x_i}^2 dQ \right)^{1/2}, \tag{15}$$

respectively; $W_{1*}^+, W_{2*}^+, H_{2*}^+$ be completions of the set of smooth in $\overline{Q}$ functions satisfying the conditions (6) in the same norms (13)-(15), respectively. It is easy to test whether the expressions (13)-(15) satisfy the norm axioms using the boundary conditions (4), (6) and Minkowsky inequality.

Let us build on the pairs $L_2(Q)$ and $W_i^+ (i = 1,2)$, and also $L_2(Q)$ and H_2^+, the negative spaces $W_i^- (i = 1,2)$ and H_2^-, respectively, as completions of the set $L_2(Q)$ in the norm

$$\|g\|_{W_i^-} = \sup_{u \in W_i^+} \frac{\left|(g,u)_{L_2(Q)}\right|}{\|u\|_{W_i^+}} \quad (i = 1,2) \tag{16}$$

and

$$\|g\|_{H_2^-} = \sup_{u \in H_2^+} \frac{\left|(g,u)_{L_2(Q)}\right|}{\|u\|_{H_2^+}}, \tag{17}$$

respectively.

In a similar way we can build the negative spaces $W_{i*}^- (i = 1,2)$ and H_{2*}^- on the pairs $L_2(Q)$ and $W_{i*}^+ (i = 1,2)$, and also $L_2(Q)$ and H_{2*}^+.

The following imbeddings are valid:

$$W_1^+ \subset W_2^+ \subset H_2^+ \subset L_2(Q) \subset H_2^- \subset W_2^- \subset W_1^-, \tag{18}$$

$$W_{1*}^+ \subset W_{2*}^+ \subset H_{2*}^+ \subset L_2(Q) \subset H_{2*}^- \subset W_{2*}^- \subset W_{1*}^-. \tag{19}$$

Let us introduce some additional notations. Let W_3^+, W_4^+, H_4^+ be completions of the set $D(L_1)$ in the norms

$$\|u\|_{W_3^+}=\left(\int_Q u_{tt}^2+\sum_{i=1}^n u_{x_itt}^2+\sum_{ij=1}^n u_{x_ix_jt}^2 dQ\right)^{1/2}, \tag{20}$$

$$\|u\|_{W_4^+}=\left(\int_Q u_t^2+\sum_{ij=1}^n u_{x_ix_jt}^2 dQ\right)^{1/2}, \tag{21}$$

$$\|u\|_{H_4^+}=\left(\int_Q u^2+\sum_{ij=1}^n u_{x_ix_j}^2 dQ\right)^{1/2}, \tag{22}$$

respectively; $W_{1^*}^+, W_{2^*}^+, H_{2^*}^+$ be completions of the set of smooth in $\overline{Q}$ functions satisfying the conditions (12) in the norms (20)-(22).

Let us introduce the negative spaces $W_i^-(i=3,4)$ and H_4^- , but in this case we construct their on the pairs $W_2^{0,1}$ and $W_i^+(i=3,4)$, and also $W_2^{0,1}$ and H_4^+ , as completions of the set $W_2^{0,1}$ in the norms

$$\|g\|_{W_i^-}=\sup_{u\in W_i^+}\frac{\left|(g,u)_{W_2^{0,1}}\right|}{\|u\|_{W_i^+}} \quad (i=3,4) \tag{23}$$

and

$$\|g\|_{H_4^-}=\sup_{u\in H_4^+}\frac{\left|(g,u)_{W_2^{0,1}}\right|}{\|u\|_{H_4^+}}. \tag{24}$$

In a similar way we can construct the spaces $W_{i^*}^-(i=3,4)$ and $H_{4^*}^-$. Now, the following imbeddings are valid:

$$W_3^+\subset W_4^+\subset H_4^+\subset W_2^{0,1}\subset H_4^-\subset W_4^-\subset W_3^-, \tag{25}$$

$$W_{3^*}^+\subset W_{4^*}^+\subset H_{4^*}^+\subset W_2^{0,1}\subset H_{4^*}^-\subset W_{4^*}^-\subset W_{3^*}^-. \tag{26}$$

2.2 A priori inequalities in negative norms

Lemma 1. *For arbitrary smooth in $\overline{Q}$ functions $u(t,x)$ and $v(t,x)$ satisfying the conditions (4) and (6), respectively, the following inequalities hold true:*

$$\|Lu\|_{W_{1^*}^-} \le c\|u\|_{W_1^+} \quad \text{and} \quad \|L^*v\|_{W_1^-} \le c\|v\|_{W_{1^*}^+}, \quad (c>0), \tag{27}$$

where the constant c does not depend on the functions $u(t,x)$ and $v(t,x)$.

Proof. By definition of the negative norm, we have

$$\|Lu\|_{W_{1^*}^-} = \sup_{v \in W_{1^*}^+} \frac{\left|(Lu,v)_{L_2(Q)}\right|}{\|v\|_{W_{1^*}^+}}.$$

Consider the numerator. Applying to it the formula of integration by parts, the Ostrogradsky-Gauss formula and taking into account the conditions (4) and (6), we obtain

$$(Lu,v)_{L_2(Q)} = (u_{tt},v)_{L_2(Q)} + \\ + \sum_{i,j=1}^{n}\left(\left(A_{ij}u_{x_i t}, v_{x_j}\right)_{L_2(Q)} + \left(B_{ij}u_{x_i}, v_{x_j}\right)_{L_2(Q)}\right) + \\ + (Cu_t, v)_{L_2(Q)} + (Du, v)_{L_2(Q)}$$

Applying to the right-hand side the Cauchy inequality ,we have

$$\left|(Lu,v)_{L_2(Q)}\right| \le \|u_{tt}\|_{L_2(Q)}\|v\|_{L_2(Q)} + \\ + \sum_{i,j=1}^{n}\left(\left\|A_{ij}u_{x_i t}\right\|_{L_2(Q)}\left\|v_{x_j}\right\|_{L_2(Q)} + \left\|B_{ij}u_{x_i}\right\|_{L_2(Q)}\left\|v_{x_j}\right\|_{L_2(Q)}\right) + \\ + \|Cu_t\|_{L_2(Q)}\|v\|_{L_2(Q)} + \|Du\|_{L_2(Q)}\|v\|_{L_2(Q)}$$

Taking into account that $A_{ij}(x)$, $B_{ij}(x)$, $C(x)$, $D(x)$ are continuous in $\overline{\Omega}$, and since, they are bounded, and granting the inequalities (16) and (21), it is easy to see that every term in the right-hand side does not exceed $\overline{c}\|u\|_{W_1^+}\|v\|_{W_{1^*}^+}\ (\overline{c}>0)$, and as far as the number of these terms is finite then

$$\left|(Lu,v)_{L_2(Q)}\right|\le c\|u\|_{W_1^+}\|v\|_{W_{1^*}^+}.$$

This completes the proof of the theorem. The proof of the second inequality is the same.

In a similar way it is possible to prove the following three lemmas.

Lemma 2. *For arbitrary smooth in* $\overline{Q}$ *functions* $u(t,x)$ *and* $v(t,x)$ *satisfying the conditions (4) and (6), respectively, the following inequalities hold true:*

$$\|Lu\|_{W_{2^*}^-}\le c\|u\|_{W_2^+} \text{ and } \left\|L^*v\right\|_{W_2^-}\le c\|v\|_{W_{2^*}^+},\quad (c>0), \tag{28}$$

where a constant c *does not depend on the function* $u(t,x)$ *and* $v(t,x)$.

Lemma 3. *For arbitrary smooth in* $\overline{Q}$ *functions* $u(t,x)$ *and* $v(t,x)$ *satisfying the conditions (8) and (12), respectively, the following inequalities hold true:*

$$\|L_1u\|_{W_{3^*}^-}\le c\|u\|_{W_3^+} \text{ and } \left\|L_1^*v\right\|_{W_3^-}\le c\|v\|_{W_{3^*}^+}\quad (c>0), \tag{29}$$

where a constant c *does not depend of the functions* $u(t,x)$ *and* $v(t,x)$.

Lemma 4. *For arbitrary smooth in* $\overline{Q}$ *functions* $u(t,x)$ *and* $v(t,x)$ *satisfying conditions (8) and (12), respectively, the following inequalities hold true*

$$\|L_1u\|_{W_{4^*}^-}\le c\|u\|_{W_4^+} \text{ and } \left\|L_1^*v\right\|_{W_4^-}\le c\|v\|_{W_{4^*}^+},\ (c>0), \tag{30}$$

where constant c *does not depend on the functions* $u(t,x)$ *and* $v(t,x)$.

Note that Lemmas 1-4 give the possibility to consider some extensions of the operators $L(\cdot)$ ($L^*(\cdot)$), $L_2(\cdot)$ ($L_2^*(\cdot)$) and $L_2(\cdot)$ ($L_2^*(\cdot)$). For example, the first inequality of Lemma 1 gives the possibility to extend with respect to continuity the operator $L(\cdot)$ and consider it as an operator mapping W_{1}^{+} into W_{1*}^{-}. We save the previous notations for the extended operators $L(\cdot)$ ($L^*(\cdot)$), $L_1(\cdot)$ ($L_1^*(\cdot)$) and $L_2(\cdot)$ ($L_2^*(\cdot)$) as far as they will be specified by context. Note that the inequalities in Lemmas 1-4 hold true for the extended operators also but on the whole space W^+. The proof of this fact follows from the passing to limit in the inequalities for smooth functions.

Lemma 5. *Let $u(t,x)$ be an arbitrary smooth in $\overline{Q}$ function satisfying the condition (4), and $v = I_t^1 u$ is the integral operator defined by the following expression:*

$$v(t,x) = \int_T^t \varphi_1(\xi) \int_T^\xi \psi_1(\eta) u(\eta,x)\,d\eta\,d\xi, \tag{31}$$

where

$$\varphi_1(\xi) = \exp\left(\frac{-0{,}5\alpha_A\alpha_B(0{,}5l_2\alpha_A\alpha_B + l_2\xi + l_1)}{\sigma(\xi)} \right),$$

$$\psi_1(\xi) = \frac{1}{\varphi_1(\xi)\sigma(\xi)},$$

$$\sigma(\xi) = \exp\left(2(\alpha_A\alpha_B)^{-1}\xi\right),\ l_1 = \sup_{x\in\Omega}(D(x)) + 1,$$

$$l_2 = 2\sup_{x\in\Omega}(D(x)) \cdot \exp\left(2(\alpha_A\alpha_B)^{-1}T\right) + 2.$$

Then,

$$\left(Lu, I_t^1 u\right)_{L_2(Q)} = (Lu, v)_{L_2(Q)} \geq c\|v\|^2_{W_{1*}^{+}}. \tag{32}$$

Proof. At first, note that $v = I_t^1 u \in W_{1*}^{+}$. Indeed, the initial conditions in (6)

$$v|_{t=T} = \frac{\partial v}{\partial t}|_{t=T} = 0 \tag{33}$$

are valid at the expense of the form of the operator $I_t^1 u$. Concerning the boundary conditions in (6)

$$\frac{\partial_{L^*} v}{\partial \vec{n}}|_{x \in \partial\Omega} = 0,$$

it should been noted that the function $v = I_t^1 u$ not necessarily satisfies this condition, as far as in completion of the set of smooth functions in the norm (13) this condition becomes not valid. Let us express $u(t, x)$ through $v(t, x)$ using (13).

$$u(t,x) = D_1 v = \exp\left(2(\alpha_A \alpha_B)^{-1} t\right) v_{tt} - (l_1 + l_2 t) v_t = \\ = \sigma(t) v_{tt} - (l_1 + l_2 t) v_t$$

Consider the left-hand side (32). Applying the formula of integration by parts, the Ostrogradsky-Gauss formula and the conditions (4), (33), which are satisfied by the functions $u(t, x)$ and $v(t, x)$ we have

$$(Lu, v)_{L_2(Q)} = (u_{tt}, v)_{L_2(Q)} + (A(u_t) + B(u), v)_{L_2(Q)} + \\ + (Cu_t, v)_{L_2(Q)} + (Du, v)_{L_2(Q)} = (u, v_{tt})_{L_2(Q)} + (A(u_t) + \\ + B(u), v)_{L_2(Q)} - (u, Cv_t)_{L_2(Q)} + (u, Dv)_{L_2(Q)}. \tag{34}$$

Consider every of the last terms separately.

1. Then,

$$(u, v_{tt})_{L_2(Q)} = (D_1 v, v_{tt})_{L_2(Q)} = \\ = (\sigma(t) v_{tt}, v_{tt})_{L_2(Q)} - ((l_2 t + l_1) v_t, v_{tt})_{L_2(Q)}$$

Applying the formula of integration by parts to the second term, we obtain

$$((l_2t+l_1)v_t,v_{tt})_{L_2(Q)}=\int_Q(l_2t+l_1)v_tv_{tt}dQ=$$

$$=\int_Q\left((l_2t+l_1)v_t^2\right)_t dQ-\int_Q(l_2t+l_1)v_tv_{tt}dQ-\int_Q l_2v_t^2dQ$$

Whence using the Ostrogradsky-Gauss formula and the initial conditions, we obtain

$$((l_2t+l_1)v_t,v_{tt})_{L_2(Q)}=-\frac{1}{2}\int_\Omega l_1v_t^2\big|_{t=0}d\Omega-\frac{1}{2}\int_Q l_2v_t^2dQ.$$

Thus,

$$(u,v_{tt})_{L_2(Q)}=\int_Q\sigma(t)v_{tt}^2dQ+\frac{1}{2}\int_\Omega l_1v_t^2\big|_{t=0}d\Omega+\frac{1}{2}\int_Q l_2v_t^2dQ.$$

Granting the conditions imposed on l_1 and l_2, we have

$$(u,v_{tt})_{L_2(Q)}=\int_Q v_{tt}^2dQ+\int_Q(\sigma(t)D+1)v_t^2dQ+\frac{1}{2}\int_\Omega Dv_t^2\big|_{t=0}d\Omega..$$

2. Let us apply the formula of integration by parts to the second term in (34):

$$(A(u_t)+B(u),v)_{L_2(Q)}=-\int_Q\sum_{i,j=1}^n\left((A_{ij}u_{x_it}+B_{ij}u_{x_i})v\right)_{x_j}dQ+$$

$$+\int_Q\sum_{i,j=1}^n(A_{ij}u_{x_it}+B_{ij}u_{x_i})v_{x_j}dQ$$

Using the Ostrogradsky-Gauss formula, we obtain

$$(A(u_t)+B(u),v)_{L_2(Q)}=-\int_0^T\int_{\partial\Omega}\frac{\partial_L u}{\partial\vec{n}}\cdot v\,d\partial\Omega dt+$$

$$+\int_Q\sum_{i,j=1}^n A_{ij}u_{x_it}v_{x_j}dQ+\int_Q\sum_{i,j=1}^n B_{ij}u_{x_i}v_{x_j}dQ$$

As far as $u(t,x)$ satisfies the conditions (4), then the first integral equals to zero. Let us execute some transformations with the second integral. We shall apply to it the formula of integration by parts and the Ostrogradsky-Gauss formula:

$$\int_Q \sum_{i,j=1}^{n} A_{ij} u_{x_i t} v_{x_j} \, dQ = \int_\Omega \sum_{i,j=1}^{n} A_{ij} u_{x_i} v_{x_j} \Big|_{t=0}^{t=T} d\Omega - \int_Q \sum_{i,j=1}^{n} A_{ij} u_{x_i} v_{x_j t} dQ .$$

Since $u(0,x)=0$ and $v(T,x)=0$ then the first integral equals to zero, and therefore

$$\left(A(u_t)+B(u),v\right)_{L_2(Q)} = -\int_Q \sum_{i,j=1}^{n} A_{ij} u_{x_i} v_{x_j t} dQ + \tag{35}$$

$$+\int_Q \sum_{i,j=1}^{n} B_{ij} u_{x_i} v_{x_j} dQ$$

Consider every of the last integrals separately.

a) Granting that $u(t,x) = D_1 v$, we have

$$-\int_Q \sum_{i,j=1}^{n} A_{ij} u_{x_i} v_{x_j t} dQ = -\int_Q \sum_{i,j=1}^{n} \sigma(t) A_{ij} v_{x_i tt} v_{x_j t} dQ +$$

$$+\int_Q \sum_{i,j=1}^{n} (l_1 + l_2 t) A_{ij} v_{x_i t} v_{x_j t} dQ$$

Applying to the first integral the formula of integration by parts, we obtain:

$$-\int_Q \sum_{i,j=1}^{n} A_{ij} u_{x_i} v_{x_j t} dQ = -\frac{1}{2}\int_Q \left(\sum_{ij=1}^{n} \sigma(t) A_{ij} v_{x_i t} v_{x_j t} \right)_t dQ +$$

$$+\int_Q \sum_{i,j=1}^{n} \sigma'(t) A_{ij} v_{x_i t} v_{x_j t} dQ + \int_Q \sum_{i,j=1}^{n} (l_1 + l_2 t) A_{ij} v_{x_i t} v_{x_j t} dQ$$

Passing in the first term to the integral on surface and taking into account the conditions of the uniform ellipticity of $A(\cdot)$, and also the relations $\sigma'(t) = 2\left(\alpha_A \alpha_B\right)^{-1} \sigma(t)$, we have

$$-\int_Q \sum_{ij=1}^{n} A_{ij} u_{x_i} v_{x_j t} dQ = \frac{1}{2} \int_\Omega \sum_{i,j=1}^{n} A_{ij} v_{x_i t} v_{x_j t} \Big|_{t=0} d\Omega +$$

$$+\int_Q \sum_{i,j=1}^{n} \left((\alpha_A \alpha_B)^{-1} \sigma(t) + l_1 + l_2 t \right) A_{ij} v_{x_i t} v_{x_j t} dQ \geq$$

$$\geq \frac{\alpha_A}{2} \int_\Omega \sum_{i=1}^{n} v_{x_i t}^2 \Big|_{t=0} d\Omega + \int_Q \sum_{i=1}^{n} \left(\alpha_B^{-1} \sigma(t) + l_1 \alpha_A \right) \sum_{i=1}^{n} v_{x_i t}^2 dQ$$

b) Consider the second term in (35):

$$\int_Q \sum_{i,j=1}^{n} B_{ij} u_{x_i} v_{x_j} dQ = \int_Q \sum_{ij=1}^{n} \sigma(t) B_{ij} v_{x_i tt} v_{x_j} dQ -$$

$$-\int_Q \sum_{i,j=1}^{n} (l_1 + l_2 t) B_{ij} v_{x_i t} v_{x_j} dQ$$

Applying to the first integral the formula of integration by parts, we have:

$$\int_Q \sum_{i,j=1}^{n} B_{ij} u_{x_i} v_{x_j} dQ = \int_Q \left(\sum_{i,j=1}^{n} \sigma(t) B_{ij} v_{x_i t} v_{x_j} \right)_t dQ -$$

$$-\int_Q \sum_{i,j=1}^{n} \sigma'(t) B_{ij} v_{x_i t} v_{x_j} dQ -$$

$$-\int_Q \sum_{i,j=1}^{n} \sigma(t) B_{ij} v_{x_i t} v_{x_j t} dQ - \int_Q \sum_{i,j=1}^{n} (l_1 + l_2 t) B_{ij} v_{x_i t} v_{x_j} dQ$$

In the first term we pass to the surface integral, join the second term with the fourth, apply the formula of integration by parts, and take into account the condition for the third term

$$\alpha_B^{-1} \sum_{i=1}^{n} \xi_i^2 \geq \sum_{i,j=1}^{n} B_{ij}(x) \xi_i \xi_j .$$

Then we obtain

$$\int_Q \sum_{i,j=1}^n B_{ij} u_{x_i} v_{x_j} dQ \geq -\int_\Omega \sum_{i,j=1}^n B_{ij} v_{x_i t} v_{x_j} \big|_{t=0} d\Omega -$$

$$-\int_Q \sum_{i,j=1}^n \left(2(\alpha_A \alpha_B)^{-1} \sigma(t) + l_1 + l_2 t\right) B_{ij} v_{x_i t} v_{x_j} dQ -$$

$$-\int_Q \sum_{i=1}^n \alpha_B^{-1} \sigma(t) v_{x_i t}^2 dQ = -\int_\Omega \sum_{i,j=1}^n B_{ij} v_{x_i t} v_{x_j} \big|_{t=0} d\Omega +$$

$$+\frac{1}{2} \int_\Omega \sum_{i,j=1}^n 2(\alpha_A \alpha_B)^{-1} + l_1 B_{ij} v_{x_i} v_{x_j} \big|_{t=0} d\Omega +$$

$$+\frac{1}{2} \int_Q \sum_{i,j=1}^n \left(2(\alpha_A \alpha_B)^{-1} \sigma'(t) + l_2\right) B_{ij} v_{x_i} v_{x_j} dQ -$$

$$-\int_Q \sum_{i=1}^n \alpha_B^{-1} \sigma(t) v_{x_i t}^2 dQ.$$

Taking into account the fact that the operator $B(\cdot)$ is non-negative and the values of the constants l_1 and l_2, we obtain

$$\int_Q \sum_{i,j=1}^n B_{ij} u_{x_i} v_{x_j} dQ \geq$$

$$\geq \int_\Omega \sum_{i,j=1}^n \frac{1}{2} (\alpha_A \alpha_B)^{-1} B_{ij} v_{x_i} v_{x_j} - \sum_{i,j=1}^n B_{ij} v_{x_i t} v_{x_j} \big|_{t=0} d\Omega -$$

$$-\int_Q \sum_{i=1}^n \alpha_B^{-1} \sigma(t) v_{x_i t}^2 dQ.$$

Thus, taking into account the results of the subsections a) and b), and returning to the identity (35), we obtain

$$(A(u_t)+B(u),v)_{L_2(Q)}\geq$$

$$\geq l_1\alpha_A\int\limits_Q\sum_{i=1}^n v_{x_it}^2dQ+\int\limits_\Omega\frac{\alpha_A}{2}\sum_{i=1}^n v_{x_it}^2-\sum_{i,j=1}^n B_{ij}v_{x_it}v_{x_j}+$$

$$+\sum_{i,j=1}^n\frac{1}{2}(\alpha_A\alpha_B)^{-1}B_{ij}v_{x_i}v_{x_j}\big|_{t=0}d\Omega\geq$$

$$\geq c\int\limits_Q\sum_{i=1}^n v_{x_it}^2dQ+\int\limits_\Omega\frac{\alpha_A}{2}\sum_{i=1}^n v_{,x_it}^2-\sum_{i,j=1}^n\frac{\alpha_A\alpha_B}{2}B_{ij}v_{x_it}v_{x_jt}\big|_{t=0}d\Omega+$$

$$+\int\limits_\Omega\sum_{i,j=1}^n B_{ij}\left(\frac{\alpha_A\alpha_B}{2}v_{x_it}v_{x_jt}-v_{x_it}v_{x_j}+\frac{1}{2}(\alpha_A\alpha_B)^{-1}v_{x_i}v_{x_j}\right)\bigg|_{t=0}d\Omega$$

Convince us that the second and the third terms in the right-hand side are non-negative. Indeed, as far as

$$\alpha_B^{-1}\sum_{i=1}^n\xi_i^2\geq\sum_{i,j=1}^n B_{ij}(x)\xi_i\xi_j\geq 0,$$

then

$$\frac{\alpha_A}{2}\sum_{i=1}^n v_{x_it}^2\geq\alpha_B\cdot\frac{\alpha_A}{2}\sum_{i,j=1}^n B_{ij}v_{x_it}v_{x_jt}\,.$$

Therefore, the second term is non-negative. Furthermore,

$$\sum_{i,j=1}^n B_{ij}\left(\frac{\alpha_A\alpha_B}{2}v_{x_it}v_{x_jt}-v_{x_it}v_{x_j}+\frac{1}{2}(\alpha_A\alpha_B)^{-1}v_{x_i}v_{x_j}\right)=$$

$$=\frac{1}{2\alpha_A\alpha_B}\sum_{i,j=1}^n B_{ij}(\alpha_A\alpha_Bv_{x_it}-v_{x_i})(\alpha_A\alpha_Bv_{x_jt}-v_{x_j})\geq 0$$

that proves the non-negativity of the third term.

Finally, we have

$$(A(u_t)+B(u),v)_{L_2(Q)}\geq c\int\limits_Q\sum_{i=1}^n v_{x_it}^2dQ\,.$$

Consider the third and the fourth terms in (34).

3. For the third term we have the following estimation:

$$(u,-Cv_t)_{L_2(Q)} = (D_1 v,-Cv_t)_{L_2(Q)} = (\sigma(t)v_{tt},-Cv_t)_{L_2(Q)} +$$
$$+((l_1 + l_2 t)v_t, Cv_t)_{L_2(Q)} \geq -\int_Q \sigma(t)C(x)v_{tt}v_t dQ.$$

Applying to the integral in the right-hand side the formula of the integration by parts and the Ostrogradsky-Gauss formula, we obtain

$$-\int_Q \sigma(t)C(x)v_{tt}v_t dQ = \frac{1}{2}\int_\Omega C(x)v_t^2\big|_{t=0} d\Omega +$$
$$+\frac{1}{2}\int_Q \sigma'(t)C(x)v_t^2 dQ \geq 0.$$

Thus,

$$(u,-Cv_t)_{L_2(Q)} \geq 0.$$

4. The fourth term is estimated in the following way:

$$(u,D(x)v)_{L_2(Q)} = (D_1 v, D(x)v)_{L_2(Q)} =$$
$$= \int_Q \sigma(t)v_{tt}D(x)v dQ - \int_Q (l_1 + l_2 t)v_t D(x)v dQ.$$

Integrate by parts the first term in the right-hand side and pass to the surface integral:

$$(u,D(x)v)_{L_2(Q)} = -\int_\Omega D(x)vv_t\big|_{t=0} d\Omega -$$
$$-\int_Q \sigma(t)D(t)v_t^2 dQ - \int_Q (\sigma'(t) + l_1 + l_2 t)D(x)vv_t dQ$$

Apply again the formula of integration by parts to the last term:

$$-\int_Q (\sigma'(t) + l_1 + l_2 t)D(x)vv_t dQ =$$
$$= -\frac{1}{2}\int_Q ((\sigma'(t) + l_1 + l_2 t)D(x)v^2)_t dQ +$$

$$+\frac{1}{2}\int_Q (\sigma''(t)+l_2)D(x)v^2 dQ=$$

$$=\frac{1}{2}\int_\Omega (2(\alpha_A\alpha_B)^{-1}\sigma(0)+l_1)D(x)v^2\big|_{t=0}d\Omega+$$

$$+\frac{1}{2}\int_Q (\sigma''(t)+l_2)D(x)v^2 dQ \ge \frac{1}{2}\int_\Omega D(x)v^2\big|_{t=0}d\Omega$$

Hence,

$$(u,D(x)v)_{L_2(Q)} \ge \int_\Omega -D(x)vv_t+\frac{1}{2}D(x)v^2\big|_{t=0}d\Omega-$$

$$-\int_Q \sigma(t)D(x)v_t^2 dQ.$$

Thus, every of the four terms in (34) are considered. Taking into account the transforms carried out, we have

$$(Lu,v)_{L_2(Q)} \ge c\int_Q v_{tt}^2+\sum_{i=1}^n v_{x_i t}^2 dQ+$$

$$+\int_\Omega \frac{1}{2}D(x)v_t^2-D(x)vv_t+\frac{1}{2}D(x)v^2\big|_{t=0}d\Omega \ge c\|v\|^2_{W^+_{1^*}}+$$

$$+\int_\Omega \frac{1}{2}D(x)(v_t-v)^2\big|_{t=0}d\Omega \ge c\|v\|^2_{W^+_{1^*}}$$

This proves the lemma..

Corollary. *For an arbitrary function $u(t,x)\in W_1^+$ the following inequality holds true:*

$$c\|u\|_{L_2(Q)} \le \|Lu\|_{W^-_{1^*}}.$$

Proof. The lemma implies that for an arbitrary smooth function $u(t,x)$ satisfying the condition (4) the following inequality holds true:

$$(Lu,v)_{L_2(Q)} \ge c\|v\|^2_{W^+_{1^*}}.$$

Apply to the left-hand side the Schwartz inequality:

$$\|Lu\|_{W_{1^*}^-}\|v\|_{W_{1^*}^+} \geq (Lu,v)_{L_2(Q)} \geq c\|v\|^2_{W_{1^*}^+}.$$

Reducing by $\|v\|_{W_{1^*}^+}$, we obtain the inequality

$$\|Lu\|_{W_{1^*}^-} \geq c\|v\|_{W_{1^*}^+}.$$

Now we must to justify the following inequality:

$$\|v\|_{W_{1^*}^+} = \left\|I_t^{'1}u\right\|_{W_{1^*}^+} \geq \|u\|_{L_2(Q)},$$

that proves the required inequality for smooth functions $u(t,x)$. Taking into account that operator $L(\cdot):W_1^+ \to W_{1^*}^-$ is continuous and passing to the limit, we prove the required inequality for an arbitrary function $u(t,x) \in W_1^+$.

Lemma 6. *Let* $v(t,x)$ *is an arbitrary smooth in* $\overline{Q}$ *function satisfying the conditions (6), and* $u = I_t^{1^*} v$ *is the integral operator defined by the following expression:*

$$u(t,x) = \int_0^t \varphi_{1^*}(\xi) \int_0^\xi \psi_{1^*}(\eta) v(\eta,x)\, d\eta\, d\xi,$$

where

$$\varphi_{1^*}(\xi) = \exp\left(-0.5\alpha_A\alpha_B\left(0.5 l_{2^*}\alpha_A\alpha_B + l_{1^*} - l_{2^*}(\xi - T)\right)\sigma\right)(\xi),$$

$$\psi_{1^*}\xi = \frac{\sigma(\xi)}{\varphi_{1^*}(\xi)},\ l_{1^*} = \sup_{x\in\Omega}\left(D(x)\right)+1,\ l_{2^*} = 2l_{1^*}.$$

Then,

$$\left(L^* v, I_t^{1^*} v\right)_{L_2(Q)} = \left(L^* v, u\right)_{L_2(Q)} \geq c\|u\|^2_{W_1^+}.$$

The proof of this lemma is similar to the previous one, therefore we shall write only calculations.

The inverse operator to $u = I_t^{1^*} v$ is

$$v(t,x) = D_{1^*}u = \exp\left(-2(\alpha_A\alpha_B)^{-1}t\right)u_{tt} + \left(\left(l_{1^*} - l_{2^*}(t-T)\right)\right)u_t =$$
$$= \sigma^{-1}u_{tt} + \left(l_{1^*} - l_{2^*}(t-T)\right)u_t.$$

Applying to $\left(L^*v, I_t^{1^*}v\right)_{L_2(Q)}$ the formula of integration by parts, we have

$$\left(L^*v, u\right)_{L_2(Q)} = \left(v_{tt}, u\right)_{L_2(Q)} + \left(-A(v_t) + B(v), u\right)_{L_2(Q)} -$$
$$- Cv_t, u_{L_2(Q)} + \left(Dv, u\right)_{L_2(Q)} = \left(v, u_{tt}\right)_{L_2(Q)} +$$
$$+ \left(-A(v_t) + B(v), u\right)'_{L_2(Q)} + \left(v, Cu_t\right)_{L_2(Q)} + \left(v, Du\right)_{L_2(Q)}.$$

Consider every of the four terms of the right-side hand separately .

1. The first term is estimated in the following way:

$$\left(v, u_{tt}\right)_{L_2(Q)} = \int_Q \sigma^{-1}u_{tt}^2 dQ + \frac{1}{2}\int_\Omega l_{1^*}u_t^2\Big|_{t=T} d\Omega +$$
$$+ \frac{1}{2}\int_Q l_{2^*}u_t^2 dQ \geq c\int_Q u_{tt}^2 dQ +$$
$$+ \frac{1}{2}\int_Q (D(x)+1)u_t^{2'} dQ + \frac{1}{2}\int_Q \sigma^{-1}(T)D(x)u_t^2\Big|_{t=T} dQ.$$

2. Now, let us estimate the second term.

$$\left(-A(v_t) + B(v), u\right)_{L_2(Q)} = -\int_Q \sum_{i,j=1}^n A_{ij}v_{x_i t}u_{x_j} dQ +$$
$$+ \int_Q \sum_{ij=1}^n B_{ij}v_{x_i}u_{x_j} dQ = \int_Q \sum_{ij=1}^n \sigma^{-1}A_{ij}u_{x_i tt}u_{x_j t} dQ +$$
$$+ \int_Q \sum_{ij=1}^n (l_{1^*} - l_{2^*}(t-T))A_{ij}u_{x_i t}u_{x_j t} dQ +$$

$$+\int_Q \sum_{ij=1}^{n} \sigma^{-1} B_{ij} u_{x_i tt} u_{x_j} dQ +$$

$$+\int_Q \sum_{ij=1}^{n} (l_{1^*} - l_{2^*}(t-T)) B_{ij} u_{x_i t} u_{x_j} dQ$$

Integrating by parts, we obtain

$$\left(-A(v_t) + B(v), u\right)_{L_2(Q)} = \frac{1}{2}\int_\Omega \sum_{i,j=1}^{n} \sigma^{-1}(T) A_{ij} u_{x_i t} u_{x_j t}\Big|_{t=T} d\Omega +$$

$$+\frac{1}{2}\int_Q \sum_{i,j=1}^{n} 2(\alpha_A \alpha_B \sigma)^{-1} A_{ij} u_{x_i t} u_{x_j t} dQ +$$

$$+\int_Q \sum_{i,j=1}^{n} \left(l_{1^*} - l_{2^*}(t-T)\right) A_{ij} u_{x_i t} u_{x_j t} dQ +$$

$$+\int_\Omega \sum_{ij=1}^{n} \sigma^{-1}(T) B_{ij} u_{x_i t} u_{x_j}\Big|_{t=T} d\Omega - \int_Q \sum_{i,j=1}^{n} \sigma^{-1} B_{ij} u_{x_i t} u_{x_j t} dQ +$$

$$+\int_Q \sum_{ij=1}^{n} \left(2(\alpha_A \alpha_B \sigma)^{-1} + l_{1^*} - l_{2^*}(t-T)\right) B_{ij} u_{x_i t} u_{x_j} dQ.$$

Taking into account the relations

$$\alpha_A^{-1} \sum_{i=1}^{n} \xi_i^2 \geq \sum_{i,j=1}^{n} A_{ij}(x) \xi_i \xi_j \geq \alpha_A \sum_{i=1}^{n} \xi_i^2,$$

$$\alpha_B^{-1} \sum_{i=1}^{n} \xi_i^2 \geq \sum_{ij=1}^{n} B_{ij}(x) \xi_i \xi_j \geq 0,$$

and integrating by parts, we have

$$\left(-A(v_t)+B(v),u\right)_{L_2(Q)}\geq$$
$$\geq\frac{\alpha_A\alpha_B}{2}\int_\Omega\sum_{ij=1}^n\sigma^{-1}(T)B_{ij}u_{x_it}u_{x_jt}\big|_{t=T}d\Omega+$$
$$+\int_Q\sum_{ij=1}^n\sigma^{-1}B_{ij}u_{x_it}u_{x_jt}dQ+\int_Q\sum_{ij=1}^n\left(l_{1^*}-l_{2^*}(t-T)\right)A_{ij}u_{x_it}u_{x_jt}dQ+$$
$$+\int_\Omega\sum_{ij=1}^n\sigma^{-1}(T)B_{ij}u_{x_it}u_{x_j}\big|_{t=T}d\Omega-\int_Q\sum_{ij=1}^n\sigma^{-1}B_{ij}u_{x_it}u_{x_jt}dQ+$$
$$+\frac{1}{2}\int_\Omega\sum_{ij=1}^n\left(2(\alpha_A\alpha_B\sigma(T))^{-1}+l_{1^*}\right)B_{ij}u_{x_i}u_{x_j}\big|_{t=T}d\Omega+$$
$$+\frac{1}{2}\int_Q\sum_{ij=1}^n\left(4(\alpha_A\alpha_B)^{-2}\sigma^{-1}+l_{2^*}\right)B_{ij}u_{x_i}u_{x_j}dQ.$$

Making obvious reductions, we obtain

$$\left(-A(v_t)+B(v),u\right)_{L_2(Q)}\geq\int_Q\sum_{i=1}^n\alpha_A l_{1^*}u_{x_it}^2dQ+$$
$$+\int_\Omega\sum_{ij=1}^n\sigma^{-1}(T)B_{ij}\left(\frac{\alpha_A\alpha_B}{2}u_{x_it}u_{x_jt}+\right.$$
$$\left.u_{x_it}u_{x_j}+\frac{1}{2(\alpha_A\alpha_B)}u_{x_i}u_{x_j}\right)\bigg|_{t=T}d\Omega\geq-\int_Q\sum_{i=1}^n u_{x_it}^2dQ.$$

3. The third term is estimated in the following way:

$$\left(v,Cu_t\right)_{L_2(Q)}=\int_Q\sigma^{-1}Cu_tu_{tt}dQ+\int_Q\left(l_{1^*}-l_{2^*}(t-T)\right)Cu_t^2dQ\geq$$
$$\geq\frac{1}{2}\int_\Omega\sigma^{-1}(T)Cu_t^2\big|_{t=T}d\Omega+\frac{1}{2}\int_Q2(\alpha_A\alpha_B\sigma)^{-1}Cu_t^2dQ\geq0$$
$$\left(v,Du\right)_{L_2(Q)}=\int_Q\sigma^{-1}Duu_{tt}dQ+\int_Q\left(l_{1^*}-l_{2^*}(t-T)\right)Du_t\,udQ=$$

$$= \int_{\Omega} \sigma^{-1}(T) Du_t u\big|_{t=T} d\Omega - \int_{Q} \sigma^{-1} Du_t^2 dQ +$$

$$+ \int_{Q} \left(2(\alpha_A \alpha_B \sigma)^{-1} + l_{1^*} - l_{2^*}(t-T)\right) Du_t u dQ.$$

Integrating by parts, we have

$$(v, Du)_{L_2(Q)} \geq \int_{\Omega} \sigma^{-1}(T) Du_t u\big|_{t=T} d\Omega - \int_{Q} \sigma^{-1} Du_t^2 dQ +$$

$$+ \frac{1}{2}\int_{\Omega} \left(2(\alpha_A \alpha_B \sigma)(T)\right)^{-1} + l_{1^*}\Big) Du^2\big|_{t=T} d\Omega +$$

$$+ \frac{1}{2}\int_{Q} \left(4(\alpha_A \alpha_B)^{-2} \sigma^{-1} + l_{2^*}\right) Du^2 dQ \geq$$

$$\geq \int_{\Omega} \sigma^{-1}(T) Du_t u + \frac{1}{2}\sigma^{-1}(T) Du^2\big|_{t=T} d\Omega - \int_{Q} Du_t^2 dQ.$$

Thus,

$$\left(L^* v, u\right)_{L_2(Q)} \geq c \int_{Q} u_{tt}^2 + \sum_{i=1}^{n} u_{x_i t}^2 dQ +$$

$$+ \int_{\Omega} \frac{1}{2}\sigma^{-1}(T) D(u_t^2 + 2u_t u + u^2)\big|_{t=T} d\Omega \geq c\|u\|_{W_1^+}^2.$$

Corollary. *For an arbitrary function* $v(t,x) \in W_{1^*}^+$ *the following inequality holds true*

$$c\|v\|_{L_2(Q)} \leq \left\|L^* v\right\|_{W_1^-}.$$

The proof is similar to the proof of the corollary of Lemma 5.

Thus, granting Lemma 1 and the corollaries of Lemmas 5 and 6, we have

$$c^{-1}\|u\|_{L_2(Q)} \leq \|Lu\|_{W_{1^*}^-} \leq c\|u\|_{W_1^+}, \tag{36a}$$

$$c^{-1}\|v\|_{L_2(Q)} \leq \left\|L^* v\right\|_{W_1^-} \leq c\|v\|_{W_{1^*}^+}. \tag{36b}$$

Definition 3. *A generalized solution of Problem 1 is such function* $u(t,x) \in W_1^+$ *that there exists a sequence of smooth in* $\overline{Q}$ *functions* $u_i(t,x)$ *satisfying the conditions (4) and*

$$\|u_i - u\|_{W_1^+} \xrightarrow[i\to\infty]{} 0, \qquad \|Lu_i - f\|_{W_r^-} \xrightarrow[i\to\infty]{} 0.$$

Definition 4. *A generalized solution of Problem 1 is such function* $u(t,x) \in L_2(Q)$ *that there exists a sequence of smooth in* $\overline{Q}$ *functions* $u_i(t,x)$ *satisfying the conditions (4) and*

$$\|u_i - u\|_{L_2(Q)} \xrightarrow[i\to\infty]{} 0, \qquad \|Lu_i - f\|_{W_r^-} \xrightarrow[i\to\infty]{} 0.$$

Lemma 7. *Let* $u(t,x)$ *be an arbitrary smooth in* $\overline{Q}$ *function satisfying the conditions (4), and* $v = I_t^2 u$ *is the integral operator defined by the following expression:*

$$v(t,x) = -\int_T^t e^{-\eta} u(\eta,x)\,d\eta. \tag{37}$$

Then,

$$(Lu, I_t^2 u)_{L_2(Q)} = (Lu, v)_{L_2(Q)} \geq c\|v\|^2_{W_{2^*}^+}.$$

Proof. As in the proof of Lemma 5 note that $v = I_t^2 u \in W_{2^*}^+$. Indeed, the initial condition $v(T,x) = 0$ holds true at the expense of the form of the operator $v = I_t^2 u$. Remaining conditions (6) do not hold true after completion of smooth functions in the norm of the space $W_{2^*}^+$.

Consider

$$(Lu, v)_{L_2(Q)} = I_1 + I_2 + I_3 + I_4,$$

where

$$I_1 = (u_{tt}, v)_{L_2(Q)},$$

$$I_2 = \big(A(u_t) + B(u), v\big)_{L_2(Q)},$$

$$I_3 = \big(Cu_t, v\big)_{L_2(Q)},$$

$$I_4 = (Du, v)_{L_2(Q)}.$$

Consider every of these term separately.

1.Applying the formula of integration by parts, we obtain

$$I_1 = \int_Q u_{tt} v dQ = \int_Q (u_t v)_t dQ - \int_Q u_t v_t dQ.$$

To calculate the first integral we use the Ostrogradsky-Gauss formula and also the conditions $u(0, x) = v(T, x) = 0$. In the second integral we take into account that (37) implies that $v_t = -e^{-t} u$.

$$I_1 = \int_\Omega e^{-t} u_t v \Big|_{t=0}^{t=T} d\Omega + \int_Q e^{-t} u_t u dQ = \int_Q e^{-t} u_t u dQ.$$

Integrating the last expression by parts, we obtain

$$I_1 = \int_Q e^{-t} u_t u dQ = \frac{1}{2}\int_Q \left(e^{-t} u^2\right)_t dQ - \frac{1}{2}\int_Q \left(e^{-t}\right)_t u^2 dQ =$$

$$= \frac{1}{2}\int_\Omega e^{-T} u^2 \big|_{t=T} d\Omega + \frac{1}{2}\int_Q e^{-t} u^2 dQ \geq c\int_Q u^2 dQ.$$

2. Applying the formula of integration by parts, the Ostrogradsky-Gauss formula and taking into account the condition $\dfrac{\partial_L u}{\partial \vec{n}}\Big|_{x\in\partial\Omega} = 0$, we have

$$I_2 = -\int_Q \left(\sum_{i,j=1}^{n} \left(A_{ij} u_{x_i t}\right)_{x_j} + \sum_{i,j=1}^{n} \left(B_{ij} u_{x_i}\right)_{x_j} \right) v dQ =$$

$$= -\int_Q \sum_{i,j=1}^{n} \left(\left(A_{ij} u_{x_i t} + B_{ij} u_{x_i}\right) v\right)_{x_j} dQ +$$

$$+\int_Q \sum_{i,j=1}^n A_{ij} u_{x_i t} v_{x_j}\, dQ + \int_Q \sum_{i,j=1}^n B_{ij} u_{x_i} v_{x_j}\, dQ =$$

$$= -\int_0^T \int_{\partial\Omega} \frac{\partial_L u}{\partial \vec{n}} v\, dt\, d\partial\Omega + \int_Q \sum_{i,j=1}^n A_{ij} u_{x_i t} v_{x_j}\, dQ +$$

$$+\int_Q \sum_{i,j=1}^n B_{ij} u_{x_i} v_{x_j}\, dQ = \int_Q \sum_{i,j=1}^n A_{ij} u_{x_i t} v_{x_j}\, dQ + \int_Q \sum_{i,j=1}^n B_{ij} u_{x_i} v_{x_j}\, dQ.$$

Consider both terms separately.

a) To calculate the first integral, we use the formula of integration by parts. We obtain that

$$\int_Q \sum_{i,j=1}^n A_{ij} u_{x_i t} v_{x_j}\, dQ = \int_Q \left(\sum_{i,j=1}^n A_{ij} u_{x_i} v_{x_j} \right)_t dQ - \int_Q \sum_{i,j=1}^n A_{ij} u_{x_i} v_{x_j t}\, dQ.$$

Passing to the integral on surface in the first term and taking into account the conditions $u(0,x) = v(T,x) = 0$, we have that this integral equals to zero. To calculate the second integral, we use the condition $v_{x_j t} = -e^{-t} u_{x_j}$ and also take into account the uniform ellipticity of the operator $\dot{A}(\cdot)$:

$$-\int_Q A_{ij} u_{x_i} v_{x_j t}\, dQ = \int_Q \sum_{i,j=1}^n e^{-t} A_{ij} u_{x_i} u_{x_j}\, dQ \geq \alpha_A \int_Q e^{-t} \sum_{i=1}^n u_{x_i}^2\, dQ.$$

b) Consider the second term. It is clear that the following identity holds true

$$\int_Q \sum_{i,j=1}^n B_{ij} u_{x_i} v_{x_j}\, dQ = \int_Q \sum_{i,j=1}^n B_{ij} \frac{-1}{e^{-t}} \left(-\int_T^t e^{-\eta} u_{x_i} \right)_t v_{x_j}\, dQ =$$

$$= -\int_Q \sum_{i,j=1}^n e^t B_{ij} v_{x_i t} v_{x_j}\, dQ.$$

Let us prove that the last expression in non-negative. To do this, we apply the operation of integration by parts and take into account the conditions

$$\alpha_B^{-1}\sum_{i=1}^{n}\xi_i^2 \geq \sum_{i,j=1}^{n} B_{ij}(x)\xi_i\xi_j \geq 0.$$

Hence,

$$-\int_Q \sum_{i,j=1}^{n} e^t B_{ij} v_{x_i t} v_{x_j}\, dQ = -\frac{1}{2}\int_Q \left(\sum_{i,j=1}^{n} e^t B_{ij} v_{x_i} v_{x_j}\right)_t dQ +$$

$$+\frac{1}{2}\int_Q \sum_{i,j=1}^{n} e^t B_{ij} v_{x_i} v_{x_j}\, dQ =$$

$$= \frac{1}{2}\int_\Omega \sum_{i,j=1}^{n} B_{ij} v_{x_i} v_{x_j}\Big|_{t=0} d\Omega + \frac{1}{2}\int_Q \sum_{i,j=1}^{n} e^t B_{ij} v_{x_i} v_{x_j}\, dQ \geq 0.$$

Thus,

$$I_2 \geq \alpha_A \int_Q e^{-t} \sum_{i=1}^{n} u_{x_i}^2\, dQ \geq c \int_Q \sum_{i=1}^{n} u_{x_i}^2\, dQ.$$

Consider I_3 and I_4.

3. Similarly to the previous reasoning we have

$$I_3 = \left(Cu_t, v\right)_{L_2(Q)} = \int_Q Cu_t v\, dQ = \int_Q (Cuv)_t\, dQ -$$

$$-\int_Q Cuv_t\, dQ = \int_Q e^{-t} Cu^2\, dQ \geq 0.$$

4. Analogously,

$$I_4 = (Du, v)_{L_2(Q)} = \int_Q D(x) uv\, dQ =$$

$$= \int_Q -\frac{1}{e^{-t}} D(x)\left(-\int_T^t e^{-\eta} u\, d\eta\right)_t v\, dQ = -\int_Q e^t D(x) v_t v\, dQ =$$

$$\frac{1}{2}\int_\Omega Dv^2\Big|_{t=0} d\Omega + \frac{1}{2}\int_Q e^t D(x) v^2\, dQ \geq 0.$$

Thus, summarizing all written above, we have

$$(Lu,v)_{L_2(Q)} \geq c\int_Q u^2 + \sum_{i=1}^{n} u_{x_i}^2 \, dQ \geq c\|u\|_{H_2^+}^2 .$$

Let us prove that $\|u\|_{H_2^+}^2 \geq \|v\|_{W_{2^*}^+}^2$. Indeed,

$$\|v\|_{W_{2^*}^+}^2 = \int_Q v_t^2 + \sum_{i=1}^{n} v_{x_i t}^2 dQ = \int_Q \left(\int_T^t e^{-\eta} u d\eta \right)_t^2 +$$

$$+ \sum_{i=1}^{n} \left(\int_T^t e^{-\eta} u d\eta \right)_{x_i t}^2 dQ = \int_Q e^{-2t} \left(u^2 + \sum_{i=1}^{n} u_{x_i}^2 \right) dQ \leq \|u\|_{H_2^+}^2 . \tag{38}$$

Hence, $(Lu,v)_{L_2(Q)} \geq c\|v\|_{W_{2^*}^+}^2$. The lemma is proved.

Corollary. *For an arbitrary function* $u(t,x) \in W_2^+$ *the following inequality holds true*

$$c\|u\|_{H_2^+} \leq \|Lu\|_{W_{2^*}^-} .$$

Proof. Applying to the left-hand side of the inequality the Schwartz inequality, we have:

$$\|Lu\|_{W_{2^*}^-} \|v\|_{W_{2^*}^+} \geq (Lu,v)_{L_2(Q)} \geq c\|v\|_{W_{2^*}^+}^2 .$$

Reducing by $\|v\|_{W_{2^*}^+}$, we obtain

$$\|Lu\|_{W_{2^*}^-} \geq c\|v\|_{W_{2^*}^+} .$$

Taking into account (38), we obtain

$$\|v\|_{W_{2^*}^+}^2 = \int_Q e^{-2t} \left(u^2 + \sum_{i=1}^{n} u_{x_i}^2 \right) dQ \geq c\|u\|_{H_2^+}^2 ,$$

that proves the required inequality for smooth functions $u(t,x)$. Taking into account that the operator $L(\cdot): W_2^+ \to W_{2^*}^-$ is continuous and passing to the limit we find that the required inequality holds true.

Lemma 8. *Let* $v(t,x)$ *be an arbitrary smooth function satisfying the conditions (6), and* $u = I_t^{2^*} v$ *is the integral operator defined by the following expression*

$$u(t,x) = \int_0^t e^{\eta} v(\eta, x) d\eta .$$

Then,

$$\left(L^* v, I_t^{2^*} v\right)_{L_2(Q)} = \left(L^* v, u\right)_{L_2(Q)} \geq c\|u\|^2_{W_2^+} .$$

The proof of this lemma is similar to the previous, therefore we represent it in brief version. Let us write

$$\left(Lu, v\right)_{L_2(Q)} = I_1 + I_2 + I_3 + I_4 ,$$

where

$$I_1 = (v_{tt}, u)_{L_2(Q)} = \frac{1}{2}\int_\Omega v^2\big|_{t=0} d\Omega + \frac{1}{2}\int_Q e^t v^2 dQ \geq c\int_Q v^2 dQ ,$$

$$I_2 = (-A(v_t) + B(v), u)_{L_2(Q)} = \int_Q e^t \sum_{i,j=1}^n A_{ij} v_{x_i} v_{x_j} dQ +$$

$$+ \frac{1}{2}\int_Q \sum_{i,j=1}^n e^{-t} B_{ij} u_{x_i} u_{x_j} dQ +$$

$$+ \frac{1}{2}\int_\Omega \sum_{i,j=1}^n e^{-T} B_{ij} u_{x_i} u_{x_j}\big|_{t=T} d\Omega \geq c\int_Q \sum_{i=1}^n v_{x_i}^2 dQ,$$

$$I_3 = (-Cv_t, u)_{L_2(Q)} = \int_Q e^t C v^2 dQ \geq 0 ,$$

$$I_4 = (Dv, u)_{L_2(Q)} = \frac{1}{2}\int_\Omega e^{-T} D u^2\big|_{t=T} d\Omega + \frac{1}{2}\int_Q e^{-t} D(x) u^2 dQ \geq 0 .$$

Thus,

$$(L^* v, u)_{L_2(Q)} \geq c\|v\|^2_{H^+_{2^*}} .$$

Taking into account the form of the operator $u = I_t^{2^*} v$, we obtain the required inequality.

Corollary. *For an arbitrary function* $v(t,x) \in W_{2^*}^+$ *the following inequality holds true*

$$c\|v\|_{H_{2^*}^+} \le \|L^* v\|_{W_2^-}.$$

The proof of the lemma is similar to the proof of Lemma 7.

Thus, by Lemma 2 and Corollaries of Lemmas 7 and 8, the following inequalities hold true:

$$c^{-1}\|u\|_{H_2^+} \le \|Lu\|_{W_{2^*}^-} \le c\|u\|_{W_2^+}, \tag{39à}$$

$$c^{-1}\|v\|_{H_{2^*}^+} \le \|L^* v\|_{W_2^-} \le c\|v\|_{W_{2^*}^+}. \tag{39b}$$

Definition 5. *A generalized solution of Problem 1 is such function* $u(t,x) \in W_2^+$ *that there exists a sequence of smooth in* $\overline{Q}$ *functions* $u_i(t,x)$ *satisfying the conditions (4) and*

$$\|u_i - u\|_{W_2^+} \xrightarrow[i\to\infty]{} 0, \quad \|Lu_i - f\|_{W_{2^*}^-} \xrightarrow[i\to\infty]{} 0.$$

Definition 6. *A generalized solution of Problem 1 is such function* $u(t,x) \in H_2^+$ *that there exists a sequence of smooth in* $\overline{Q}$ *functions* $u_i(t,x)$ *satisfying the conditions (4) and*

$$\|u_i - u\|_{H_2^+} \xrightarrow[i\to\infty]{} 0, \quad \|Lu_i - f\|_{W_{2^*}^-} \xrightarrow[i\to\infty]{} 0.$$

In a similar way the definitions for Problem 2 are formulated.

Lemma 9. *Let* $u(t,x)$ *is an arbitrary smooth function satisfying the conditions (8), and* $v = I_t^3 u$ *is the integral operator defined by the following expression*

$$v(t,x) = \int_T^t \varphi_3(\xi) \int_T^\xi \psi_3(\eta) u(\eta,x)\, d\eta\, d\xi,$$

where

$$\varphi_3(\xi) = \exp\left(-\frac{1}{4}(l_2' + 2l_1' + 2l_2'\xi)e^{-2\xi}\right),\ \psi_3(\xi) = \frac{1}{\varphi_3(\xi)e^{2\xi}},$$

$$l_1' = k_1 e^{2T} + k_3 + 1,\ \ l_2' = 2k_3 e^{2T} + 1.$$

Then,

$$\left(L_2 u, I_t^3 u\right)_{W_2^{0,1}} = (L_2 u, v)_{\overset{\circ}{W}{}_2^{0,1}} \geq c\|v\|^2_{W^+_{3^*}}.$$

P r o o f. It is easily to see that $v = I_t^3 u \in W^+_{3^*}$, as far as all the conditions (12) hold true. Consider

$$(L_2 u, v)_{W_2^{0,1}} = (u, L_2^* v)_{W_2^{0,1}} = (D_3 v, L_2^* v)_{W_2^{0,1}},$$

where $u = D_3 v$ is the inverse operator to $v = I_t^3 u$ and the operator:

$$D_3 v = e^{2t} v_{tt} - (l_1' + l_2' t) v_t.$$

Then,

$$(L_2 u, v)_{W_2^{0,1}} = \left(D_3 v, L_2^* v\right)_{L_2(Q)} + \int_Q \sum_{i,j=1}^n A_{ij} (D_3 v)_{x_i} \left(L_2^* v\right)_{x_j} dQ. \quad (40)$$

The first term is considered in a similar way as in Lemma 5:

$$\left(D_3 v, L_2^* v\right)_{L_2(Q)} = I_1 + I_2 + I_3 + I_4 + I_5,$$

where

$$I_1 = \left(D_3 v, v_{tt}\right)_{L_2(Q)} = \int_Q e^{2t} v_{tt}^2 + \frac{1}{2} l_2' v_t^2 dQ + \frac{1}{2}\int_\Omega l_1' v_t^2\Big|_{t=0} d\Omega,$$

$$I_2 = \left(D_3 v, -A(v_t)\right)_{L_2(Q)} = \frac{1}{2}\int_\Omega \sum_{i,j=1}^n A_{ij} v_{x_i t} v_{x_j t}\Big|_{t=0} d\Omega +$$

$$+ \int_Q \left(e^{2t} + l_1' + l_2' t\right) \sum_{i,j=1}^n A_{ij} v_{x_i t} v_{x_j t} dQ,$$

$$I_3 = (D_3 v, A(k_1 v))_{L_2(Q)} =$$

$$= -k_1 \int_\Omega \sum_{i,j=1}^n A_{ij} v_{x_i t} v_{x_j} - \left(1 + \frac{l_1'}{2}\right) \sum_{i,j=1}^n A_{ij} v_{x_i} v_{x_j} \big|_{t=0} d\Omega +$$

$$+ \int_Q - k_1 e^{2t} \sum_{i,j=1}^n A_{ij} v_{x_i t} v_{x_j t} + k_1 \left(2e^{2t} + \frac{1}{2} l_2'\right) \sum_{i,j=1}^n A_{ij} v_{x_i} v_{x_j} dQ,$$

$$I_4 = (D_3 v, -k_2 v_t)_{L_2(Q)} =$$

$$= \frac{1}{2} \int_\Omega k_2 v_t^2 \big|_{t=0} d\Omega + \int_Q k_2 \left(e^{2t} + l_1' + l_2' t\right) v_t^2 dQ,$$

$$I_5 = (D_3 v, k_3 v_t)_{L_2(Q)} = -\int_\Omega k_3 v_t v \big|_{t=0} d\Omega - \int_Q k_3 e^{2t} v_t^2 dQ +$$

$$+ \int_Q k_3 \left(2e^{2t} + \frac{1}{2} l_2'\right) v^2 dQ + \int_\Omega k_3 \left(1 + \frac{1}{2} l_1'\right) v^2 \big|_{t=0} d\Omega.$$

Consider the second term in (40)

$$\int_Q \sum_{i,j=1}^n A_{ij} (D_3 v)_{x_i} (L_2^* v)_{x_j} dQ = I_6 + I_7 + I_8 + I_9 + I_{10},$$

where

$$I_6 = \int_Q \sum_{i,j=1}^n A_{ij} (D_3 v)_{x_i} (v_{tt})_{x_j} dQ =$$

$$= \int_Q \sum_{i,j=1}^n A_{ij} e^{2t} v_{ttx_i} v_{ttx_j} dQ - \int_Q \sum_{i,j=1}^n A_{ij} (l_1' + l_2' t) v_{tx_i} v_{ttx_j} dQ.$$

Integrating the second term by parts, we have

$$I_6 = \int_Q \sum_{i,j=1}^n A_{ij} e^{2t} v_{ttx_i} v_{ttx_j} dQ + \frac{1}{2} \int_Q \sum_{i,j=1}^n A_{ij} (l_1') v_{tx_i} v_{tx_j} \big|_{t=0} d\Omega +$$

$$+ \frac{1}{2} \int_Q \sum_{i,j=1}^n A_{ij} (l_{21}') v_{tx_i} v_{tx_j} dQ.$$

$$I_7 = \int_Q \sum_{i,j=1}^{n} A_{ij}(D_3 v)_{x_i} (-A(v_t))_{x_j} dQ =$$

$$= \int_Q \sum_{i,j=1}^{n} \left(A_{ij}(D_3 v)_{x_i} (-A(v_t))\right)_{x_j} dQ +$$

$$+ \int_Q \sum_{i,j=1}^{n} \left(A_{ij}(D_3 v)_{x_i}\right)_{x_j} (A(v_t)) dQ.$$

Note that granting that $u = D_3 v$ and the conditions (8), we have $\frac{\partial D_3 v}{\partial \bar{\mu}_A}\Big|_{x \in \partial\Omega} = 0$. Therefore, applying to the first integral the Ostrogradsky-Gauss formula, we obtain that it equals to zero. Thus,

$$I_7 = -\int_Q A(D_3 v) A(v_t) dQ = \frac{1}{2}\int_\Omega (A(v_t))^2\Big|_{t=0} d\Omega +$$

$$+ \frac{1}{2}\int_Q \left(e^{2t} + l_1' + l_2' t\right)(A(v_t))^2 dQ,$$

$$I_8 = \int_Q \sum_{i,j=1}^{n} A_{ij}(D_3 v)_{x_i} (k_1 v)_{x_j} dQ.$$

Similarly to the previous case, we have

$$I_8 = \int_Q A(D_3 v)(A(k_1 v)) dQ = \int_Q k_1 e^{2t} A(v_{tt})(A(v)) dQ -$$

$$- \int_Q k_1 (l_1' + l_2' t) A(v_t)(A(v)) dQ.$$

Applying to the first integral the operation of integration by parts, we obtain

$$I_8 = -\int_Q \sum_{ij=1}^{n} k_1 A(v_t)(A(v))\Big|_{t=0} d\Omega -$$

$$\int_Q k_1\left(2e^{2t} + l_1' + l_2' t\right) A(v_t)(A(v)) dQ - \int_Q k_1 e^{2t}\left(A(v_t)\right)^2 dQ.$$

Let us integrate by parts the second term again:

$$I_8 = \int_\Omega -k_1 A(v_t)\big(A(v)\big) + k_1\left(1+\frac{1}{2}l_1'\right)\big(A(v)\big)^2\Big|_{t=0} d\Omega +$$

$$+\int_Q -k_1 e^{2t}\big(A(v_t)\big)^2 + k_1\left(2e^{2t}+\frac{1}{2}l_2'\right)\big(A(v)\big)^2 dQ,$$

$$I_9 = \int_Q \sum_{i,j=1}^{n} A_{ij}(D_3 v)_{x_i}(-k_2 v_t)_{x_j} dQ = \frac{1}{2}\int_Q k_2 \sum_{i,j=1}^{n} A_{ij} v_{tx_i} v_{tx_j}\Big|_{t=0} d\Omega +$$

$$+\int_Q k_2\left(e^{2t}+l_1'+l_2't\right)\sum_{i,j=1}^{n} A_{ij} v_{tx_i} v_{tx_j} dQ,$$

$$I_{10} = \int_Q \sum_{i,j=1}^{n} A_{ij}(D_3 v)_{x_i}(k_3 v_t)_{x_j} dQ = \int_Q \sum_{i,j=1}^{n} A_{ij}\left(e^{2t}v_{tt}\right)_{x_i}(k_3 v)_{x_j} dQ -$$

$$-\int_Q k_3(l_1'+l_2't)\sum_{i,j=1}^{n} A_{ij} v_{tx_i} v_{x_j} dQ.$$

Integrating by parts the first term, we obtain:

$$I_{10} = -\int_Q k_3 \sum_{i,j=1}^{n} A_{ij} v_{tx_i} v_{tx_j}\Big|_{t=0} d\Omega - \int_Q k_3 e^{2t} \sum_{i,j=1}^{n} A_{ij} v_{tx_i} v_{tx_j} dQ -$$

$$-\int_Q k_3\left(e^{2t}+l_1'+l_2't\right)\sum_{i,j=1}^{n} A_{ij} v_{tx_i} v_{tx_j} dQ.$$

Whence, we have

$$I_{10} = \int_{\Omega} - k_3 \sum_{i,j=1}^{n} A_{ij} v_{tx_i} v_{x_j} + k_3\left(1 + \frac{l_1'}{2}\right) \sum_{i,j=1}^{n} A_{ij} v_{x_i} v_{x_j} \big|_{t=0} d\Omega +$$
$$+ \int_{Q} - k_3 e^{2t} \sum_{i,j=1}^{n} A_{ij} v_{tx_i} v_{tx_j} + k_3\left(2e^{2t} + \frac{1}{2} l_2'\right) \sum_{i,j=1}^{n} A_{ij} v_{x_i} v_{x_j} dQ.$$

Thus, granting that the operator $A(\cdot)$ is uniformly elliptic, we finally obtain:

$$(L_2 u, v)_{W_2^{0,1}} \geq \int_{Q} e^{2t} v_{tt}^2 + \left(\frac{1}{2} l_2' - k_3 e^{2t}\right) v_t^2 +$$
$$+ \left(l_1' - k_1 e^{2t} + \frac{1}{2} l' - k_3 e^{2t}\right) \sum_{ij=1}^{n} A_{ij} v_{x_i t} v_{x_j t} +$$
$$+ e^{2t} \sum_{i,j=1}^{n} A_{ij} v_{x_i tt} v_{x_j tt} + \left(l_1' - k_1 e^{2t}\right)\left(A(v_t)\right)^2 dQ +$$
$$+ \int_{\Omega} \left(\frac{1}{2} l' v_t^2 - k_3 v v_t + \frac{1}{2} k_3 v^2\right) +$$
$$+ \left(\frac{1}{2} \sum_{i,j=1}^{n} A_{ij} v_{x_i t} v_{x_j t} - k_1 \sum_{i,j=1}^{n} A_{ij} v_{x_i t} v_{x_j} + k_1 \frac{1}{2} l_1' \sum_{i,j=1}^{n} A_{ij} v_{x_i} v_{x_j}\right) +$$
$$+ \left(\frac{1}{2} l_1' \sum_{i,j=1}^{n} A_{ij} v_{x_i t} v_{x_j t} - k_3 \sum_{ij=1}^{n} A_{ij} v_{x_i t} v_{x_j} + k_3 \frac{1}{2} \sum_{i,j=1}^{n} A_{ij} v_{x_i} v_{x_j}\right) +$$
$$+ \left(\frac{1}{2} \left(A(v_t)\right)^2 - k_1 A(v_t) A(v) + k_1 \frac{1}{2} l' \left(A(v)\right)^2\right)\Big|_{t=0} d\Omega.$$

Taking into account the values of the constants l_1', l_2' and carrying out the obvious estimations, we obtain

$$(L_2u,v)_{W_2^{0,1}} \geq c\int_Q v_{tt}^2 + v_t^2 + \sum_{i,j=1}^n A_{ij}v_{x_itt}v_{x_jtt} + (A(v_t))^2\,dQ +$$

$$+\int_\Omega \frac{1}{2}k_3\left(v_t^2 - 2vv_t + v^2\right)+$$

$$+\frac{1}{2}\sum_{i,j=1}^n A_{ij}\left(v_{x_it}v_{x_jt} - 2k_1 v_{x_it}v_{x_j} + k_1^2 v_{x_i}v_{x_j}\right)+$$

$$+k_3\frac{1}{2}\sum_{i,j=1}^n A_{ij}\left(v_{x_it}v_{x_jt} - 2v_{x_it}v_{x_j} + v_{x_i}v_{x_j}\right)+$$

$$+\frac{1}{2}\left((A(v_t))^2 - 2k_1A(v_t)A(v) + k_1^2(A(v))^2\right)\Big|_{t=0}\,d\Omega.$$

Taking into account the uniform ellipticity of the operator $A(\cdot)$, we obtain

$$(L_2u,v)_{W_2^{0,1}} \geq c\int_Q v_{tt}^2 + \sum_{i=1}^n v_{x_itt}^2\,dQ + c\|A(v_t)\|_{L_2(Q)}^2 + c\|v_t\|_{L_2(Q)}^2.$$

Note that for the elliptic operators the inequality of coercitivity holds true [74]:

$$\|Au\|_{W_2^s(G)}^2 + p\|u\|_{L_2(G)}^2 \geq c\|u\|_{W_2^{s+2}(G)}^2,$$

Taking into account the inequality of coercitivity with $s = 0$, we have

$$(L_2u,v)_{W_2^{0,1}} \geq c\int_Q v_{tt}^2 + \sum_{i=1}^n v_{x_itt}^2 + \sum_{i,j=1}^n v_{x_ix_jt}^2\,dQ = c\|v\|_{W_{3*}^+}^2.$$

Corollary. *For an arbitrary* $u \in W_3^+$ *the following inequality holds true*

$$c\|u\|_{W_2^{0,1}} \leq \|L_2u\|_{W_{3*}^-}.$$

The proof of is similar to the previous proofs of the corollaries of the lemmas.

Lemma 10. *Let* $v(t,x)$ *is an arbitrary smooth in* $\overline{Q}$ *function from* $W_{3^*}^{+}$, *and* $u = I_t^{3^*} v$ *is the integral operator*

$$u(t,x) = I_t^{3^*} v = \int_0^t \varphi_{3^*}(\xi) \int_0^\xi \psi_{3^*}(\eta) v(\eta,x) d\eta d\xi, \tag{41}$$

where

$$\varphi_{3^*}(\xi) = \exp\left(-\frac{1}{2k_1}\left(l'_{1^*} - l'_{2^*}(\xi - T) + \frac{l'_{2^*}}{2k_1}\right) e^{2k_1\xi}\right),$$

$$\psi_{3^*}(\xi) = \frac{e^{2k_1\xi}}{\varphi_{3^*}(\xi)},$$

the constants l'_{1^*} *and* l'_{2^*} *are defined as* $l'_{1^*} = k_3 + 1$ *and* $l'_{2^*} = 2l'_{1^*}$.

Then,

$$\left(L_2^* v, I_t^{3^*} v\right)_{W_2^{0,1}} \geq c\|u\|^2_{W_3^+}.$$

Proof. Let us express from $v(t,x)$ through $u(t,x)$ from (41).

$$v(t,x) = D_{3^*} u = e^{-2k_1 t} u_{tt} + \left(\bar{l}_{1^*} - \bar{l}_{2^*}(t-T)\right) u_t .$$

Consider

$$\left(L_2^* v, u\right)_{W_2^{0,1}} = \left(v, L_2 u\right)_{W_2^{0,1}} = \left(D_{3^*} u, L_2 u\right)_{W_2^{0,1}} = I_1 + I_2 + \ldots + I_{10},$$

where

$$I_1 = \left(D_{3^*} u, u_{tt}\right)_{W_2^{0,1}} = \int_Q e^{-2k_1 t} u_{tt}^2 + \frac{\bar{l}_{2^*}}{2} u_t^2 dQ + \int_\Omega \frac{\bar{l}_{1^*}}{2} u_t^2\Big|_{t=T} d\Omega \geq$$

$$\geq c\int_Q u_{tt}^2 + u_t^2 dQ + \int_Q k_3 u_t^2 dQ + \int_\Omega \frac{k_3 e^{-2k_1 T}}{2} u_t^2\Big|_{t=T} d\Omega,$$

$$I_2=\left(D_{3^*}u,A(u_t)\right)_{L_2}=\int_Q k_1\sum_{ij=1}^{n}e^{-2k_1t}A_{ij}u_{tx_i}u_{tx_j}dQ+$$

$$+\int_\Omega\frac{e^{-2k_1T}}{2}\sum_{ij=1}^{n}A_{ij}u_{tx_i}u_{tx_j}\Big|_{t=T}d\Omega+\int_Q\left(\bar{l}_{1^*}-\bar{l}_{2^*}(t-T)\right)\sum_{ij=1}^{n}A_{ij}u_{tx_i}u_{tx_j}dQ,$$

$$I_3=\left(D_{3^*}u,A(k_1u)\right)_{L_2}=\int_\Omega k_1\sum_{ij=1}^{n}e^{-2k_1T}A_{ij}u_{x_i}u_{tx_j}\Big|_{t=T}d\Omega-$$

$$-\int_Q k_1\sum_{ij=1}^{n}e^{-2k_1t}A_{ij}u_{tx_i}u_{tx_j}dQ+$$

$$+\int_\Omega\left(k_1^2e^{-2k_1T}+k_1\frac{\bar{l}_{1^*}}{2}\right)\sum_{ij=1}^{n}A_{ij}u_{x_i}u_{x_j}\Big|_{t=T}d\Omega+$$

$$+\int_Q k_1\left(2k_1^2e^{-2k_1t}+\frac{\bar{l}_{2^*}}{2}\right)\sum_{ij=1}^{n}A_{ij}u_{x_i}u_{x_j}dQ\geq$$

$$\geq\int_\Omega k_1\sum_{ij=1}^{n}e^{-2k_1T}A_{ij}u_{x_i}u_{tx_j}+\frac{k_1^2e^{-2k_1T}}{2}\sum_{ij=1}^{n}A_{ij}u_{x_i}u_{x_j}\Big|_{t=T}d\Omega-$$

$$-\int_Q k_1\sum_{ij=1}^{n}e^{-2k_1t}A_{ij}u_{tx_i}u_{tx_j}dQ,$$

$$I_4=\left(D_{3^*}u,k_2u_t\right)_{L_2}=\frac{1}{2}\int_\Omega k_2e^{-2k_1T}u_t^2\Big|_{t=T}d\Omega+$$

$$+\int_Q k_2\left(k_1e^{-2k_1t}+\left(\bar{l}_{1^*}-\bar{l}_{2^*}(t-T)\right)\right)u_t^2dQ\geq 0,$$

$$I_5=\left(D_{3^*}u,k_3u\right)_{L_2}=\int_\Omega k_3e^{-2k_1T}uu_t\Big|_{t=T}d\Omega-\int_Q k_3e^{-2k_1T}u_t^2dQ+$$

$$+\int_\Omega k_3\left(k_1e^{-2k_1T}+\frac{\bar{l}_{1^*}}{2}\right)u^2\Big|_{t=T}d\Omega+\int_Q k_3\left(2k_1^2e^{-2k_1t}+\frac{\bar{l}_{2^*}}{2}\right)u^2dQ\geq$$

$$\geq \int_{\Omega} k_3 e^{-2k_1 T}\left(uu_t + \frac{u^2}{2}\right)\Big|_{t=T} d\Omega - \int_{Q} k_3 u_t^2 dQ,$$

$$I_6 = \int_{Q} \sum_{ij=1}^{n} A_{ij}\left(D_{3^*}u\right)_{x_i}\left(u_{tt}\right)_{x_j} dQ =$$

$$= \int_{Q} e^{-2k_1 t} \sum_{ij=1}^{n} A_{ij} u_{x_i tt} u_{x_j tt} + \frac{\bar{l}_{2^*}}{2} \sum_{ij=1}^{n} A_{ij} u_{tx_i} u_{tx_j} dQ +$$

$$+ \int_{\Omega} \frac{\bar{l}_{1^*}}{2} \sum_{ij=1}^{n} A_{ij} u_{x_i t} u_{x_j t}\Big|_{t=T} d\Omega \geq c \int_{Q} \sum_{ij=1}^{n} A_{ij} u_{x_i tt} u_{x_j tt} dQ +$$

$$+ \int_{Q} k_3 \sum_{ij=1}^{n} A_{ij} u_{tx_i} u_{tx_j} dQ + \int_{\Omega} \frac{k_3}{2} e^{-2k_1 T} \sum_{ij=1}^{n} A_{ij} u_{tx_i} u_{tx_j}\Big|_{t=T} d\Omega,$$

$$I_7 = \int_{Q} \sum_{ij=1}^{n} A_{ij}\left(D_{3^*}u\right)_{x_i}\left(A(u_t)\right)_{x_j} dQ =$$

$$= \int_{Q} \left(k_1 e^{-2k_1 t} + \bar{l}_{1^*} - \bar{l}_{2^*}(t-T)\right)\left(A(u_t)\right)^2 dQ +$$

$$+ \int_{\Omega} \frac{1}{2} e^{-2k_1 T} \left(A(u_t)\right)^2\Big|_{t=T} d\Omega,$$

$$I_8 = \int_{Q} \sum_{ij=1}^{n} A_{ij}\left(D_{3^*}u\right)_{x_i}\left(A(k_1 u)\right)_{x_j} = \int_{\Omega} k_1 e^{-2k_1 T} A(u) A(u_t)\Big|_{t=T} d\Omega -$$

$$- \int_{Q} k_1 e^{-2k_1 t}\left(A(u_t)\right)^2 dQ + \int_{\Omega}\left(k_1^2 e^{-2k_1 T} + k_1 \frac{\bar{l}_{1^*}}{2}\right)\left(A(u_t)\right)^2\Big|_{t=T} d\Omega +$$

$$+ \int_{Q} k_1 \left(2k_1^2 e^{-2k_1 t} + \frac{\bar{l}_{2^*}}{2}\right)\left(A(u)\right)^2 dQ \geq$$

$$\geq \int_\Omega e^{-2k_1T}\left(k_1 A(u)A(u_t) + \frac{k_1^2}{2}(A(u_t))^2 \right)\Big|_{t=T} d\Omega -$$

$$-\int_Q k_1 e^{-2k_1t}(A(u_t))^2 dQ,$$

$$I_9 = \int_Q \sum_{ij=1}^n A_{ij}(D_{3*}u)_{x_i}(k_2 u_t)_{x_j} dQ =$$

$$= \frac{1}{2}\int_\Omega k_2 e^{-2k_1T}\sum_{ij=1}^m A_{ij}u_{x_it}u_{x_jt}\Big|_{t=T} d\Omega +$$

$$+\int_Q k_2\left(k_1 e^{-2k_1t} + (\bar{l}_{1*} - \bar{l}_{2*}(t-T))\right)\sum_{ij=1}^n A_{ij}u_{x_it}u_{x_jt} dQ \geq 0,$$

$$I_{10} = \int_Q \sum_{ij=1}^n A_{ij}(D_{3*}u)_{x_i}(k_3 u)_{x_j} dQ =$$

$$= \int_\Omega k_3 e^{-2k_1T}\sum_{ij=1}^n A_{ij}u_{x_i}u_{x_jt}\Big|_{t=T} d\Omega - \int_Q k_3 e^{-2k_1t}\sum_{ij=1}^n A_{ij}u_{x_it}u_{x_jt} dQ +$$

$$+\int_\Omega k_3\left(k_1 e^{-2k_1t} + \frac{\bar{l}_{1*}}{2}\right)\sum_{ij=1}^n A_{ij}u_{x_i}u_{x_j}\Big|_{t=T} d\Omega +$$

$$+\int_Q k_3\left(2k_1^2 e^{-2k_1t} + \frac{\bar{l}_{2*}}{2}\right)\sum_{ij=1}^n A_{ij}u_{x_i}u_{x_j} dQ \geq$$

$$\geq \int_\Omega k_3 e^{-2k_1T}\left(\sum_{ij=1}^n A_{ij}u_{x_i}u_{x_jt} + \frac{1}{2}\sum_{ij=1}^n A_{ij}u_{x_i}u_{x_j} \right)\Big|_{t=T} d\Omega -$$

$$-\int_Q k_3 \sum_{ij=1}^n A_{ij}u_{x_it}u_{x_jt} dQ.$$

Finally, we have

$$\left(L_2^* v,u\right)_{W_2^{0,1}} \geq c\int_Q u_{tt}^2 + \sum_{i,j=1}^n A_{ij}u_{x_itt}u_{x_jtt}dQ +$$
$$+c\left(\|A(u_t)\|_{L_2}^2 + \|u_t\|_{L_2}^2\right)+$$
$$+\int_\Omega \frac{1}{2}k_3 e^{-2k_1T}\left(u_t^2 + 2uu_t + u^2\right)+$$
$$+\frac{1}{2}e^{-2k_1T}\left((A(u_t))^2 + 2k_1A(u_t)A(u) + k_1^2(A(u_t))^2\right)+$$
$$+\frac{1}{2}k_3e^{-2k_1T}\left(\sum_{i,j=1}^n A_{ij}\left(u_{x_it}+u_{x_i}\right)\left(u_{x_jt}+u_{x_j}\right)\right)+$$
$$+\frac{1}{2}e^{-2k_1T}\left(\sum_{i,j=1}^n A_{ij}\left(u_{x_it}+k_1u_{x_i}\right)\left(u_{x_jt}+k_1u_{x_j}\right)\right)\Bigg|_{t=T}d\Omega.$$

Applying the inequality of coercitivity as in Lemma 9, we obtain that the lemma holds true.

Corollary. *For an arbitrary function $v \in W_{3*}^+$ the following inequality holds true:*

$$c\|v\|_{W_2^{0,1}} \leq \left\|L_2^* v\right\|_{W_3^-}.$$

The proof is similar to the previous ones.

Thus, by Lemma 3 and the corollaries of Lemmas 9 and 10, we find that the following inequalities hold true

$$c^{-1}\|u\|_{W_2^{0,1}} \leq \|L_2u\|_{W_{3*}^-} \leq c\|u\|_{W_3^+}, \tag{42a}$$

$$c^{-1}\|v\|_{W_2^{0,1}} \leq \left\|L_2^*v\right\|_{W_3^-} \leq c\|v\|_{W_{3*}^+}, \tag{42b}$$

for an arbitrary functions $u \in W_3^+$ and $v \in W_{3*}^+$.

Definition 7. *A generalized solution of Problem 3 is such function $u(t,x) \in W_4^+$ that there exists a sequence of smooth in $\overline{Q}$ functions $u_i(t,x)$ satisfying the conditions (8) and*

$$\|u_i - u\|_{W_3^+} \xrightarrow[i\to\infty]{} 0, \quad \|Lu_i - f\|_{W_{3^*}^-} \xrightarrow[i\to\infty]{} 0$$

Definition 8. *A generalized solution of Problem 3 is such function* $u(t,x) \in W_2^{0,1}$ *that there exists a sequence of smooth in* $\overline{Q}$ *functions* $u_i(t,x)$ *satisfying the conditions (8) and*

$$\|u_i - u\|_{W_2^{0,1}} \xrightarrow[i\to\infty]{} 0, \quad \|Lu_i - f\|_{W_{3^*}^-} \xrightarrow[i\to\infty]{} 0$$

In a similar way the definitions for the adjoint Problem 4 are introduced.

Lemma 11. *Let* $u(t,x)$ *be an arbitrary smooth in* $\overline{Q}$ *function satisfying the conditions (8), and* $v = I_t^2 u$ *is the integral operator defined by the following expression:*

$$v(t,x) = -\int_T^t e^{-\eta} u(\eta, x)\, d\eta .$$

Then

$$(L_2 u, I_t^4 u)_{W_2^{0,1}} = (L_2 u, v)_{W_2^{0,1}} \geq c\|v\|_{W_{4^*}^+}^2 .$$

The proof is similar to the previous ones.

Corollary. *For an arbitrary function* $u \in W_4^+$ *the following inequality holds true*

$$c\|u\|_{H_4^+} \leq \|L_2 u\|_{W_{4^*}^-}$$

The proof is similar to the previous ones.

Lemma 12. *Let* $v(t,x)$ *be an arbitrary smooth in* $\overline{Q}$ *function satisfying the conditions (12), and* $u = I_t^{4^*} v$ *is the integral operator defined by the following expression*

$$u(t,x) = \int_0^t e^{\eta} v(\eta, x)\, d\eta .$$

Then

$$\left(L_2^* v, I_t^{2^*} v\right)_{W_2^{0,1}} = \left(L^*{}_2 v, u\right)_{W_2^{0,1}} \geq c\|u\|^2_{W_4^+}.$$

The proof is similar to the previous Lemma.

Corollary. *For an arbitrary function* $v(t,x) \in W^+_{4^*}$ *the following inequality holds true*

$$c\|v\|_{H^+_{4^*}} \leq \|L^* v\|_{W_4^-}.$$

By Lemma 4 and the corollaries of Lemmas 11 and 12 the following inequalities hold true

$$c^{-1}\|u\|_{H_4^+} \leq \|L_2 u\|_{W^-_{4^*}} \leq c\|u\|_{W_4^+}, \tag{43a}$$

$$c^{-1}\|v\|_{H^+_{4^*}} \leq \|L_2^* u\|_{W_4^-} \leq c\|u\|_{W^+_{4^*}}, \tag{43b}$$

Definition 9. *A generalized solution of Problem 3 is such function* $u(t,x) \in W_4^+$ *that there exists a sequence of smooth in* $\overline{Q}$ *functions* $u_i(t,x)$ *satisfying the conditions (8) and*

$$\|u_i - u\|_{W_4^+} \xrightarrow[i\to\infty]{} 0, \qquad \|Lu_i - f\|_{W^-_{4^*}} \xrightarrow[i\to\infty]{} 0$$

Definition 10. *A generalized solution of Problem 3 is such function* $u(t,x) \in H_4^+$ *that there exists a sequence of smooth in* $\overline{Q}$ *functions* $u_i(t,x)$ *satisfying the conditions (8) and*

$$\|u_i - u\|_{H_4^+} \xrightarrow[i\to\infty]{} 0, \qquad \|Lu_i - f\|_{W^-_{4^*}} \xrightarrow[i\to\infty]{} 0.$$

The definitions of Problem 4 are formulated in a similar way.

Using the results obtained above and applying the general theorems from Chapter 1, we obtain the following

Theorem 1. *For an arbitrary functions* $f \in L_2(Q)$ *there exists a unique solution* $u(t,x)$ *of Problem 1 in the sense of Definition 3.*

Theorem 2. *For an arbitrary function* $f \in W^-_{1^*}$ *there exists a unique solution* $u(t,x)$ *of Problem 1 in the sense of Definition 4.*

Theorem 3. *For an arbitrary function* $f \in H^{-}_{2^*}$ ($f \in W^{-}_{2^*}$*, respectively) there exists a unique solution* $u(t, x)$ *of Problem 1 in the sense of Definition 5 (Definition 6, respectively).*

Theorem 4. *For an arbitrary function* $f \in W^{0,1}_{2}$ ($f \in W^{-}_{3^*}$ *, respectively) there exists a unique solution* $u(t, x)$ *of Problem 3 in the sense of Definition 7 (Definition 8, respectively).*

Theorem 5. *For an arbitrary function* $f \in H^{-}_{4^*}$ ($f \in W^{-}_{4^*}$*, respectively) there exists a unique solution* $u(t, x)$ *of Problem 3 in the sense of Definition 9 (Definition 10, respectively).*

The proofs of these theorems follow from general Theorems 1.1.1 and 1.1.3

Similar theorems hold true for Problem 2 and 4, also.

3. ANALOGIES OF GALERKIN METHOD FOR PSEUDO-HYPERBOLIC SYSTEMS

Consider Galerkin's method for second boundary value problems for pseudo-hyperbolic systems, which was studied in paragraph 2 of the chapter. All denotations of the following correspond to paragraph 2 ones.

Let the right-hand side $f(t, x)$ of equation (2.1) be a smooth in $\overline{Q}$ function, that satisfies the condition: $f\big|_{t=0} = 0$. Consider approximate solution of problem (2.1) in the following form
where $g_i(t)$ is a solution of the Cauchy problem for the set of linear ordinary differential equations with constant coefficients.

$$\sum_{i=1}^{s}\left(\frac{d^2 g_i(t)}{dt^2}\left(\omega_i,\omega_j\right)_{L_2(\Omega)}+\right.$$

$$+\frac{dg_i(t)}{dt}\left(\sum_{kl=1}^{n}\left(A_{kl}\frac{\partial\omega_i}{\partial x_k},\frac{\partial\omega_j}{\partial x_l}\right)_{L_2(\Omega)}+\left(C\omega_i,\omega_j\right)_{L_2(\Omega)}\right)+$$

$$\left.+g_i(t)\left(\sum_{kl=1}^{n}\left(B_{kl}\frac{\partial\omega_i}{\partial x_k},\frac{\partial\omega_j}{\partial x_l}\right)_{L_2(\Omega)}+\left(D\omega_i,\omega_j\right)_{L_2(\Omega)}\right)\right)=\left(f,\omega_j\right)_{L_2(\Omega)},$$

$$j=\overline{1,s}, \tag{2}$$

$$g_i(0)=\frac{dg_i(0)}{dt}=0,\quad i=\overline{1,s}, \tag{3}$$

where $\{\omega_i(x)\}$ is a sequence of smooth in $\overline{\Omega}$ functions, that the set $\{\varphi(t)\omega_i(x)\}_{i=1}^{\infty}$ is total in $W_{1^*}^{+}$, $\varphi(t)$ is a smooth function: $\varphi(T)=\varphi_t(T)=0$.

Denote by H' the completion of set of smooth in $\overline{Q}$ functions that satisfy the conditions (2.4), in the norm

$$\|u\|_{H'}=\left(\int_Q u_{tt}^2+\sum_{i=1}^{n}u_{ttx_i}^2\,dQ\right)^{1/2}.$$

Lemma 2. *Consider the differential operator* $D_{1^*}(\cdot)$

$$D_{1^*}u=\exp(-2(\alpha_A\alpha_B)^{-1}t)u_{tt}+$$
$$+(l_{1^*}-l_{2^*}(t-T))u_t=a(t)u_{tt}+b(t)u_t$$

then for an arbitrary function $u(t,x)\in H'$ *the following inequality is true*

$$c_1\|u\|_{W_1^+}^2\le\left\langle Lu,D_{1^*}u\right\rangle_1\le c_2\|u\|_{H'}^2,$$

where by the bilinear form $\left\langle Lu,D_{1^*}u_1\right\rangle$ *we mean*

Denote by H' the completion of set of smooth in $\overline{Q}$ functions that satisfy the conditions (2.4), in the norm

$$\|u\|_{H'} = \left(\int_Q u_{tt}^2 + \sum_{i=1}^{n} u_{ttx_i}^2 dQ \right)^{1/2}.$$

Lemma 2. *Consider the differential operator* $D_{1^*}(\cdot)$

$$D_{1^*}u = \exp(-2(\alpha_A \alpha_B)^{-1} t) u_{tt} +$$
$$+ (l_{1^*} - l_{2^*}(t-T)) u_t = a(t) u_{tt} + b(t) u_t$$

then for an arbitrary function $u(t,x) \in H'$ *the following inequality is true*

$$c_1 \|u\|_{W_1^+}^2 \le \left\langle Lu, D_{1^*}u \right\rangle_1 \le c_2 \|u\|_{H'}^2,$$

where by the bilinear form $\left\langle Lu, D_{1^*}u_1 \right\rangle$ *we mean*

$$\left\langle Lu, D_{1^*}u \right\rangle_1 = (u_{tt} + Cu_t' + Du, D_{1^*}u)_{L_2(Q)} +$$
$$+ \sum_{k,l=1}^{n} (A_{kl} u_{x_k t} + B_{kl} u_{x_k}, D_{1^*} u_{x_l})_{L_2(Q)}.$$

Proof. Consider an arbitrary smooth in $\overline{Q}$ function $u(t,x)$, that satisfies the conditions (2.4). Applying he Schwarz inequality, we have

$$\left\langle Lu, D_{1^*}u \right\rangle_1 \le \|u_{tt}\|_{L_2(Q)} \|D_{1^*}u\|_{L_2(Q)} +$$
$$+ \|Cu_t\|_{L_2(Q)} \|D_{1^*}u\|_{L_2(Q)} + \|Du\|_{L_2(Q)} \|D_{1^*}u\|_{L_2(Q)} +$$
$$+ \sum_{k,l=1}^{n} \left(\|A_{kl} u_{x_k t}\|_{L_2(Q)} + \|B_{kl} u_{x_k}\|_{L_2(Q)} \right) \|D_{1^*} u_{x_l}\|_{L_2(Q)}.$$

Taking into account the continuity of the functions $C(x), D(x), A_{kl}(x), B_{kl}(x)$ in the domain $\overline{Q}$, and $a(t), b(t)$ in $[0,T]$, we have that each norm in the right-hand side not exceed the $M\|u\|_{H'}$, that is what had to be proved for the right-hand side of the desired inequality. To prove the left-hand side, consider the following expression $\left\langle Lu, D_{1^*}u \right\rangle_1$, where u is a smooth function. Transforming the $\left\langle Lu, D_{1^*}u \right\rangle_1$ in the much the same way as in Lemma 2.6, we have the desired inequality for smooth functions. To prove the inequality for all functions u form H' it is necessary to pass to the limit. Let $u(t,x)$ be a fuñction from H' and u_k be a sequence of smooth functions, that satisfy the conditions (2.4) and $\|u-u_k\|_{H'} \xrightarrow[k\to\infty]{} 0$. Since

$$\|u\|_{L_2(Q)} \le c_1 \|u_t\|_{L_2(Q)} \le c_2 \|u_{tt}\|_{L_2(Q)} \le c_3 \|u\|_{H'},$$
$$\|u_{x_k}\|_{L_2(Q)} \le c_1 \|u_{x_k t}\|_{L_2(Q)} \le c_2 \|u_{x_k tt}\|_{L_2(Q)} \le c_3 \|u\|_{H'},$$

then the sequences $\{u_k\}, \left\{\frac{\partial u_k}{\partial t}\right\}, \left\{\frac{\partial^2 u_k}{\partial t^2}\right\}, \left\{\frac{\partial u_k}{\partial x_l}\right\}, \left\{\frac{\partial^2 u_k}{\partial x_l \partial t}\right\}, \left\{\frac{\partial^3 u_k}{\partial x_l \partial t^2}\right\}$

converge to $u, \frac{\partial u}{\partial t}, \frac{\partial^2 u}{\partial t^2}, \frac{\partial u}{\partial x_l}, \frac{\partial^2 u}{\partial x_l \partial t}, \frac{\partial^3 u}{\partial x_l \partial t^2}$, respectively, in the space $L_2(Q)$ (by the derivatives we mean derivatives of the distributions). It is clear, that $\|u_k\|_{W_1^+} \xrightarrow[k\to\infty]{} \|u\|_{W_1^+}$, thus

$$\left|\left\langle Lu_k, D_{1^*}u_k \right\rangle_1 - \left\langle Lu, D_{1^*}u \right\rangle_1\right| \le$$
$$\le \left|\left\langle Lu_k - Lu, D_{1^*}u \right\rangle_1\right| + \left|\left\langle Lu_k, D_{1^*}u_k - D_{1^*}u \right\rangle_1\right| \xrightarrow[k\to\infty]{} 0.$$

It is what had to be proved.

Lemma 3. *Consider the differential operator* $D_{2^*}(\cdot)$ *(*$D_{2^*}u = e^{-t}u_t$*), then for an arbitrary function* $u(t,x) \in H'$ *the following inequality is true*

$$c_1 \|u\|^2_{W_2^+} \le \left\langle Lu, D_{2^*}u \right\rangle_1 \le c_2 \|u\|^2_{H'}.$$

P r o o f. Let $u(t,x)$ be a smooth in $\overline{Q}$ function, that satisfies the conditions (2.4). Consider

$$\langle Lu, D_{2^*}u\rangle_1 = I_1 + \ldots + I_5,$$

where

$$I_1 = (u_{tt}, e^{-t}u_t)_{L_2(Q)} = \frac{1}{2}\int_\Omega e^{-T}u_t^2\big|_{t=T}d\Omega + \frac{1}{2}\int_Q e^{-t}u_t^2 dQ \geq c\int_Q u_t^2 dQ,$$

$$I_2 = (Cu_t, e^{-t}u_t)_{L_2(Q)} \geq 0,$$

$$I_3 = (Du, e^{-t}u_t)_{L_2(Q)} = \frac{1}{2}\int_\Omega e^{-T}Du^2\big|_{t=T}d\Omega + \frac{1}{2}\int_Q e^{-t}Du^2 dQ \geq 0,$$

$$I_4 = \sum_{k,l=1}^n (A_{kl}u_{x_kt}, e^{-t}u_{x_lt})_{L_2(Q)} \geq \alpha_A\int_Q e^{-t}\sum_{i=1}^n u_{x_it}^2 dQ \geq c\int_Q\sum_{i=1}^n u_{x_it}^2 dQ,$$

$$I_5 = \sum_{k,l=1}^n (B_{kl}u_{x_k}, e^{-t}u_{x_lt})_{L_2(Q)} = \frac{1}{2}\int_\Omega e^{-T}\sum_{k,l=1}^n B_{kl}u_{x_k}u_{x_l}\big|_{t=T}d\Omega +$$

$$+\frac{1}{2}\int_Q e^{-t}\sum_{k,l=1}^n B_{kl}u_{x_k}u_{x_l}dQ \geq 0.$$

Thus,

$$\langle Lu, D_{2^*}u\rangle_1 \geq c\int_Q u_t^2 + \sum_{i=1}^n u_{x_it}^2 dQ \geq c_1\|u\|_{W_2^+}^2.$$

The verity of the inequality for all functions $u(t,x) \in H'$ we obtain by passing to the limit. The right-hand side of the inequality is proved in the same way as the previous one.

L e m m a 4. *For the functions* $u_s(t,x)$, *which have been defined in (1), the following inequality is true*

$$\|u_s\|_{H'} \leq c\left\|\frac{\partial f}{\partial t}\right\|_{H_{2^*}^-}.$$

Proof. Let $Tu_s = e^{-t}\dfrac{\partial^2 u_s}{\partial t^2}$. Prove that $\left\langle L\left(\dfrac{\partial u_s}{\partial t}\right), Tu_s\right\rangle_1 \geq c\|u_s\|^2_{H'}$.

Consider

$$\left\langle L\left(\frac{\partial u_s}{\partial t}\right), Tu_s\right\rangle_1 = I_1 + \ldots + I_5,$$

where

$$I_1 = (u_{s\,ttt}, e^{-t}u_{stt})_{L_2(Q)} =$$

$$= \frac{1}{2}\int_\Omega e^{-T}u_{s\,tt}^2\Big|_{t=T} d\Omega - \frac{1}{2}\int_\Omega u_{s\,tt}^2\Big|_{t=0} d\Omega + \frac{1}{2}\int_Q e^{-t}u_{s\,tt}^2 dQ,$$

$$I_2 = (Cu_{stt}, e^{-t}u_{stt})_{L_2(Q)} \geq 0,$$

$$I_3 = (Du_{st}, e^{-t}u_{s\,tt})_{L_2(Q)} =$$

$$= \frac{1}{2}\int_\Omega e^{-T}Du_{s\,t}^2\Big|_{t=T} d\Omega + \frac{1}{2}\int_Q e^{-t}Du_{s\,t}^2 dQ \geq 0,$$

$$I_4 = \sum_{k,l=1}^n (A_{kl}u_{s\,x_k tt}, e^{-t}u_{s\,x_l tt})_{L_2(Q)} \geq$$

$$\geq \alpha_A \int_Q e^{-t}\sum_{i=1}^n u_{s\,x_i tt}^2 dQ \geq c\int_Q \sum_{i=1}^n u_{s\,x_i tt}^2 dQ,$$

$$I_5 = \sum_{kl=1}^n (B_{kl}u_{s\,x_k t}, e^{-t}u_{s\,x_l tt})_{L_2(Q)} = \frac{1}{2}\int_\Omega e^{-T}\sum_{kl=1}^n B_{kl}u_{s\,tx_k}u_{s\,tx_l}\Big|_{t=T} d\Omega +$$

$$+\frac{1}{2}\int_Q e^{-t}\sum_{kl=1}^n B_{kl}u_{s\,tx_k}u_{s\,tx_l} dQ \geq 0.$$

Thus,

$$\langle Lu_{st}, Tu_s \rangle_1 \geq -\frac{1}{2}\int_\Omega u_{stt}^2\big|_{t=0} d\Omega + c\int_Q u_{stt}^2 + \sum_{i=1}^n u_{sx_itt}^2 dQ.$$

Prove, that $\int_\Omega u_{stt}^2\big|_{t=0} d\Omega = 0$. Multiplying the both right and left hand sides of (2) by $\frac{d^2 g_j}{dt^2}$ and summing up over j from 1 to s, we obtain

$$\left(u_{stt} + Cu_{st} + D_s u, u_{stt}\right)_{L_2(\Omega)} +$$

$$+\sum_{kl=1}^n \left(A_{kl}u_{sx_kt} + B_{kl}u_{sx_k}, u_{stt}\right)_{L_2(\Omega)} = \left(f, u_{stt}\right)_{L_2(\Omega)}.$$

Substitute $t = 0$ into the relation and take into account that $u_s(0) = \frac{du_s(0)}{dt} = f\big|_{t=0} = 0$, whence $\int_\Omega u_{stt}^2\big|_{t=0} d\Omega = 0$. Thus,

$$\langle Lu_{st}, Tu_s \rangle_1 \geq c\int_Q u_{stt}^2 + \sum_{i=1}^n u_{sx_itt}^2 dQ \geq \|u_s\|_{H'}^2.$$

Consider the relation (2) again. Differentiating the relation (2) with respect to t, multiplying by $e^{-t}\frac{d^2 g_j}{dt^2}$, summing up over j from 1 to s and integrating with respect to t from 0 to T, we have

$$\left\langle L\left(\frac{\partial u_s}{\partial t}\right), Tu_s\right\rangle_1 = \left(f_t, Tu_s\right)_{L_2(Q)} \leq \|f_t\|_{H_{2*}^-} \|Tu_s\|_{H_{2*}^+}.$$

From the previous inequality we find

$$c\|u_s\|_{H'}^2 \leq \|f_t\|_{H_{2*}^-} \|Tu_s\|_{H_{2*}^+}.$$

It remains to mark that $\|Tu_s\|_{H_{2*}^+} \leq \|u_s\|_{H'}$, which is what had to be proved.

By Lemma 4, we obtain that the sequence $\{u_s\}_{s=1}^{\infty}$ is bounded in the space H', thus there exists a function u^* and a subsequence $\{u_{s_k}\}_{k=1}^{\infty}$: $u_{s_k} \xrightarrow[k\to\infty]{} u^*$ weakly in H'. It is clear that the sequences of norm

$$\left\|\frac{\partial u_{s_k}}{\partial t}\right\|_{L_2(Q)}, \left\|\frac{\partial^2 u_{s_k}}{\partial t^2}\right\|_{L_2(Q)}, \left\|\frac{\partial u_{s_k}}{\partial x_i}\right\|_{L_2(Q)}, \left\|\frac{\partial^2 u_{s_k}}{\partial x_i \partial t}\right\|_{L_2(Q)}$$

are bounded (since $\{u_{s_k}\}_{k=1}^{\infty}$ is bounded in H'), and then there exists subsequence (which is denoted by u_{s_k} again), that

$$\frac{\partial u_{s_k}}{\partial t} \xrightarrow[k\to\infty]{} u_{,1}, \quad \frac{\partial^2 u_{s_k}}{\partial t^2} \xrightarrow[k\to\infty]{} u_{,2},$$

$$\frac{\partial u_{s_k}}{\partial x_i} \xrightarrow[k\to\infty]{} u_{,3}, \quad \frac{\partial^2 u_{s_k}}{\partial x_i \partial t} \xrightarrow[k\to\infty]{} u_{,4}$$

weakly in $L_2(Q)$. Since $u_{s_k} \xrightarrow[k\to\infty]{} u^*$ weakly, it is easy to prove that

$$u_{,1} = \frac{\partial u^*}{\partial t}, \quad u_{,2} = \frac{\partial^2 u^*}{\partial t^2}, \quad u_{,3} = \frac{\partial u^*}{\partial x_i}, \quad u_{,4} = \frac{\partial^2 u^*}{\partial x_i \partial t}, \tag{4}$$

where derivatives of the function u^* are understood in the sense of distributions.

Theorem 2. *For all smooth in the domain $\overline{Q}$ function $f(t,x)$: $f\big|_{t=0} = 0$ the approximate sequence $\{u_s\}_{s=1}^{\infty}$ converges to the solution of Problem 2.1 in the sense of Definition 2.3 in the norm of space W_1^+ and $\|Lu_s - f\|_{W_1^-} \xrightarrow[s\to\infty]{} 0$.*

P r o o f. Multiplying the relation (2) by

$$\exp(-2(\alpha_A\alpha_B)^{-1}t)\frac{d^2 g_j}{dt^2}+\left(l_{1^*}-l_{2^*}(t-T)\right)\frac{dg_j}{dt},$$

summing up over j from 1 to s and integrating with respect to t from 0 to T, we obtain

$$\left\langle Lu_s, D_{1^*}u_s\right\rangle_1 = \left(f, D_{1^*}u_s\right)_{L_2(Q)},$$

thus,

$$Lu_{s_k}, D_{1^*}u_{s_k 1} = \left(f, D_{1^*}u_{s_k}\right)_{L_2(Q)},$$

where the sequence $\left\{u_{s_k}\right\}_{k=1}^{\infty}$ converges weakly to u^* in the space H'. Taking into account the relation (4), we have that

$$D_{1^*}u_{s_k} \xrightarrow[k\to\infty]{} D_{1^*}u^* \text{ weakly in } L_2(Q),$$

so

$$\lim_{k\to\infty}\left\langle Lu_{s_k}, D_{1^*}u_{s_k}\right\rangle_1 = \lim_{k\to\infty}\left(f, D_{1^*}u_{s_k}\right)_{L_2(Q)} = \left(f, D_{1^*}u^*\right)_{L_2(Q)}. \quad (5)$$

From Lemma 2 we find

$$c\left\|u_{s_k}-u^*\right\|^2_{W_1^+} \le \left\langle L\left(u_{s_k}-u^*\right), D_{1^*}\left(u_{s_k}-u^*\right)\right\rangle_1 =$$

$$= \left\langle Lu_{s_k}, D_{1^*}u_{s_k}\right\rangle_1 - \left\langle Lu_{s_k}, D_{1^*}u^*\right\rangle_1 - \quad (6)$$

$$- \left\langle Lu^*, D_{1^*}u_{s_k}\right\rangle_1 + \left\langle Lu^*, D_{1^*}u^*\right\rangle_1.$$

Prove that $\left\langle Lu^*, D_{1^*}u^*\right\rangle_1 = \left(f, D_{1^*}u^*\right)_{L_2(Q)}$

Multiplying the equality (2) by a function $\varphi_j(t)$: $\varphi_j(T) = \frac{\partial\varphi_j}{\partial t}(T) = 0$, summing up over j form 1 to p and integrating with respect to t from 0 to T, we have

$$\left\langle Lu_{s_k}, \sum_{j=1}^{p}\varphi_j\omega_j\right\rangle_1 = \left(f, \sum_{j=1}^{p}\varphi_j\omega_j\right)_{L_2(Q)}, \qquad p=\overline{1, s_k}.$$

Let $v_p = \sum_{j=1}^{p} \varphi_j \omega_j$, then

$$\langle Lu_{s_k}, v_p \rangle_1 = (f, v_p)_{L_2(Q)}, \quad p = \overline{1, s_k}.$$

From the equalities (4), we obtain

$$\langle Lu_{s_k}, v_p \rangle_1 \xrightarrow[k\to\infty]{} \langle Lu^*, v_p \rangle_1.$$

Therefore,

$$\langle Lu^*, v_p \rangle_1 = (f, v_p)_{L_2(Q)}.$$

By virtue of the totality of the system $\{\varphi(t)\omega_i(x)\}_{i=1}^{\infty}$ in the space $H_{2^*}^+$, there exists sequence $v'_p \xrightarrow[p\to\infty]{} D_{1^*}u^*$ in space $H_{2^*}^+$, then

$$\langle Lu^*, v'_p \rangle_1 \xrightarrow[p\to\infty]{} \langle Lu^*, D_{1^*}u^* \rangle_1,$$

$$(f, v'_p)_{L_2(Q)} \xrightarrow[p\to\infty]{} (f, D_{1^*}u^*)_{L_2(Q)}.$$

That is why $\langle Lu^*, D_{1^*}u^* \rangle_1 = (f, D_{1^*}u^*)_{L_2(Q)}$. Returning to (6), we have

$$c\left\|u_{s_k} - u^*\right\|_{W_1^+}^2 \le \langle Lu_{s_k}, D_{1^*}u_{s_k} \rangle_1 - \langle Lu_{s_k}, D_{1^*}u^* \rangle_1 - \langle Lu^*, D_{1^*}u_{s_k} \rangle_1 + (f, D_{1^*}u^*)_{L_2(Q)}. \tag{7}$$

From (4), we find

$$\langle Lu_{s_k}, D_{1^*}u^* \rangle_1 \xrightarrow[k\to\infty]{} \langle Lu^*, D_{1^*}u^* \rangle_1 = (f, D_{1^*}u^*)_{L_2(Q)},$$

$$\langle Lu^*, D_{1^*}u_{s_k} \rangle_1 \xrightarrow[k\to\infty]{} \langle Lu^*, D_{1^*}u^* \rangle_1 = (f, D_{1^*}u^*)_{L_2(Q)}.$$

Passing in the (7) to the limit and taking into account (5)

$$c\lim_{k\to\infty}\left\|u_{s_k} - u^*\right\|_{W_1^+}^2 \le \lim_{k\to\infty}\langle Lu_{s_k}, D_{1^*}u_{s_k} \rangle_1 - \lim_{k\to\infty}(f, D_{1^*}u^*)_{L_2(Q)} = 0.$$

Therefore, there exists a function u^* and a sequence $\{u_{s_k}\}_{k=1}^{\infty}$, which converges to the function u^* in the norm of space W_1^+. Prove that the function u^* is a solution of Problem 2.1. Applying the inequality of Lemma 2.1, we have

$$\left\| Lu_{s_k} - Lu_{s_p} \right\|_{W_{1^*}^-} \leq c \left\| u_{s_k} - u_{s_p} \right\|_{W_1^+} \xrightarrow[k,p\to\infty]{} 0. \tag{8}$$

So that, the sequence Lu_{s_k} is fundamental in the complete space $W_{1^*}^-$, thus there exists an element $\hat{f} \in W_{1^*}^-$: $\left\| Lu_{s_k} - \hat{f} \right\|_{W_{1^*}^-} \xrightarrow[k\to\infty]{} 0.$

Prove that $f = \hat{f}$ in the space $W_{1^*}^-$. Multiply the equality (2) by a function $h(t) : h(T) = h'(T) = 0$, and let $\psi_j = h(t)\omega_j(x)$. Integrating the both right and left-hand sides of the equality (8) with respect to t from 0 to T, we have:

$$\left\langle Lu_{s_k}, \psi_j \right\rangle_1 = \left(f, \psi_j\right)_{L_2(Q)}.$$

Pass to the limit as $k \to \infty$:

$$\left\langle \hat{f}, \psi_j \right\rangle_{W_{1^*}} = \left(f, \psi_j\right)_{L_2(Q)}.$$

Since the totality of the set $\left\{\psi_j\right\}_{j=1}^{\infty}$ in the space $W_{1^*}^+$, we obtain $f = \hat{f}$. Using the convergence of the sequence $\left\{u_{s_k}\right\}_{k=1}^{\infty}$ and (8), it's easy to prove that the function u^* is a solution of Problem 2.1 with the right-hand side $f(t, x)$ in the sense of Definition 2.3. It remains to mark that by Theorem 2.1 the solution is unique, thus it is not necessary to choose the subsequence $\left\{u_{s_k}\right\}_{k=1}^{\infty}$. If there is an accumulation point of the sequence $\left\{u_s\right\}_{s=1}^{\infty}$ in fact that differ from u^*, then by the same reasoning Problem 2.1 has another solution.

Theorem 3. *For an arbitrary smooth in $\overline{Q}$ function $f(t, x) : f\big|_{t=0} = 0$ the approximate sequence $\left\{u_s\right\}_{s=1}^{\infty}$ converges to* *the solution of Problems 2.1 in the sense of Definition 2.5 in the norm of space* W_2^+ *and* $\|Lu_s - f\|_{W_{2^*}^-} \xrightarrow[s\to\infty]{} 0$.

Proof is analogous to the reasoning of previous theorem.

Now, consider the case $f \in L_2(Q)$. Consider the approximate sequence in the form of the relations (1)-(3).

Lemma 5. *The following inequality is valid*

$$\|u_s\|_{W_1^+} \le c\|f\|_{L_2}.$$

Proof. Reasoning as in the proof of Lemma 2.6, we prove that for a smooth in $\overline{Q}$ function $u(t,x)$, which satisfy conditions $u|_{t=0} = \frac{\partial u}{\partial t}|_{t=0} = 0$ the following inequality $c\|u\|_{W_1^+}^2 \le \langle Lu, D_{1^*}u\rangle_1$ is valid. Choose the sequence of smooth on $[0,T]$ functions $g_i^k(t)$: $g_i^k(0) = \frac{dg_i^k(0)}{dt} = 0$, that converges to the solution of the set of the differential equations in the space $H_2^2[0,T]$, where $H_2^2[0,T]$ is the Sobolev space of functions $v(t)$: $v(0) = v'(0) = 0$.

Prove that

$$\|u_s - \tilde{u}_s^k\|_{W_1^+} \xrightarrow[k\to\infty]{} 0,$$

where $\tilde{u}_s^k = \sum_{i=1}^{s} g_i^k(t)\omega_i(x)$.

Actually,

$$\left\|u_s - \tilde{u}_s^k\right\|_{W_1^+}^2 = \int_Q \left(\sum_{i=1}^{s} \left(g_i^k(t) - g_i(t)\right)_{tt} \omega_i(x) \right)^2 +$$

$$+ \sum_{j=1}^{n} \left(\sum_{i=1}^{s} \left(g_i^k(t) - g_i(t)\right)_t \frac{\partial \omega_i}{\partial x_j} \right)^2 dQ \le$$

$$\le c \sum_{i=1}^{s} \left(\int_Q \left(\left(g_i^k(t) - g_i(t)\right)_{tt} \omega_i(x) \right)^2 + \right.$$

$$\left. + \sum_{j=1}^{n} \left(\left(g_i^k(t) - g_i(t)\right)_t \frac{\partial \omega_i}{\partial x_j} \right)^2 dQ \right) =$$

$$= c \sum_{i=1}^{s} \left(\int_0^T \left(\left(g_i^k(t) - g_i(t)\right)_{tt} \right)^2 . dt \int_\Omega \left(\omega_i(x)\right)^2 d\Omega + \right.$$

$$\left. + c \sum_{j=1}^{n} \int_0^T \left(\left(g_i^k(t) - g_i(t)\right)_t \right)^2 dt \int_\Omega \left(\frac{\partial \omega_i}{\partial x_j} \right)^2 d\Omega \right) \le$$

$$\le c_1 \sum_{i=1}^{s} \left\| g_i^k - g_i \right\|_{\overset{\circ}{H}{}_2^2[0,T]}^2 \xrightarrow[k \to \infty]{} 0.$$

Therefore, $\left\|u_s - \tilde{u}_s^k\right\|_{W_1^+} \xrightarrow[k \to \infty]{} 0$.

Since the function $\tilde{u}_s^k = \sum_{i=1}^{s} g_i^k(t)\omega_i(x)$ is smooth in $\overline{Q}$, then Lemma 2 is valid. Thus, $c\left\|\tilde{u}_s^k\right\|_{W_1^+}^2 \le \left\langle L\tilde{u}_s^k, D_{1*}\tilde{u}_s^k \right\rangle_1$. We have

$$\begin{aligned} \left\langle Lu_s, D_{1*}u_s \right\rangle_1 &= \left\langle L\left(u_s - \tilde{u}_s^k\right), D_{1*}u_s \right\rangle_1 + \\ &+ \left\langle L\tilde{u}_s^k, D_{1*}\left(u_s - \tilde{u}_s^k\right) \right\rangle_1 + \left\langle L\tilde{u}_s^k, D_{1*}\tilde{u}_s^k \right\rangle_1 \ge \\ &\ge \left\langle L\left(u_s - \tilde{u}_s^k\right), D_{1*}u_s \right\rangle_1 + \left\langle L\tilde{u}_s^k, D_{1*}\left(u_s - \tilde{u}_s^k\right) \right\rangle_1 + c\left\|\tilde{u}_s^k\right\|_{W_1^+}^2 . \end{aligned} \tag{9}$$

It is easy to show that

$$\left\langle L\left(u_s - \tilde{u}_s^k\right), D_{1^\bullet} u_s \right\rangle_1 \xrightarrow[k\to\infty]{} 0,$$
$$\left\langle L\tilde{u}_s^k, D_{1^\bullet}\left(u_s - \tilde{u}_s^k\right)\right\rangle \xrightarrow[k\to\infty]{} 0,$$

thus passing to the limit in the inequality (9) as $k \to \infty$, we obtain that

$$\left\langle Lu_s, D_{1^\bullet} u_s \right\rangle_1 \geq c\left\|u_s\right\|^2_{W_1^+}.$$

Now, we return to the relations (2). Multiplying the relations (2) by

$$\exp(-2(\alpha_A\alpha_B)^{-1}t)\frac{d^2 g_j}{dt^2} + (l_{1^\bullet} - l_{2^\bullet}(t-T))\frac{dg_j}{dt},$$

summing up over j from 1 to s and integrating with respect to t from 0 to T, we obtain

$$\left\langle Lu_s, D_{1^\bullet} u_s \right\rangle_1 = \left(f, D_{1^\bullet} u_s\right)_{L_2(Q)} \leq \|f\|_{L_2(Q)}\left\|D_{1^\bullet} u_s\right\|_{L_2(Q)}.$$

Whence,

$$c\left\|u_s\right\|^2_{W_1^+} \leq \|f\|_{L_2(Q)}\left\|D_{1^\bullet} u_s\right\|_{L_2(Q)}.$$

To prove the lemma, it suffices to mark that $\left\|D_{1^\bullet} u_s\right\|_{L_2(Q)} \leq c\left\|u_s\right\|_{W_1^+}$. The lemma is proved.

Theorem 4. *For an arbitrary function $f(t,x) \in L_2(Q)$ the approximate sequence $\{u_s\}_{i=1}^{\infty}$ converges to the solution of Problem 2.1 in the sense of Definition 2.3 in the norm of space W_1^+ and $\left\|Lu_s - f\right\|_{W_{1^*}^-} \xrightarrow[s\to\infty]{} 0$.*

Proof. Choose the sequence of smooth in $\overline{Q}$ functions f_m: $f_m|_{t=0} = 0$ such that $\left\|f_m - f\right\|_{L_2(Q)} \xrightarrow[m\to\infty]{} 0$. Let u_s^m be the approximate sequence (1) of the problem with the right-hand side f_m. By Theorem 2, the sequence u_s^m converges to the solution $\overline{u}^m$ as $s \to \infty$. Prove that the sequence $\overline{u}^m$ is fundamental in the space W_1^+. In fact, by Lemma 5 we have

$$\left\|\overline{u}^m-\overline{u}^n\right\|_{W_1^+}\le\left\|\overline{u}^m-u_s^m\right\|_{W_1^+}+\left\|u_s^m-u_s^n\right\|_{W_1^+}+\left\|u_s^n-\overline{u}^n\right\|_{W_1^+}\le$$
$$\le\left\|\overline{u}^m-u_s^m\right\|_{W_1^+}+c\left\|f_m-f_n\right\|_{L_2(Q)}+\left\|u_s^n-\overline{u}^n\right\|_{W_1^+}.$$

By Theorem 2,

$$\left\|u_s^n-\overline{u}^n\right\|_{W_1^+}\xrightarrow[s\to\infty]{}0$$

and

$$\left\|u_s^m-\overline{u}^m\right\|_{W_1^+}\xrightarrow[s\to\infty]{}0.$$

From the other hand, since $\left\|f_m-f\right\|_{L_2(Q)}\xrightarrow[m\to\infty]{}0$, the sequence f_m is fundamental, whence $\left\|\overline{u}^m-\overline{u}^n\right\|_{W_1^+}\xrightarrow[m,n\to\infty]{}0$. Therefore, there exists such element $\overline{u}\in L_2(Q)$, that $\left\|\overline{u}^m-\overline{u}\right\|_{W_1^+}\xrightarrow[m\to\infty]{}0$.

Consider

$$\left\|u_s-\overline{u}\right\|_{W_1^+}\le\left\|u_s-u_s^m\right\|_{W_1^+}+\left\|u_s^m-\overline{u}^m\right\|_{W_1^+}+\left\|\overline{u}^m-\overline{u}\right\|_{W_1^+}.$$

By Lemma 4

$$\left\|u_s-u_s^m\right\|_{W_1^+}\le c\left\|f-f_m\right\|_{L_2(Q)}\xrightarrow[m\to\infty]{}0.$$

Thus,

$$\left\|u_s-\overline{u}\right\|_{W_1^+}\le c\left\|f-f_m\right\|_{L_2(Q)}+\left\|u_s^m-\overline{u}^m\right\|_{W_1^+}+\left\|\overline{u}^m-\overline{u}\right\|_{W_1^+}.$$

Approaching to the limit as $s\to\infty$, we have

$$\overline{\lim_{s\to\infty}}\left\|u_s-\overline{u}\right\|_{W_1^+}\le c\left\|f-f_m\right\|_{L_2(Q)}+\left\|\overline{u}^m-\overline{u}\right\|_{W_1^+}.$$

Making $m\to\infty$, we have

$$\lim_{s\to\infty}\left\|u_s-\overline{u}\right\|_{W_1^+}=0. \tag{10}$$

Prove that $\overline{u}(t,x)$ is a solution of Problem 2.1 with the right-hand side $f(t,x)$ in the sense of Definition 2.3:

$$\left\|Lu_s-f\right\|_{W_{1^*}^-}\le\left\|Lu_s-Lu_s^m\right\|_{W_{1^*}^-}+\left\|Lu_s^m-f_m\right\|_{W_{1^*}^-}+\left\|f_m-f\right\|_{W_{1^*}^-}.$$

Prove that every summand in the right-hand side vanishes. By Lemmas 2.6 and 5, we have

$$\left\|Lu_s - Lu_s^m\right\|_{W_{1^*}^-} \le c\left\|u_s - u_s^m\right\|_{W_1^+} \le c^2\left\|f - f_m\right\|_{L_2(Q)} \xrightarrow[m\to\infty]{} 0 .$$

By Theorem 2, $\left\|Lu_s^m - f_m\right\|_{W_{1^*}^-} \xrightarrow[s\to\infty]{} 0$. Thus,

$$\overline{\lim_{s\to\infty}}\left\|Lu_s - f\right\|_{W_{1^*}^-} \le (c^2+1)\left\|f_m - f\right\|_{W_{1^*}^-} \xrightarrow[m\to\infty]{} 0 . \quad (11)$$

Applying (10) and (11), it is easy to prove that $\bar{u}$ is a solution of Problem 2.1 by Definition 2.3. Now the theorem is proved.

Let $f \in H_{2^*}^-$ and the approximate sequence of solution $\{u_s\}$ is defined in (1)-(3) (in the case the integrals are defined in the sense of distribution theory).

Lemma 6. The following inequalities is valid

$$\left\|u_s\right\|_{W_2^+} \le c\left\|f\right\|_{H_{2^*}^-} .$$

The lemma can be proved in much the same way as Lemma 5 (it is necessary to substitute the space W_1^+ for the space W_2^+, the operator D_{1^*} for D_{2^*}, and to apply the Schwarz inequality to the right-hand side of the following expression

$$\left\langle Lu_s, D_{2^*}u_s\right\rangle_1 = \left\langle f, D_{2^*}u_s\right\rangle_{H_{2^*}} .$$

Theorem 5. *For every function $f \in H_{2^*}^-$ the approximate sequence $\{u_s\}_{s=1}^{\infty}$ converges to the solution of Problem 2.1 in the sense of Definition 2.5 in the norm of the space W_2^+ and* $\left\|Lu_s - f\right\|_{W_{2^*}^-} \xrightarrow[s\to\infty]{} 0$.

The Proof is completely analogous to that of Theorem 4.

Now, let $f(t,x)$ be an element of the negative space $W_{1^*}^-$, f_m be a sequence of functions from $L_2(Q)$ such that $\|f_m - f\|_{W_{1^*}^-} \xrightarrow[m\to\infty]{} 0$.

Consider the approximate sequence $u_{s,m}(t,x)$ of the form

$$u_{s,m}(t,x) = \sum_{i=1}^{s} g_{i,m}(t)\omega_i(x), \tag{12}$$

where functions $g_{i,m}(t)$ are solutions of the following system

$$\begin{aligned}
&\sum_{i=1}^{s}\left(\frac{d^2 g_{i,m}(t)}{dt^2}(\omega_i,\omega_j)_{L_2(\Omega)} + \right.\\
&+\frac{dg_{i,m}(t)}{dt}\left(\sum_{k,l=1}^{n}\left(A_{kl}\frac{\partial\omega_i}{\partial x_k},\frac{\partial\omega_j}{\partial x_l}\right)_{L_2(\Omega)} + (C\omega_i,\omega_j)_{L_2(\Omega)}\right)+\\
&\left.+g_{i,m}(t)\left(\sum_{k,l=1}^{n}\left(B_{kl}\frac{\partial\omega_i}{\partial x_k},\frac{\partial\omega_j}{\partial x_l}\right)_{L_2(\Omega)} + (D\omega_i,\omega_j)_{L_2(\Omega)}\right)\right)=\\
&=(f_m,\omega_j)_{L_2(\Omega)}, j=\overline{1,s}
\end{aligned} \tag{13}$$

$$g_{i,m}(0) = \frac{dg_{i,m}(0)}{dt} = 0, \qquad i=\overline{1,s}, \tag{14}$$

Theorem 6. *Let ε_m be an arbitrary number sequence $\varepsilon_m > 0, \varepsilon_m \xrightarrow[m\to\infty]{} 0$. Therefore, for each integer $s(m)$: $\|Lu_{s(m),m} - f_m\|_{W_{1^*}^-} < \varepsilon_m$, (which exists necessarily), and for each right-hand side $f \in W_{1^*}^-$, the approximate sequence $u_{s(m),m}$ converges to the solution of Problem 2.1 in the sense of Definition 2.6 in the norm of space $L_2(Q)$.*

Proof. By Theorem 4 the approximate sequence $u_{s,m}(t,x)$ converges to the solution of the following equation $Lu = f_m$ with $s \to \infty$, then there exists such function $\overline{u}^m \in W_1^+$, that

$$\left\| u_{s,m} - \overline{u}^m \right\|_{W_1^+} \xrightarrow[s\to\infty]{} 0, \quad \left\| Lu_{s,m} - f_m \right\|_{W_{1^*}^-} \xrightarrow[s\to\infty]{} 0.$$

Prove that the sequence $\overline{u}^m \in W_1^+$ is fundamental in the space $L_2(Q)$.

$$\left\| \overline{u}^m - \overline{u}^n \right\|_{L_2(Q)} \le \left\| \overline{u}^m - u_{s,m} \right\|_{L_2(Q)} + \left\| u_{s,m} - u_{s,n} \right\|_{L_2(Q)} + \left\| u_{s,n} - \overline{u}^n \right\|_{L_2(Q)}.$$

Using Lemma 2.5, we have

$$\left\| \overline{u}^m - \overline{u}^n \right\|_{L_2(Q)} \le c\left\| \overline{u}^m - u_{s,m} \right\|_{W_1^+} + c\left\| Lu_{s,m} - Lu_{s,n} \right\|_{W_{1^*}^-} + c\left\| u_{s,n} - \overline{u}^n \right\|_{W_1^+} \le c\left\| \overline{u}^m - u_{s,m} \right\|_{W_1^+} + c\left\| Lu_{s,m} - f_m \right\|_{W_{1^*}^-} + c\left\| f_m - f_n \right\|_{W_{1^*}^-} + c\left\| f_n - Lu_{s,n} \right\|_{W_{1^*}^-} + c\left\| u_{s,n} - \overline{u}^n \right\|_{W_1^+}.$$

Pass to the limit as $s \to \infty$. Then,

$$\left\| \overline{u}^m - \overline{u}^n \right\|_{L_2(Q)} \le c\left\| f_m - f_n \right\|_{W_{1^*}^-} \xrightarrow[m,n\to\infty]{} 0,$$

and there exists a function $\overline{u}(t,x) \in L_2(Q)$: $\left\| \overline{u}^m - \overline{u} \right\|_{L_2(Q)} \xrightarrow[m\to\infty]{} 0$.

Consider

$$\left\| Lu_{s,m} - f \right\|_{W_{1^*}^-} \le \left\| Lu_{s,m} - f_m \right\|_{W_{1^*}^-} + \left\| f_m - f \right\|_{W_{1^*}^-}.$$

Choose $s = s(m)$ such that $\left\| Lu_{s(m),m} - f_m \right\|_{W_{1^*}^-} \le \varepsilon_m$, (by Theorem 4, the number $s = s(m)$ exists). Thus,

$$\left\| Lu_{s(m),m} - f \right\|_{W_{1^*}^-} \le \varepsilon_m + \left\| f_m - f \right\|_{W_{1^*}^-} \xrightarrow[m\to\infty]{} 0,$$

and

$$\left\|u_{s(m),m} - \overline{u}\right\|_{L_2(Q)} \le \left\|u_{s(m),m} - \overline{u}^m\right\|_{L_2(Q)} + \left\|\overline{u}^m - \overline{u}\right\|_{L_2(Q)} \le$$

$$\le c\left\|Lu_{s(m),m} - f_m\right\|_{W_{1^*}^-} + \left\|\overline{u}^m - \overline{u}\right\|_{L_2(Q)} \le$$

$$\le c\varepsilon_m + \left\|\overline{u}^m - \overline{u}\right\|_{L_2(Q)} \xrightarrow[m\to\infty]{} 0,$$

which is what had to be proved.

If $f(t, x)$ be an element of the negative space $W_{2^*}^-$, then choose an arbitrary sequence of functions $f_m \in H_{2^*}^-$: $\left\|f_m - f\right\|_{W_{2^*}^-} \xrightarrow[m\to\infty]{} 0$. Approximate sequence $u_{s,m}(t, x)$ is defined in by the relations (12)-(14). In this case the following analogous of Theorem 6 can be proved.

Theorem 7. *Let* ε_m *be an arbitrary number sequence* $\varepsilon_m > 0, \varepsilon_m \xrightarrow[m\to\infty]{} 0$. *Therefore, for each integer* $s(m)$ *:* $\left\|Lu_{s(m),m} - f_m\right\|_{W_{2^*}^-} < \varepsilon_m$, *(which exists necessarily), and for each right-hand side* $f \in W_{2^*}^-$, *the approximate sequence* $u_{s(m),m}$ *converges to the solution of Problem 2.1 in the sense of Definition 2.6 in the norm of space* H_2^+.

Let the right-hand side $f(t, x)$ of the equation (2.1) is a smooth in $\overline{Q}$ function, that satisfies the following condition $f\big|_{t=0} = 0$. Approximate solution can be found in the form

$$u_s(t,x) = \sum_{i=1}^{s} g_i(t)\omega_i(x), \tag{15}$$

where the function $g_i(t)$ is a solution of the Cauchy problem for the set of differential equations with constant coefficients:

$$\sum_{i=1}^{s}\left(a_{ij}\frac{d^2 g_i(t)}{dt^2} + b_{ij}\frac{dg_i(t)}{dt} + c_{ij}g_i(t)\right) = \\ = \left(f, \omega_j + A(\omega_j)\right)_{L_2(\Omega)}, j = \overline{1,s}, \tag{16}$$

$$g_i(0) = \frac{dg_i(0)}{dt} = 0, \qquad i = \overline{1,s}, \tag{17}$$

where

$$a_{ij} = \left(\omega_i, \omega_j\right)_{L_2(\Omega)} + \sum_{k,l=1}^{n}\left(A_{kl}\frac{\partial\omega_i}{\partial x_k}, \frac{\partial\omega_j}{\partial x_l}\right)_{L_2(\Omega)},$$

$$b_{ij} = k_2\left(\omega_i, \omega_j\right)_{L_2(\Omega)} + k_2\sum_{k,l=1}^{n}\left(A_{kl}\frac{\partial\omega_i}{\partial x_k}, \frac{\partial\omega_j}{\partial x_l}\right)_{L_2(\Omega)} + \\ + \sum_{k,l=1}^{n}\left(A_{kl}\frac{\partial\omega_i}{\partial x_k}, \frac{\partial\omega_j}{\partial x_l}\right)_{L_2(\Omega)} + \left(A(\omega_i), A(\omega_j)\right)_{L_2(\Omega)}, \tag{18}$$

$$c_{ij} = k_3\left(\omega_i, \omega_j\right)_{L_2(\Omega)} + k_3\sum_{k,l=1}^{n}\left(A_{kl}\frac{\partial\omega_i}{\partial x_k}, \frac{\partial\omega_j}{\partial x_l}\right)_{L_2(\Omega)} + \\ + k_1\sum_{k,l=1}^{n}\left(A_{kl}\frac{\partial\omega_i}{\partial x_k}, \frac{\partial\omega_j}{\partial x_l}\right)_{L_2(\Omega)} + k_1\left(A(\omega_i), A(\omega_j)\right)_{L_2(\Omega)},$$

where $\{\omega_i(x)\}$ is a sequence of smooth in $\overline{\Omega}$ functions, which satisfy the condition $\frac{\partial\omega_i}{\partial\overline{\mu}_A}\Big|_{x\in\partial\Omega} = 0$, and the set $\{\varphi(t)\omega_i(x)\}_{i=1}^{\infty}$ is total in $W_{3^*}^{+}$, where $\varphi(t)$ is an arbitrary smooth function: $\varphi(T) = \varphi_t(T) = 0$.

Denote by H'' the completion of the set of smooth in $\overline{Q}$ functions, which satisfy the condition (2.4), in the norm

$$\|u\|_{H''} = \left(\int_Q u_{tt}^2 + \sum_{i=1}^{n} u_{ttx_i}^2 + \sum_{i,j=1}^{n} u_{ttx_ix_j}^2\, dQ\right)^{1/2}.$$

Lemma 7. *Let* $D_{3^*}(\cdot)$ *be the following differential operator:*

$$D_{3^*}u = \exp(-2k_1 t)\, u_{tt} + (l_{1^*} - l_{2^*}(t-T))\, u_t ,$$

then for every function $u(t,x) \in H''$ *the following inequality is valid*

$$c_1 \|u\|^2_{W_3^+} \leq \left\langle L_2 u, D_{3^*} u \right\rangle_2 \leq c_2 \|u\|^2_{H^*},$$

where by $\left\langle L_2 u, D_{3^*} u \right\rangle_2$ *we mean*

$$\left\langle L_2 u, D_{3^*} u \right\rangle_2 = (u_{tt} + k_2 u_t + k_3 u, D_{3^*} u)_{W_2^{0,1}} +$$
$$+ \left(A(u_t) + k_1 A(u), D_{3^*} u\right)_{L_2(Q)} + \left(A(u_t) + k_1 A(u), A\left(D_{3^*} u\right)\right)_{L_2(Q)}.$$

Proof. Let $u(t,x)$ be a smooth in $\overline{Q}$ function, which satisfy the condition (2.8). Then, the right-hand side of the required inequality we obtain by applying partial integration. To prove the left-hand side, it is necessary to repeat the transformation of Lemma 2.10 with the expression $\left\langle Lu, D_{3^*} u \right\rangle_2$. The required inequality in the space H'' can be obtained by passing to the limit.

Lemma 8. *Let* $D_{4^*}(\cdot)$ *be the following differential operator:* $D_{4^*}u = e^{-t} u_t$, *then for every function* $u(t,x) \in H''$ *the following inequality is valid*

$$c_1 \|u\|^2_{W_4^+} \leq \left\langle Lu_2, D_{4^*} u \right\rangle_2 \leq c_2 \|u\|^2_{H^*}.$$

Proof is analogous to that of previous lemma.

Lemma 9. *For functions* $u_s(t,x)$, *which were defined in (15)-(17), the inequality is true*

$$\|u_s\|_{H^*} \leq c \left\| \frac{\partial f}{\partial t} \right\|_{H^-_{4^*}}.$$

Proof. Prove that

$$\left\langle L_2\left(\frac{\partial u_s}{\partial t}\right), Tu_s \right\rangle_2 \geq c\|u_s\|^2_{H''}.$$

Consider

$$\left\langle L_2\left(\frac{\partial u_s}{\partial t}\right), Tu_s \right\rangle_2 = I_1 + \ldots + I_5,$$

where

$$I_1 = (u_{sttt}, e^{-t}u_{stt})_{L_2(Q)} + \sum_{k,l=1}^{'n}\left(A_{kl}u_{sx_ltt}, e^{-t}u_{sx_ktt}\right)_{L_2(Q)}.$$

Transform each summand as I_1 and I_4 in Lemma 4. Hence,

$$I_1 = (u_{sttt}, e^{-t}u_{stt})_{L_2(Q)} + \sum_{k,l=1}^{n}\left(A_{kl}u_{sx_ltt}, e^{-t}u_{sx_ktt}\right)_{L_2(Q)} \geq$$

$$\geq c\int_Q u^2_{stt} + \sum_{i=1}^{n} u^2_{sx_ktt}dQ - \frac{1}{2}\int_\Omega u^2_{stt}\big|_{t=0}d\Omega,$$

$$I_2 = (k_2 u_{stt} + k_3 u_{st} + k_1 A(u_{st}), e^{-t}u_{stt})_{L_2(Q)} \geq 0,$$

$$I_3 = \sum_{k,l=1}^{n}(A_{kl}u_{sx_kttt}, e^{-t}u_{sx_ltt})_{L_2(Q)} = \frac{1}{2}\int_\Omega e^{-T}\sum_{k,l=1}^{n}A_{kl}u_{sx_ktt}u_{sx_ltt}\big|_{t=T}d\Omega -$$

$$-\frac{1}{2}\int_\Omega\sum_{k,l=1}^{n}A_{kl}u_{sx_ktt}u_{sx_ltt}\big|_{t=0}d\Omega + \frac{1}{2}\int_Q e^{-t}\sum_{k,l=1}^{n}A_{kl}u_{sx_ktt}u_{sx_ltt}dQ \geq$$

$$\geq -\frac{1}{2}\int_\Omega\sum_{k,l=1}^{n}A_{kl}u_{sx_ktt}u_{sx_ltt}\big|_{t=0}d\Omega,$$

$$I_4 = \int_Q e^{-t}k_2\sum_{k,l=1}^{n}A_{kl}u_{sx_ktt}u_{sx_ltt}dQ + \int_Q e^{-t}k_3\sum_{k,l=1}^{n}A_{kl}u_{sx_kt}u_{sx_ltt}dQ \geq$$

$$\geq \frac{1}{2}\int_\Omega e^{-T}k_3\sum_{k,l=1}^{n}A_{kl}u_{sx_kt}u_{sx_lt}\big|_{t=T}d\Omega +$$

$$+\frac{1}{2}\int_Q e^{-t}k_3\sum_{k,l=1}^{n}A_{kl}u_{sx_kt}u_{sx_lt}dQ \geq 0,$$

$$I_5 = \int_Q \left(A(u_{s_{tt}}) + k_1 A(u_{s_t}) \right) \left(A\left(e^{-t} u_{s_{tt}}\right) \right) dQ = \int_Q e^{-t} \left(A(u_{s_{tt}}) \right)^2 dQ +$$

$$+ \int_Q e^{-t} k_1 A(u_{s_t}) A\left(u_{s_{tt}}\right) dQ = \int_Q e^{-t} \left(A(u_{s_{tt}}) \right)^2 dQ +$$

$$+ \frac{1}{2} \int_\Omega e^{-T} k_1 \left(A(u_{s_t}) \right)^2 \Big|_{t=T} d\Omega + \frac{1}{2} \int_Q e^{-t} k_1 \left(A(u_{s_t}) \right)^2 dQ \geq$$

$$\geq c \int_Q e^{-t} \left(A(u_{s_{tt}}) \right)^2 dQ.$$

Therefore, we have

$$\left\langle L_2\left(\frac{\partial u_s}{\partial t} \right), Tu_s \right\rangle_2 \geq c \int_Q u_{s\,tt}^2 + \left(A\left(u_{s_{tt}}\right)\right)^2 dQ -$$

$$- \frac{1}{2} \int_\Omega u_{s\,tt}^2 + \sum_{i,j=1}^n A_{ij} u_{sttx_i} u_{sttx_j} \Big|_{t=0} d\Omega.$$

In a way analogous to Lemma 4, we prove that

$$\int_\Omega u_{s\,tt}^2 + \sum_{i,j=1}^n A_{ij} u_{sttx_i} u_{sttx_j} \Big|_{t=0} d\Omega = 0.$$

Taking into account the inequality of coercivity, we obtain

$$\left\langle L_2\left(\frac{\partial u_s}{\partial t} \right), Tu_s \right\rangle_2 \geq c \int_Q u_{s\,tt}^2 + \sum_{i=1}^n u_{s\,ttx_i}^2 + \sum_{ij=1}^n u_{s\,ttx_ix_j}^2 \, dQ \geq c \|u\|_{H''}^2.$$

Differentiating the equality (16) with respect to t, multiplying on $e^{-t} \dfrac{\partial^2 g_j}{\partial t^2}$, summing up over j from 1 to s and integrating with respect to t from 0 to T, we have

$$\left\langle L_2\left(\frac{\partial u_s}{\partial t}\right), Tu_s\right\rangle_2 = (f_t, Tu_s)_H \le \|f_t\|_{H^-_{4^*}} \|Tu_s\|_{H^+_{4^*}}.$$

Whence, we find

$$c\|u_s\|^2_{H^*} \le \|f_t\|_{H^-_{4^*}} \|Tu_s\|_{H^+_{4^*}}.$$

It suffices to remark that $\|Tu_s\|_{H^+_{4^*}} \le c\|u_s\|_{H''}$.

Theorem 8. *For all smooth in $\overline{Q}$ functions $f(t,x): f|_{t=0} = 0$ the approximate sequence $\{u_s\}_{s=1}^{\infty}$ (see (15)-(17)) converges to the solution of Problem 2.3 in the sense of Definition 2.7 in the norm of space W_3^+ and $\|L_2 u_s - f\|_{W^-_{3^*}} \xrightarrow[s\to\infty]{} 0$.*

Theorem 9. *For all smooth in $\overline{Q}$ functions $f(t,x): f|_{t=0} = 0$ the approximate sequence $\{\dot{u}_s\}_{s=1}^{\infty}$ converges to the solution of Problem 2.3 in the sense of Definition 2.9 in the norm of space W_4^+ and $\|L_2 u_s - f\|_{W^-_{4^*}} \xrightarrow[s\to\infty]{} 0$.*

Let $f(t,x)$ be an arbitrary function from $W_2^{0,1}$, and approximate sequence $\{u_s\}_{s=1}^{\infty}$ is defined by the equalities (15)-(17). Using the results of Lemma 2.10 and repeating the proof of Lemma 5, the following lemma can be proved.

Lemma 10. *The following inequality is true*

$$\|u_s\|_{W_3^+} \le c\|f\|_{W_2^{0,1}}.$$

Theorem 10. *For all functions $f(t,x) \in W_2^{0,1}$ the approximate sequence $\{u_s\}_{i=1}^{\infty}$ converges to the solution of Problem 2.3 in the sense of Definition 2.7 in the norm of space W_3^+ and $\|L_2 u_s - f\|_{W^-_{3^*}} \xrightarrow[s\to\infty]{} 0$.*

In the case when the function $f(t,x)$ is from the space $H^-_{4^*}$, the analogous results of convergence of the approximate solution (15)-(17) can be obtained.

Lemma 11. *The following inequality is true*

$$\|u_s\|_{W_4^+} \le c\|f\|_{H^-_{4^*}}.$$

Theorem 11. *For all function $f(t,x) \in H^-_{4^*}$ the approximate sequence $\{u_s\}_{i=1}^{\infty}$ converges to the solution of Problem 2.3 in the sense of Definition 2.9 in the norm of the space W_4^+ and $\|L_2 u_s - f\|_{W^-_{4^*}} \xrightarrow[s\to\infty]{} 0$.*

Let $f(t,x)$ be an arbitrary function from $W^-_{3^*}$. Choose a sequence of functions f_m from $W_2^{0,1}$ such that $\|f_m - f\|_{W^-_{3^*}} \xrightarrow[m\to\infty]{} 0$, and consider an approximate sequence of the solution

$$u_{s,m}(t,x) = \sum_{i=1}^{s} g_{i,m}(t)\omega_i(x), \tag{19}$$

where $g_i(t)$ is a solution of the Cauchy problem for the set of linear differential equations with constant coefficients

$$\sum_{i=1}^{s}\left(a_{ij}\frac{d^2 g_{i,m}(t)}{dt^2} + b_{ij}\frac{dg_{i,m}(t)}{dt} + c_{ij}g_{i,m}(t)\right) = \\ = \left(f_m, \omega_j + A(\omega_j)\right)_{L_2(\Omega)}, j=\overline{1,s} \tag{20}$$

$$g_{i,m}(0) = \frac{dg_{i,m}(0)}{dt} = 0, \quad i=\overline{1,s}, \tag{21}$$

where the constants a_{ij}, b_{ij}, c_{ij} were defined in (18), the sequence $\{\omega_i(x)\}$ satisfies the same conditions as in (15)-(18).

Theorem 12. *Let ε_m be an arbitrary number sequence $\varepsilon_m > 0, \varepsilon_m \xrightarrow[m\to\infty]{} 0$. Therefore, for each integer $s(m)$.*

$\left\|L_2 u_{s(m),m} - f_m\right\|_{W_{3^*}^-} < \varepsilon_m$, *(which exists necessarily), and for each right-hand side* $f \in W_{3^*}^-$, *the approximate sequence* $u_{s(m),m}$ *converges to the solution of Problem 2.3 in the sense of Definition 2.8 in the norm of space* $W_2^{0,1}$.

Analogously for $f(t,x) \in W_{4^*}^-$, we have.

Theorem 13. *Let* ε_m' *be an arbitrary number sequence* $\varepsilon_m > 0, \varepsilon_m \xrightarrow[m\to\infty]{} 0$. *Therefore, for each integer* $s(m)$ *:* $\left\|L_2 u_{s(m),m} - f_m\right\|_{W_{4^*}^-} < \varepsilon_m$, *(which exists necessarily), and for every right-hand side* $f \in W_{4^*}^-$, *the approximate sequence* $u_{s(m),m}$ *converges to the solution of Problem 2.3 in the sense of Definition 2.10 in the norm of space* H_4^+.

4. PULSE OPTIMAL CONTROL OF PSEUDO-HYPERBOLIC SYSTEMS (THE DIRICHLET INITIAL BOUNDARY VALUE PROBLEM)

Apply the obtained results to optimization of the pseudo-hyperbolic systems. The denotations are the same as previous.

Consider the pseudo-hyperbolic equation

$$Lu = f + A(h), \tag{1}$$

$$u\big|_{t=0} = \frac{\partial u}{\partial t}\Big|_{t=0} = 0; \quad u\big|_{x\in\partial\Omega} = 0. \tag{2}$$

where $L(\cdot)$ is a one of the pseudo-hyperbolic operator $L_i(\cdot)$ (see Section 1).

Using the template theorems od Section 3.6, complete the tables for the pseudo-hyperbolic equation (the first boundary value problem).

Table 1.

N	Operator	Space N	Space $W^{-}(Q)$
1.	$L_1(\cdot)$	H^{+}_{bd}	$W^{-l}_{bd^+}$
2.	$L_1(\cdot)$	W^{+l}_{bd}	$H^{-}_{bd^+}$
3.	$L_3(\cdot)$	$L_2(Q)$	$\overline{W}^{-l}_{bd^+}$
4.	$L_3(\cdot)$	$\overline{W}^{+l}_{bd}$	$L_2(Q)$

Table 2.

N	Operator $A_i(\cdot)$	Space $W^{-}(Q)$
1.	$A_1(\cdot)$	$W^{-l}_{bd^+}, \overline{W}^{-l}_{bd^+}$
2.	$A_2(\cdot)$, $k=1$	$\overline{W}^{-l}_{bd^+}$
3.	$A_3(\cdot)$	$W^{-l}_{bd^+}, \overline{W}^{-l}_{bd^+}, H^{-}_{bd^+}$
4.	$A_4(\cdot)$	$W^{-l}_{bd^+}, \overline{W}^{-l}_{bd^+}$
5.	$A_5(\cdot)$	$W^{-l}_{bd^+}, \overline{W}^{-l}_{bd^+}, H^{-}_{bd^+}$

Table 3.

N	Operator $L(\cdot)$	Space $W^{-}(Q)$	Space $W^{+}(Q)$
1.	$L_1(\cdot)$	$W^{-l}_{bd^+}$	W^{+l}_{bd}
2.	$L_3(\cdot)$	$\overline{W}^{-l}_{bd^+}$	$\overline{W}^{+l}_{bd}$

Table 4.

N	Operator $L(\cdot)$	Space $W^{-}(Q)$	Space H^{+}
1.	$L_1(\cdot)$	$W^{-l}_{bd^+}$	H^{+}_{bd}
2.	$L_3(\cdot)$	$\overline{W}^{-l}_{bd^+}$	$L_2(Q)$

Table 5.

N	Operator $L(\cdot)$	Space $W^{-}(Q)$	Space $W^{+}(Q)$
1.	$L_1(\cdot)$	$W^{-l}_{bd^+}$	$W^{+l}_{bd^+}$
2.	$L_3(\cdot)$	$\overline{W}^{-l}_{bd^+}$	$\overline{W}^{+l}_{bd^+}$

Table 6.

N	Space $W^{-}(Q)$	Map $A_i(\cdot)$
1.	$\overline{W}^{-l}_{bd^+}$	$A_1(\cdot)$

Table 7.

N	Operator $L(\cdot)$	Space $W^{-}(Q)$
1.	$L_1(\cdot)$	$W^{-l}_{bd^+}$
2.	$L_3(\cdot)$	$\overline{W}^{-l}_{bd^+}$

Table 8.

N	Operator $A_i(\cdot)$	Space $W^-(Q)$
1.	$A_1(\cdot)$	$W^{-l}_{bd^+}, \overline{W}^{-l}_{bd^+}$
2.	$A_2(\cdot)$, $k=1$	$\overline{W}^{-l}_{bd^+}$
3.	$A_3(\cdot)$	$W^{-l}_{bd^+}, \overline{W}^{-l}_{bd^+}, H^-_{bd^+}$
4.	$A_4(\cdot)$	$W^{-l}_{bd^+}, \overline{W}^{-l}_{bd^+}$
5.	$A_5(\cdot)$	$W^{-l}_{bd^+}, \overline{W}^{-l}_{bd^+}, H^-_{bd^+}$

Table 9.

N	Exponent α	Space $W^-(Q)$	Map $A_{i,\varepsilon}(\cdot)$
1.	1/2	$W^{-l}_{bd^+}, \overline{W}^{-l}_{bd^+}$	$A_{1,\varepsilon}(\cdot)$
2.	1	$\overline{W}^{-l}_{bd^+}$	$A_{1,\varepsilon}(\cdot)$
3.	1/2	$W^{-l}_{bd^+}, \overline{W}^{-l}_{bd^+}$	$A_{2,\varepsilon}(\cdot)$
4.	1	$\overline{W}^{-l}_{bd^+}$	$A_{2,\varepsilon}(\cdot)$
5.	1/2	$W^{-l}_{bd^+}, \overline{W}^{-l}_{bd^+}, H^-_{bd^+}$	$A_{3,\varepsilon}(\cdot)$
6.	1/2	$W^{-l}_{bd^+}, \overline{W}^{-l}_{bd^+}$	$A_{4,\varepsilon}(\cdot)$
7.	1/2	$W^{-l}_{bd^+}, \overline{W}^{-l}_{bd^+}, H^-_{bd^+}$	$A_{5,\varepsilon}(\cdot)$

5. PULSE OPTIMAL CONTROL OF PSEUDO-HYPERBOLIC SYSTEMS (THE NEUMANN INITIAL BOUNDARY VALUE PROBLEM)

Apply the results of Section 3.6 to optimization of the pseudo-hyperbolic systems (the Neumann initial boundary value problem). The denotations are the same as in Section 2 of this chapter.

Consider the pseudo-hyperbolic equation

$$Lu = f + A(h), \tag{1}$$

where $L(\cdot)$ is a one of the pseudo-hyperbolic operator $L_i(\cdot)$ (see Section 1) with corresponding boundary conditions:

$$u\big|_{t=0} = \frac{\partial u}{\partial t}\Big|_{t=0} = 0\,; \quad \frac{\partial_L u}{\partial \vec{n}}\Big|_{x\in\partial\Omega} = 0 \text{ or } \frac{\partial u}{\partial \vec{\mu}_A}\Big|_{x\in\partial\Omega} = 0 \tag{2}$$

Using the template theorems of Section 3.6, complete the tables for the pseudo-hyperbolic equation (the Neumann initial boundary value problem).

Table 1.

N	Operator	Space N	Space $W^{-}(Q)$
1.	$L(\cdot)$	$L_2(Q)$	W_{1*}^{-}
2.	$L(\cdot)$	H_2^{+}	W_{2*}^{-}
3.	$L_2(\cdot)$	$W_2^{0,1}$	W_{3*}^{-}
4.	$L_2(\cdot)$	H_4^{+}	W_{4*}^{-}
5.	$L(\cdot)$	W_1^{+}	$L_2(Q)$
6.	$L(\cdot)$	W_2^{+}	H_{2*}^{-}
7.	$L_2(\cdot)$	W_3^{+}	$W_2^{0,1}$
8.	$L_2(\cdot)$	W_4^{+}	H_{4*}^{-}

Table 2.

N	Operator $A_i(\cdot)$	Space $W(Q)$
1.	$A_1(\cdot)$	$W_{1*}^-, W_{2*}^-, W_{3*}^-, W_{4*}^-$
2.	$A_2(\cdot)$, $k=1$	W_{1*}^-
3.	$A_2(\cdot)$, $k=1, \varphi_i \in W_2^1(\Omega)$	W_{1*}^-, W_{3*}^-
4.	$A_3(\cdot)$	$W_{1*}^-, W_{2*}^-, H_{2*}^-$
5.	$A_4(\cdot)$	W_{1*}^-, W_{2*}^-
6.	$A_5(\cdot)$	$W_{1*}^-, W_{2*}^-, H_{2*}^-$
7.	$A_6(\cdot)$	W_{1*}^-

Table 3.

N	Operator $L(\cdot)$	Space $W^-(Q)$	Space $W^+(Q)$
1.	$L(\cdot)$	W_{1*}^-	W_1^+
2.	$L(\cdot)$	W_{2*}^-	W_2^+
3.	$L_2(\cdot)$	W_{3*}^-	W_3^+
4.	$L_2(\cdot)$	W_{4*}^-	W_4^+

Table 4.

N	Operator $L(\cdot)$	Space $W^-(Q)$	Space H^+
1.	$L(\cdot)$	W_{1*}^-	$L_2(Q)$
2.	$L(\cdot)$	W_{2*}^-	H_2^+
3.	$L_2(\cdot)$	W_{3*}^-	$W_2^{0,1}$
4.	$L_2(\cdot)$	W_{4*}^-	H_4^+

Table 5.

N	Operator $L(\cdot)$	Space $W^-(Q)$	Space $W^+(Q)$
1.	$L(\cdot)$	W^-_{1*}	W^+_{1*}

Table 6.

N	Space $W^-(Q)$	Map $A_i(\cdot)$
1.	W^-_{1*}	$A_1(\cdot)$

Table 7.

N	Operator $L(\cdot)$	Space $W^-(Q)$
1.	$L(\cdot)$	W^-_{1*}

Table 8.

N	Operator $A_i(\cdot)$	Space $W^-(Q)$
1.	$A_1(\cdot)$	$W^-_{1*}, W^-_{2*}, W^-_{3*}, W^-_{4*}$
2.	$A_2(\cdot)$, $k=1$	W^-_{1*}
3.	$A_2(\cdot)$, $k=1, \varphi_i \in W_2^1(\Omega)$	W^-_{1*}, W^-_{3*}
4.	$A_3(\cdot)$	$W^-_{1*}, W^-_{2*}, H^-_{2*}$
5.	$A_4(\cdot)$	W^-_{1*}, W^-_{2*}
6.	$A_5(\cdot)$	$W^-_{1*}, W^-_{2*}, H^-_{2*}$
7.	$A_6(\cdot)$	W^-_{1*}

Table 9.

N	Exponent α	Space $W^{-}(Q)$	Map $A_{i,\varepsilon}(\cdot)$
1.	1/2	$W_{1^*}^{-}, W_{2^*}^{-}, W_{3^*}^{-}, W_{4^*}^{-}$	$A_{1,\varepsilon}(\cdot)$
2.	1	$W_{1^*}^{-}, W_{3^*}^{-},$	$A_{1,\varepsilon}(\cdot)$
3.	1/2	$W_{1^*}^{-}, W_{2^*}^{-}, W_{3^*}^{-}, W_{4^*}^{-}$	$A_{2,\varepsilon}(\cdot)$
4.	1	$W_{1^*}^{-}, W_{3^*}^{-},$	$A_{2,\varepsilon}(\cdot)$
5.	½	$W_{1^*}^{-}, W_{2^*}^{-}, H_{2^*}^{-}$	$A_{3,\varepsilon}(\cdot)$
6.	½	$W_{1^*}^{-}, W_{3^*}^{-},$	$A_{4,\varepsilon}(\cdot)$
7.	½	$W_{1^*}^{-}, W_{2^*}^{-}, H_{2^*}^{-}$	$A_{5,\varepsilon}(\cdot)$

Remark that if we consider the problem (1), (2) with operator $L_2(\cdot)$, other theorems can be proved, because of the theorems of the solvability of the equation $L_2 u = F$ ensure smoother solution when the right-hand side belongs to some positive spaces.

Let $\overline{W}_{3^*}^{-}$ be a negative space corresponding to the pair of spaces $W_{3^*}^{+}$, $L_2(Q)$. Consider the problem of optimal control of the system (1), (2) with the right-hand side $F \in W_{2^*}^{-}$. It requires to minimize the performance criterion

$$J(h) = \Phi(u(h)) = \sum_{i=1}^{p} \int_{Q} \alpha_i(t,x)(u(h) - u_i)^2 \, dQ \tag{4}$$

where $u_i(t,x), \alpha_i(t,x)$ is functions from $W_2^{0,1}$ and $C^1(\overline{Q})$, and $\alpha_i(t,x) \geq \varepsilon > 0$ in $\overline{Q}$.

Theorem 10. *Let the state function* $u(t,x)$ *satisfy the problem (1), (2) with the right-hand side* $f(h) \in W_{2^*}^{-}$.

Performance criterion is in the form (4). If there exists a Fréchet derivative $f_{h^*}(\cdot)$ *of the map* $f(\cdot): H \to \overline{W}_{3^*}^-$, *then the performance criterion* $J(h)$ *has a Fréchet derivative at the point* h^* *in the following form*

$$J_{h^*}(\cdot) = \langle f_{h^*}(\cdot), v \rangle_{\overline{W}_{3^*}}, \tag{5}$$

where $\langle \cdot, \cdot \rangle_{\overline{W}_{3^*}}$ *is a bilinear form on* $\overline{W}_{3^*}^- \times W_{3^*}^+$, *and* $v(t,x)$ *is a solution of the adjoint problem*

$$L_2^* v = 2\sum_{i=1}^{p} \alpha_i \left(u(h^*) - u_i\right), \tag{6}$$

$$v\big|_{t=T} = \frac{\partial v}{\partial t}\Big|_{t=T} = 0, \frac{\partial v}{\partial \mu_A}\Big|_{x \in \partial\Omega} = 0. \tag{7}$$

P r o o f. Prove, that $v(t,x) \in W_{3^*}^+$. Since $f(h) \in W_{2^*}^-$, by Theorem 2.3 the solution $u(h^*)$ belongs to H_2^+. Since the spaces H_2^+ and $W_2^{0,1}$ have the equivalent norms, then we can consider that $2\sum_{i=1}^{p} \alpha_i \left(u(h^*) - u_i\right) \in W_2^{0,1}$. Applying the analogue of Theorem 2.4 for the adjoint operator, we have $v(t,x) \in W_{3^*}^+$. Thus $J_{h^*}(\cdot) = \langle f_{h^*}(\cdot), v \rangle_{\overline{W}_{3^*}} : H \to R$ is a linear continuous functional on H. From the analogue of Theorem 2.4 we have that for all functions $\overline{u} \in W_3^+$ the following equality is true

$$\langle L_2^* v, \overline{u} \rangle_{W_{3^*}} = \left(2\sum_{i=1}^{p} \alpha_i \left(u(h^*) - u_i\right), \overline{u} \right)_{W_2^{0,1}}. \tag{8}$$

In the same way as Lemmas 2.1-2.4, we can prove that for all functions $v(t,x) \in W_{3^*}^+$

$$\left\| L_2^* v \right\|_{H_4^-} \le c \left\| v \right\|_{W_{3^*}^+},$$

thus $L_2^* v \in H_4^-$. Therefore the equality (8) is valid for all $\overline{u} \in H_4^+$. Taking account that $L_2^* v \in W_2^{0,1}$, we have

$$\left\langle L_2^* v, \overline{u} \right\rangle_{H_4} = \left(2\sum_{i=1}^{p} \alpha_i \left(u(h^*) - u_i \right), \overline{u} \right)_{W_2^{0,1}}. \tag{9}$$

In a way analogous to that was made in Theorem 1.4.3, we obtain

$$J(h^* + \Delta h) - J(h^*) =$$
$$= \int_Q \Delta u \sum_{i=1}^{p} \alpha_i(t,x) \left(2u(h^*) - 2u_i \right) dQ + \int_Q (\Delta u)^2 \sum_{i=1}^{p} \alpha_i(t,x) dQ, \tag{10}$$

where $\Delta u = u(h^* + \Delta h) - u(h^*)$. It is clear that the function Δu satisfies the equation

$$L_2 \Delta u = \Delta f,$$
$$u\big|_{t=0} = \frac{\partial u}{\partial t}\Big|_{t=0} = 0, \frac{\partial \Delta u}{\partial \mu_A}\Big|_{x\in\partial\Omega} = 0,$$

where $\Delta f = f(h^* + \Delta h) - f(h^*) \in W_{2*}^-$.

By Theorem 2.3, we have that $\Delta u \in H_2^+$. Since $v \in W_{3*}^+$ and

$$\left\| L_2^* v \right\|_{L_2} \le c \|v\|_{W_{3*}^+},$$

we have that

$$L_2^* v \in L_2(Q)$$

and

$$\left(\Delta u, L_2^* v \right)_{L_2(Q)} = \langle \Delta f, v \rangle_{W_{2*}} \tag{11}$$

Consider the equation

$$A(\Delta \overline{u}) + \Delta \overline{u} = \Delta u,$$
$$\frac{\partial \Delta \overline{u}}{\partial \mu_A}\Big|_{x\in\partial\Omega} = 0.$$

As is well known from the theory of elliptic equations, the solution of the equation exists and has the degree of the smoothness with respect to the space argument two orders higher than the right-hand side of the equation Δu. Thus,

$$\left(\Delta\overline{u}, L_2^* v\right)_{L_2(Q)} = \left(A(\Delta\overline{u}) + \Delta\overline{u}, L_2^* v\right)_{L_2(Q)}$$

and it is easy to see that

$$\left(A(\Delta\overline{u}) + \Delta\overline{u}, L_2^* v\right)_{L_2(Q)} = \left\langle \Delta\overline{u}, L_2^* v\right\rangle_{H_4}.$$

Therefore,

$$\left(\Delta u, L_2^* v\right)_{L_2(Q)} = \left\langle \Delta\overline{u}, L_2^* v\right\rangle_{H_4}.$$

Substituting $\overline{u} = \Delta\overline{u}$ into (9) and taking into account (11), we have

$$\left(2\sum_{i=1}^{p} \alpha_i \left(u(h^*) - u_i\right), \Delta\overline{u}\right)_{W_2^{0,1}} = \langle \Delta f, v\rangle_{W_{2^*}}.$$

Since $W_{3^*}^+ \subset W_{2^*}^+$, we obtain

$$\langle \Delta f, v\rangle_{\overline{W}_{3^*}} = \langle \Delta f, v\rangle_{W_{2^*}} = \left(2\sum_{i=1}^{p} \alpha_i (u(h^*) - u_i), \Delta\overline{u}\right)_{W_2^{0,1}} =$$

$$= \int_Q 2\sum_{i=1}^{p} \alpha_i (u(h^*) - u_i)\Delta\overline{u} + \sum_{k,l=1}^{n} A_{kl}\left(2\sum_{i=1}^{p} \alpha_i (u(h^*) - u_i)\right)_{x_k} \Delta\overline{u}_{x_l}\, dQ.$$

Using the condition $\left.\dfrac{\partial \Delta\overline{u}}{\partial \mu_A}\right|_{x\in\partial\Omega} = 0$, we have

$$\langle \Delta f, v\rangle_{\overline{W}_{3^*}} = \langle \Delta f, v\rangle_{W_{2^*}} =$$

$$= \int_Q 2\sum_{i=1}^{p} \alpha_i (u(h^*) - u_i)\Delta\overline{u} + \sum_{kl=1}^{n} A_{kl}\left(2\sum_{i=1}^{p} \alpha_i (u(h^*) - u_i)\right)_{x_k} \Delta\overline{u}_{x_l}\, dQ =$$

$$= \int_Q 2\sum_{i=1}^{p} \alpha_i (u(h^*) - u_i)\Delta\overline{u} + 2\sum_{i=1}^{p} \alpha_i (u(h^*) - u_i) A(\Delta\overline{u})\, dQ =$$

$$=\left(2\sum_{i=1}^{p}\alpha_i(u(h^*)-u_i), A(\Delta\overline{u})+\Delta\overline{u}\right)_{L_2(Q)} =$$

$$=\left(2\sum_{i=1}^{p}\alpha_i(u(h^*)-u_i), \Delta u\right)_{L_2(Q)} .$$

Returning to (10), we have that

$$J(h^*+\Delta h)-J(h^*)=\langle\Delta f, v\rangle_{\overline{W}_{3^*}}+\int_Q(\Delta u)^2\sum_{i=1}^{p}\alpha_i(t,x)dQ .$$

The further part of the proof repeats the reasoning of Theorem 2.1.3.

Theorem 11. *If the mapping* $f(\cdot): H \to W_{2^*}^-$ *has a continuous Fréchet derivative at a point* h^* $\left(\forall\varepsilon>0\exists\delta>0\colon \forall h\in U_{ad}\left\|h-h^*\right\|<\delta\Rightarrow\left\|F_h(\cdot)-F_{h^*}(\cdot)\right\|<\varepsilon\right)$, *then the derivative* $J_h(\cdot)$ *is continuous at the point* h^*.

Proof is analogous to that of Theorem 2.1.4.

Theorem 12. *If mapping* $f(\cdot): H \to W_{2^*}^-$ *has a Fréchet derivative in a neighbourhood of a point* h^*, *that satisfies Lipschitz condition with exponent* α, $0<\alpha\le 1$ $\left(\exists C_1>0\colon\forall h_1, h_2\right.$ *from the neighbourhood of the point* h^* $\left.\left\|f_{h_1}(\cdot)-f_{h_2}(\cdot)\right\|\le C_1\left\|h_1-h_2\right\|_H^{\alpha}\right)$, *then the Fréchet derivative* $J_h(\cdot)$ *satisfies Lipschitz condition with the same exponent* α.

Proof is analogous to that of Theorem 2.1.5.

Consider the case of another type of performance criterion. Let

$$J(h) = \Phi(u(h)) = \sum_{i=1}^{p_1} \int_Q \alpha_i(t,x)\left(u(h) - u_i\right)^2 dQ +$$
$$+ \sum_{i=p_1+1}^{p} \int_Q \alpha_i(t,x) \sum_{kl=1}^{n} A_{kl}\left(u(h) - u_i\right)_{x_k}\left(u(h) - u_i\right)_{x_l} dQ, \tag{12}$$

where $u_i(t,x)$ are known functions from H_2^+, that satisfy the condition $\frac{\partial u_i}{\partial \overline{\mu}_A}\Big|_{x\in\partial\Omega} = 0, \alpha_i(t,x) \in C^1(\overline{Q})$ and $\alpha_i(t,x) \geq \varepsilon > 0$ in $\overline{Q}$.

Consider the pseudo-hyperbolic equation

$$L_1 u \equiv \frac{\partial^2 u}{\partial t^2} + A\left(\frac{\partial u}{\partial t} + k_1 u\right) + C(x)\frac{\partial u}{\partial t} + D(x)u\,, \tag{13}$$

$$u\big|_{t=0} = \frac{\partial u}{\partial t}\Big|_{t=0} = 0, \quad \frac{\partial u}{\partial \overline{\mu}_A}\Big|_{x\in\partial\Omega} = 0\,. \tag{14}$$

Consider the case when $f(\cdot): H \to W_{2^*}^-$. By Theorem 1, the functional (12) is defined correctly and if the conditions of the theorem are valid the optimal control exists.

Theorem 13. *Consider the problem (13), (14) with right-hand side $f(\cdot): H \to W_{2^*}^-$. Performance criterion is in the form (12). If there exists a Fréchet derivative $f_{h^*}(\cdot): H \to W_{2^*}^-$ of the mapping $f(\cdot): H \to W_{2^*}^-$ at a point h^*, then there exists a Fréchet defivative of performance criterion $J(h)$ at the same point h^*, in the form*

$$J_{h^*}(\cdot) = \left\langle f_{h^*}(\cdot), v\right\rangle_{W_{2^*}}, \tag{15}$$

where $v(t,x)$ is a solution of the adjoint problem

$$L_1^* v = 2\sum_{i=1}^{p_1} \alpha_i\left(u(h^*) - u_i\right) + 2\sum_{i=p_1+1}^{p} A\left(\alpha_i\left(u(h^*) - u_i\right)\right), \tag{16}$$

$$v\big|_{t=T} = \frac{\partial v}{\partial t}\Big|_{t=T} = 0, \quad \frac{\partial v}{\partial \bar{\mu}_A}\Big|_{x\in\partial\Omega} = 0. \tag{17}$$

P r o o f. We shall give the increment Δh to the control h^*

$$J(h^* + \Delta h) - J(h^*) = \sum_{i=1}^{p_1} \int_Q \alpha_i(t,x)\left(u(h^* + \Delta h) - u_i\right)^2 dQ +$$

$$+ \sum_{i=p_1+1}^{p} \int_Q \alpha_i(t,x) \sum_{k,l=1}^{n} A_{kl}\left(u(h^* + \Delta h) - u_i\right)_{x_k} \left(u(h^* + \Delta h) - u_i\right)_{x_l} dQ -$$

$$- \sum_{i=1}^{p_1} \int_Q \alpha_i(t,x)\left(u(h^*) - u_i\right)^2 dQ -$$

$$- \sum_{i=p_1+1}^{p} \int_Q \alpha_i(t,x) \sum_{k,l=1}^{n} A_{kl}\left(u(h^*) - u_i\right)_{x_k} \left(u(h^*) - u_i\right)_{x_l} dQ =$$

$$\sum_{i=1}^{p_1} \int_Q \alpha_i(t,x)\left(u(h^* + \Delta h) - u(h^*)\right)\left(u(h^* + \Delta h) + u(h^*) - 2u_i\right) dQ +$$

$$+ \sum_{i=p_1+1}^{p} \int_Q \alpha_i(t,x) \sum_{k,l=1}^{n} A_{kl}\left(u(h^* + \Delta h) - u(h^*)\right)_{x_k} \cdot$$

$$\cdot \left(u(h^* + \Delta h) + u(h^*) - 2u_i\right)_{x_l} dQ -$$

$$- \sum_{i=p_1+1}^{p} \int_Q \alpha_i(t,x) \sum_{k,l=1}^{n} A_{kl}\left(u(h^* + \Delta h) - u_i\right)_{x_k} \left(u(h^*) - u_i\right)_{x_l} dQ +$$

$$+ \sum_{i=p_1+1}^{p} \int_Q \alpha_i(t,x) \sum_{k,l=1}^{n} A_{kl}\left(u(h^* + \Delta h) - u_i\right)_{x_l} \left(u(h^*) - u_i\right)_{x_k} dQ.$$

Since the symmetry of matrix $\left\{A_{ij}(x)\right\}_{ij=1}^{n}$, each of the last summands is equal one another.

$$J(h^* + \Delta h) - J(h^*) =$$

$$\sum_{i=1}^{p_1} \int_Q \alpha_i(t,x) \big(u(h^* + \Delta h) - u(h^*)\big) \big(u(h^* + \Delta h) + u(h^*) - 2u_i\big) dQ +$$

$$+ \sum_{i=p_1+1}^{p} \int_Q \alpha_i(t,x) \sum_{k,l=1}^{n} A_{kl} \big(u(h^* + \Delta h) - u(h^*)\big)_{x_k} \cdot$$

$$\cdot \big(u(h^* + \Delta h) + u(h^*) - 2u_i\big)_{x_l} dQ =$$

$$= \sum_{i=1}^{p_1} \int_Q \alpha_i(t,x) \Delta u 2 \big(u(h^*) - u_i\big) dQ + \sum_{i=1}^{p_1} \int_Q \alpha_i(t,x) \Delta u^2 dQ +$$

$$+ 2 \sum_{i=p_1+1}^{p} \int_Q \alpha_i(t,x) \sum_{k,l=1}^{n} A_{kl} (\Delta u)_{x_k} \big(u(h^*) - u_i\big)_{x_l} dQ +$$

$$+ \sum_{i=p_1+1}^{p} \int_Q \alpha_i(t,x) \sum_{k,l=1}^{n} A_{kl} (\Delta u)_{x_k} (\Delta u)_{x_l} dQ.$$

By the definition of bilinear form $\langle \cdot,\cdot \rangle_{H_2}$, we have

$$J(h^* + \Delta h) - J(h^*) =$$

$$= \Bigg\langle \Delta u, \Bigg(2 \sum_{i=1}^{p_1} \alpha_i(t,x) \big(u(h^*) - u_i\big) +$$

$$+ 2 \sum_{i=p_1+1}^{p} A\big(\alpha_i(t,x)\big(u(h^*) - u_i\big)\big) \Bigg) \Bigg\rangle_{H_2} +$$

$$+ \sum_{i=1}^{p_1} \int_Q \alpha_i(t,x) \Delta u^2 dQ + \sum_{i=p_1+1}^{p} \int_Q \alpha_i(t,x) \sum_{k,l=1}^{n} A_{kl} (\Delta u)_{x_k} (\Delta u)_{x_l} dQ.$$

It is clear that the function Δu satisfies the following equation

$$L_1 \Delta u = f(h^* + \Delta h) - f(h^*),$$

$$u\big|_{t=0} = \frac{\partial u}{\partial t}\Big|_{t=0} = 0, \ \frac{\partial \Delta u}{\partial \mu_A}\Big|_{x \in \partial\Omega} = 0.$$

Since $f(h^* + \Delta h) - f(h^*) \in W_{2*}^-$ then by Theorem 2.3. there exists a solution $\Delta u \in H_2^+$ of the equation such that for all functions $v \in W_{2*}^+$ we have

$$\left\langle \Delta u, L_1^* v \right\rangle_{H_2} = \left\langle f(h^* + \Delta h) - f(h^*), v \right\rangle_{W_{2*}}. \tag{18}$$

From another hand, it is clear that the right-hand side of (16) belongs to H_{2*}^-. That is why, by Theorem 2.3 there exists an unique solution of the problem (16), (17) $v(t,x) \in W_{2*}^+$ and for all functions $u \in H_2^+$

$$\left\langle L_1^* v, u \right\rangle_{H_2} =$$

$$= \left\langle 2\sum_{i=1}^{p_1} \alpha_i \left(u(h^*) - u_i\right) + 2\sum_{i=p_1+1}^{p} A\left(\alpha_i \left(u(h^*) - u_i\right)\right), u \right\rangle_{H_2}.$$

Let $u = \Delta u$ and take into account (18)

$$J(h^* + \Delta h) - J(h^*) = \left\langle f(h^* + \Delta h) - f(h^*), v \right\rangle_{W_{2*}} +$$

$$+ \sum_{i=1}^{p_1} \int_Q \alpha_i(t,x) \Delta u^2 dQ + \sum_{i=p_1+1}^{p} \int_Q \alpha_i(t,x) \sum_{k,l=1}^{n} A_{kl} (\Delta u)_{x_k} (\Delta u)_{x_l} dQ.$$

Prove that the derivative $J_{h^*}(\cdot)$ is defined by expression (15). To prove this consider

$$\left| J(h^* + \Delta h) - J(h^*) - \left\langle f_{h^*}(\Delta h), v \right\rangle_{W_{2*}} \right| \leq$$

$$\leq \left| \left\langle f(h^* + \Delta h) - f(h^*) - f_{h^*}(\Delta h), v \right\rangle_{W_{2*}} \right| +$$

$$+\left|\sum_{i=1}^{p_1}\int_Q \alpha_i(t,x)\Delta u^2 dQ\right|+$$

$$+\left|\sum_{i=p_1+1}^{p}\int_Q \alpha_i(t,x)\sum_{k,l=1}^{n} A_{kl}(\Delta u)_{x_k}(\Delta u)_{x_l} dQ\right|\le \tag{19}$$

$$\le\left\|f(h^*+\Delta h)-f(h^*)-f_{h^*}(\Delta h)\right\|_{W_{2^*}^-}\left\|v\right\|_{W_{2^*}^+}+M\left\|\Delta u\right\|_{H_2^+}^2 .$$

Since $f_{h^*}(\cdot): H\to W_{2^*}^-$ is a, Fréchet derivative of the mapping $f(\cdot): H\to W_{2^*}^-$, then $\forall\varepsilon>0\ \exists\delta_1>0$: from inequality $\|\Delta h\|_H<\delta_1$ we obtain

$$\left\|f\left(h^*+\Delta h\right)-f\left(h^*\right)-f_{h^*}(\Delta h)\right\|_{W_{2^*}^-}<\frac{\varepsilon}{2\|v\|_{W_{2^*}^+}}\|\Delta h\|_H .$$

From another hand, applying the Lemma 1.1.3, we have

$$\|\Delta u\|_{H_2^+}^2\le c\left\|f\left(h^*+\Delta h\right)-f\left(h^*\right)\right\|_{W_{2^*}^-}^2\le$$

$$\le c\left(\left\|f\left(h^*+\Delta h\right)-f\left(h^*\right)-f_{h^*}(\Delta h)\right\|_{W_{2^*}^-}+\left\|f_{h^*}(\Delta h)\right\|_{W_{2^*}^-}\right)^2\le$$

$$\le c\left(\left\|f\left(h^*+\Delta h\right)-f\left(h^*\right)-f_{h^*}(\Delta h)\right\|_{W_{2^*}^-}+\left\|f_{h^*}(\Delta h)\right\|_{W_{2^*}^-}\right)^2\le$$

$$\le 2c\left\|f\left(h^*+\Delta h\right)-f\left(h^*\right)-f_{h^*}(\Delta h)\right\|_{W_{2^*}^-}^2+2c\left\|f_{h^*}(\Delta h)\right\|_{W_{2^*}^-}^2 .$$

Taking into account the definition of number δ_1, we have that from the inequality $\|\Delta h\|_H<\delta_1$ it follows

$$M\|\Delta u\|_{H_2^+}^2\le\frac{2cM\varepsilon^2}{4\|v\|_{W_{2^*}^+}^2}\|\Delta h\|_H^2+2cM\left\|f_{h^*}(\cdot)\right\|^2\|\Delta h\|_H^2 .$$

Let

$$\delta_2 = \frac{\varepsilon}{2\left(\frac{2cM\varepsilon^2}{4\|v\|^2_{W^+_{2^*}}} + 2cM\|f_{h^*}(\cdot)\|^2\right)}.$$

If $\|\Delta h\|_H < \min\{\delta_1, \delta_2\}$, then

$$M\|\Delta u\|^2_{H^+_2} \le \frac{\varepsilon}{2}\|\Delta h\|_H .$$

Returning to (19), we have

$$\left| J(h^* + \Delta h) - J(h^*) - \left\langle f_{h^*}(\Delta h), v \right\rangle_{W_{2^*}} \right| \le \left(\frac{\varepsilon}{2} + \frac{\varepsilon}{2} \right) \|\Delta h\|_H ,$$

that is what had to be proved.

Theorem 14. *If under conditions of Theorem 13 the mapping $f(\cdot): H \to W^-_{2^*}$ has a continuous Fréchet derivative at the point h^* $(\forall \varepsilon > 0\ \exists \delta > 0$: $\forall h \in U_{ad} \|h - h^*\| < \delta \Rightarrow \|F_h(\cdot) - F_{h^*}(\cdot)\| < \varepsilon)$, then the Fréchet derivative $J_h(\cdot)$ is continuous at the point h^* also.*

Proof is analogous to that of theorem 2.1.4.

Theorem 15. *If under conditions of Theorem 13 the mapping $f(\cdot): H \to W^-_{2^*}$ has a Fréchet derivative in a bounded neighbourhood of the point h^*, that satisfies Lipschitz condition with exponent α, $0 < \alpha \le 1$, then the Fréchet derivative $J_h(\cdot)$ satisfies Lipschitz condition with exponent α also.*

Proof is analogous to that of Theorem 2.1.5.

Consider application of the theorems in the case when the right-hand side of the equation (1) is defined as $f_i(t, x, h)$ $(i = \overline{3,5})$. As in previous sections, assume that the set Q is cylindrical with respect to space variable x_1 $(Q=[0,T]\times[\bar{x}_1,\bar{\bar{x}}_1]\times\Omega')$ and $h\in U_{ad}$, where U_{ad} is a bounded, closed, and convex set in the Hilbert space of control H.

Let the right-hand side of the equation be in the form

$$f_3(t,x,h)=\sum_{i=1}^{s}\delta(x_1-x_{1,i})\otimes\varphi_i(t,x_2,\dots,x_n).$$

It is easy to show that $f_3\in H^-_{r,2^*}$, thus $f_3\in W^-_{2^*}$. Analogously to the right hand side $f_1(h)$ it is easy to prove that the mapping $f_3\colon H\to\overline{W}^-_{3^*}$ has a Fréchet derivative in the form

$$\frac{\partial f_3}{\partial x_{1,i}}=-\delta^{(1)}(x_1-x_{1,i})\otimes\varphi_i(t,x_2,\dots,x_n),\qquad \frac{\partial f_3}{\partial\varphi_i}=\delta(x_1-x_{1,i}).$$

Prove that the derivative $f_{3h}(\cdot)$ satisfies the Lipschitz condition with exponent $\frac{1}{2}$. Consider

$$\left\|f_{3h_1}(\cdot)-f_{3h_2}(\cdot)\right\|=\sup_{\Delta h\in H_3}\frac{\left\|f_{3h_1}(\Delta h)-f_{3h_2}(\Delta h)\right\|_{\overline{W}^-_{3^*}}}{\|\Delta h\|_{H_3}}=$$

$$=\sup_{\Delta h\in H_3}\sup_{v\in\overline{W}^+_{3^*}}\frac{\left|\left\langle f_{3h_1}(\Delta h)-f_{3h_2}(\Delta h),v\right\rangle_{\overline{W}_{3^*}}\right|}{\|\Delta h\|_{H_3}\|v\|_{W^+_{3^*}}}.$$

Let $v(t,x)$ be a smooth in $\overline{Q}$ function from $W^+_{3^*}$. Consider the numerator of the fraction

$$I=\left\langle f_{3h_1}(\Delta h)-f_{3h_2}(\Delta h),v\right\rangle_{\overline{W}_{3^*}}=$$

$$=\sum_{i=1}^{s}\int_0^T\int_{\Omega'}\left(v_{x_1}(t,x^1_{1,i},x_2,\dots,x_n)-v_{x_1}(t,x^2_{1,i},x_2,\dots,x_n)\right)\varphi^1_i\Delta x_{1,i}+$$

$$+v_{x_1}(t,x_{1,i}^2,x_2,\dots,x_n)\left(\varphi_i^1-\varphi_i^2\right)\Delta x_{1,i}+$$

$$+\left(v(t,x_{1,i}^1,x_2,\dots,x_n)-v(t,x_{1,i}^2,x_2,\dots,x_n)\right)\Delta\varphi_i\,dtd\Omega'=$$

$$=\sum_{i=1}^{s}\int_0^T\int_{\Omega'}\left(\int_{x_{1,i}^2}^{x_{1,i}^1}v_{x_1x_1}(t,\eta,x_2,\dots,x_n)d\eta\right)\varphi_i^1\Delta x_{1,i}+$$

$$+v_{x_1}(t,x_{1,i}^2,x_2,\dots,x_n)\left(\varphi_i^1-\varphi_i^2\right)\Delta x_{1,i}+$$

$$+\left(\int_{x_{1,i}^2}^{x_{1,i}^1}v_{x_1}(t,\eta,x_2,\dots,x_n)d\eta\right)\Delta\varphi_i\,dtd\Omega'.$$

Applying the inequality

$$\max_{x\in[a,b]}|g(x)|\le c\left(\int_a^b g^2+\left(\frac{dg}{dx}\right)^2dx\right)^{\frac{1}{2}}$$

and the Schwartz inequality, we have

$$|I|\le\sum_{i=1}^{s}\int_0^T\int_{\Omega'}\left(\int_{x_{1,i}^2}^{x_{1,i}^1}d\eta\right)^{\frac{1}{2}}\left(\int_{x_{1,i}^2}^{x_{1,i}^1}v_{x_1x_1}^2(t,\eta,x_2,\dots,x_n)d\eta\right)^{\frac{1}{2}}\left|\varphi_i^1\Delta x_{1,i}\right|+$$

$$\left(\int_{\bar{x}_1}^{\bar{\bar{x}}_1}v_{x_1}^2(t,\eta,x_2,\dots,x_n)+v_{x_1x_1}^2(t,\eta,x_2,\dots,x_n)d\eta\right)^{\frac{1}{2}}\left|\varphi_i^1-\varphi_i^2\right|\left|\Delta x_{1,i}\right|+$$

$$+\left(\int_{x_{1,i}^2}^{x_{1,i}^1}d\eta\right)^{\frac{1}{2}}\left(\int_{x_{1,i}^2}^{x_{1,i}^1}v_{x_1}^2(t,\eta,x_2,\dots,x_n)d\eta\right)^{\frac{1}{2}}\left|\Delta\varphi_i\right|dtd\Omega'.$$

Apply the Schwarz inequality again:

$$|I| \le \sum_{i=1}^{s} \left|x_{1,i}^{1} - x_{1,i}^{2}\right|^{\frac{1}{2}} \left|\Delta x_{1,i}\right| \left\|v_{x_1 x_1}\right\|_{L_2(Q)} \left\|\varphi_i^1\right\|_{L_2((0,T)\times\Omega')} +$$

$$+\left|\Delta x_{1,i}\right| \left\|\varphi_i^1 - \varphi_i^2\right\|_{L_2((0,T)\times\Omega')} \left(\int_Q v_{x_1}^2 + v_{x_1 x_1}^2 dQ\right)^{\frac{1}{2}} +$$

$$+\left|x_{1,i}^{1} - x_{1,i}^{2}\right|^{\frac{1}{2}} \left\|v_{x_1 x_1}\right\|_{L_2(Q)} \left\|\Delta\varphi_i\right\|_{L_2((0,T)\times\Omega')} \le$$

$$\le \sum_{i=1}^{s} \left(\left|x_{1,i}^{1} - x_{1,i}^{2}\right|^{\frac{1}{2}} \left\|\varphi_i^1\right\|_{L_2((0,T)\times\Omega')} + \left\|\varphi_i^1 - \varphi_i^2\right\|_{L_2((0,T)\times\Omega')} + \right.$$

$$\left. + \left|x_{1,i}^{1} - x_{1,i}^{2}\right|^{\frac{1}{2}} \right) \left\|\Delta h\right\|_{H_3} \left\|v\right\|_{W_{3^*}^{+}}.$$

Whence,

$$\left\|f_{3h_1}(\cdot) - f_{3h_2}(\cdot)\right\| \le$$

$$\le \sum_{i=1}^{s} \left(\left|x_{1,i}^{1} - x_{1,i}^{2}\right|^{\frac{1}{2}} \left\|\varphi_i^1\right\|_{L_2((0,T)\times\Omega')} + \right.$$

$$\left. + \left\|\varphi_i^1 - \varphi_i^2\right\|_{L_2((0,T)\times\Omega')} + \left|x_{1,i}^{1} - x_{1,i}^{2}\right|^{\frac{1}{2}} \right).$$

Taking account that the set U_{ad} is bounded, we have $\left\|\varphi_i^1\right\|_{L_2((0,T)\times\Omega')} < C$. Therefore, $f_{3h}(\cdot)$ satisfies the Lipschitz condition with respect to h with exponent $\frac{1}{2}$. Then, by the proved theorem the performance criterion (4) has a Fréchet derivative $J_h(\cdot)$ in the domain U_{ad}, that satisfies the Lipschitz condition with respect to h with exponent $\frac{1}{2}$ and is in the form $J_{h^*}(\cdot) = \left\langle f_{3h^*}(\cdot), v\right\rangle_{\overline{W}_{3^*}}$. If we consider the performance criterion $J(h)$ directly, we prove that

$$\left\| J_{(x_{1,i}^1,\varphi_i^1)}(\cdot) - J_{(x_{1,i}^1,\varphi_i^2)}(\cdot) \right\| \le c \sum_{i=1}^{s} \left\| \varphi_i^1 - \varphi_i^2 \right\|_{L_2((0,T)\times\Omega')},$$

In same manner we can prove that the mapping $f_4(h)$ has a Fréchet derivative:

$$\frac{\partial f_4}{\partial t_i} = -\delta^{(1)}(t-t_i) \otimes \sum_{j=1}^{p} \delta(x_1 - x_{1,j}) \otimes \varphi_{ij}(x_2,\dots,x_n),$$

$$\frac{\partial f_4}{\partial x_{1,j}} = -\delta^{(1)}(x_1 - x_{1,j}) \otimes \sum_{i=1}^{s} \delta(t-t_i) \otimes \varphi_{ij}(x_2,\dots,x_n),$$

$$\frac{\partial f_4}{\partial \varphi_{ij}} = \delta(t-t_i) \otimes \delta(x_1 - x_{1,j}),$$

the derivative $f_{4h}(\cdot)$ satisfies the Lipschitz condition with the exponent $\frac{1}{2}$, and the performance criterion $J(h)$ has a Fréchet derivative, that satisfies the Lipschitz condition with the same exponent also.

Consider the other right-hand side

$$f_5(t,x,h) = \sum_{i=1}^{s} \delta(x_1 - a_i(t)) \otimes \varphi_i(t,x_2,\dots,x_n).$$

Prove that the mapping $f_5(\cdot)$ has a Fréchet derivative in the form

$$\frac{\partial f_5}{\partial a_i} = -\delta^{(1)}(x_1 - a_i(t)) \otimes \varphi_i(t,x_2,\dots,x_n),$$

$$\frac{\partial f_5}{\partial \varphi_i} = \delta(x_1 - a_i(t)).$$

Actually, we have

$$-v_{x_1}(t,a_i(t),x_2,\dots)\big)\Delta a_i(t)\varphi_i +$$

$$+\big(v(t,q(t)+\Delta a_i(t),x_2,\dots) - v(t,q(t),x_2,\dots)\big)\Delta\varphi_i\, d\Omega' dt.$$

Whence,

$$I = \left\langle f_5(h+\Delta h) - f_5(h) - (\text{grad} f_5, \Delta h), v \right\rangle_{\overline{W}_{3^*}} =$$

$$= \sum_{i=1}^{s} \int_0^T \int_{\Omega'} \left(v_{x_1}(t, a_i(t) + \theta \Delta a_i(t), x_2, \ldots) - \right.$$

$$|I| \leq \sum_{i=1}^{s} \int_0^T \int_{\Omega'} |\Delta a_i(t)|^{\frac{3}{2}} \left(\int_{\overline{x}_1}^{\overline{\overline{x}}_1} v_{x_1x_1}^2(t,x) d\xi \right)^{\frac{1}{2}} \varphi_i +$$

$$+ |\Delta a_i(t)|^{\frac{1}{2}} \left(\int_{\overline{\overline{x}}_1}^{\overline{\overline{\overline{x}}}_1} v_{x_1}^2(t,x) d\xi \right)^{\frac{1}{2}} \Delta\varphi_i d\Omega' dt.$$

Using the inequality $\Delta a(t) \in W_2^1(0,T)$, we obtain $\max_{t\in[0,T]} |\Delta a(t)| \leq c \|\Delta a\|_{W_2^1(0,T)}$. Thus,

$$|I| \leq C \sum_{i=1}^{s} \left(\int_0^T \int_{\Omega'} \int_{\overline{x}_1}^{\overline{\overline{x}}_1} v_{x_1x_1}^2(t,x) d\xi d\Omega' dt \right)^{\frac{1}{2}} \|\Delta a_i(t)\|_{W_2^1(0,T)}^{\frac{3}{2}} \|\varphi_i\|_{L_2((0,T)\times\Omega')} +$$

$$+ C \sum_{i=1}^{s} \left(\int_0^T \int_{\Omega'} \int_{\overline{x}_1}^{\overline{\overline{\overline{x}}}_1} v_{x_1}^2(t,x) d\xi d\Omega' dt \right)^{\frac{1}{2}} \|\Delta a_i(t)\|_{W_2^1(0,T)}^{\frac{1}{2}} \|\Delta\varphi_i\|_{L_2((0,T)\times\Omega')} \leq$$

$$\leq C_1 \|v\|_{W_{3^*}^+} \|\Delta h\|_{H_5}^{\frac{3}{2}}.$$

In the same manner, we prove that the mapping $f_{5h}(\cdot)$ satisfies the Lipschitz condition with the exponent $\frac{1}{2}$.

Remark that there is not any Fréchet derivative of the performance criterion (12) when the right-hand side of the equation (13) is function $f_i(\cdot)$ $(i = \overline{1,5})$, because the mapping $f_i(\cdot)$ has not any Fréchet derivative in the space $W_{2^*}^-$. To solve the problem, one can apply the method of regularization of control.

Chapter 8

SYSTEMS WITH HYPERBOLIC OPERATOR COEFFICIENTS

In many applied problems of science and engineering such as movement control, information transfer, radiolocation and object discovering it is necessary to solve boundary problems for differential equations with operator coefficients [141]:

$$Lu = u_{tt} + Bu = f(t,x).$$

Such problems with certain restrictions imposed for right-hand side were investigated for the cases of some operator coefficients B in Banach spaces in the papers [78, 141].

In the cases when the right-hand side is a distribution of finite order the analogous problem was investigated in [62-64, 142, 143, 144]. The Cauchy problem when B is a generator of a semi-group was studied in [30, 78, 145]. If B is self-adjoint and positive definite operator, mixed problems for systems of differential equations containing the first and the second derivatives were been solved with the analogy of the Galerkin method in [146].

1. HYPERBOLIC SYSTEM WITH OPERATOR COEFFICIENT

Let us consider a system governed by the differential equation

$$Lu = u_{tt} + Bu = f(t,x). \tag{1}$$

Let B be a hyperbolic operator defined by the expression

$$Bu = \sum_{i,j=1}^{n} \left(A_{ij} u_{x_j}\right)_{x_i} - Du_{yy},$$

where $A_{ij} = A_{ji}(x)$, $\{A_{ij}\}_{i,j=1}^{n}$ are functions continuously differentiable in some domain $\overline{\Omega}$, which is determined later and supposed to be sufficiently smooth in R^n, $D(x) > 0$ is continuous in $\overline{\Omega}$ function. The expression $\sum_{i,j=1}^{n}\left(A_{ij}u_{x_j}\right)_{x_i}$ is supposed to be uniformly elliptic in $\overline{\Omega}$, i.e.

$$\sum_{i,j=1}^{n} A_{ij}\xi_i\xi_j \geq \alpha\sum_{i=1}^{n}\xi_i^2\,,\ \alpha = \text{const} > 0\,,\ \xi_i \in R\,,\ i = \overline{1,n}\,.$$

The functioning of the system (1) is investigated in a tube $Q = (0,T)\times\Omega$, $\Omega \subset R^{n+1}$ is a domain in the space of the variables $(x_1,\ldots,x_n,y)$, $y \geq 0$, bounded by the characteristic surface of the equation (1) $\varphi(x,y) = 0$, so that the values of the component n_y of the outward normal is positive. The reader can be convinced in the existence of such domains himself.

Let us introduce the following denotations: W_{bd}^{+} is a completion of the set of smooth in $\overline{Q}$ functions satisfying the conditions

$$u\big|_{t=0} = u\big|_{t=T} = 0\,,\quad u\big|_{y=0} = u_y\big|_{y=0} = 0\,, \tag{2}$$

in the norm

$$\|u\|_{W_{bd}^{+}}^{2} = \int_{Q}\left(u_t^2 + \sum_{i,j=1}^{n} A_{ij}u_{x_i}u_{x_j} + Du_y^2\right)dQ\,.$$

$W_{bd^{+}}^{+}$ is a completion in the same norm of the set of smooth functions satisfying the conditions

$$v\big|_{t=0} = v\big|_{t=T} = 0\,,\quad v\big|_{\varphi(x,y)=0} = 0\,,$$

W_{bd}^{-}, $W_{bd^{+}}^{-}$ are associated negative spaces constructed by the space of square integrable functions $L_2(Q)$ and W_{bd}^{+}, $W_{bd^{+}}^{+}$, respectively.

Lemma 1. *For any functions $u \in W_{bd}^{+}$ the following relation holds true*

$$\|Lu\|_{W_{bd^{+}}^{-}} \leq c\|u\|_{W_{bd}^{+}}\,.$$

Proof. Let us prove the lemma for smooth functions $u(t,x,y)$ satisfying the conditions (2). By definition of the negative norm

$$\|Lu\|_{W^-_{bd^+}} = \sup_{\substack{v\neq 0 \\ v\in W^+_{bd^+}}} \frac{\left|\langle Lu, v\rangle_{W_{bd^+}}\right|}{\|v\|_{W^+_{bd^+}}} = \sup_{\substack{v\neq 0 \\ v\in W^+_{bd^+}}} \frac{\left|(Lu, v)_{L_2(Q)}\right|}{\|v\|_{W^+_{bd^+}}}, \quad (3)$$

as far as for smooth functions $u(t,x,y)$ the bilinear form $\langle\cdot,\cdot\rangle_{W_{bd^+}}$ coincides with an inner product $'\varphi(x,y)=0$ in $L_2(Q)$. We shall suppose that $v(t,x,y)$ is a smooth in $\overline{Q}$ function satisfying the conditions

$$v|_{t=0} = v|_{t=T} = 0, \quad v|_{\varphi(x,y)=0} = 0.$$

Let us consider the numerator in the right-hand side of (3)

$$\left|(Lu, v)_{L_2(Q)}\right| = \left|\int_Q \left(u_{tt} + \sum_{i,j=1}^{n} \left(A_{ij} u_{x_j}\right)_{x_i} - D u_{yy} \right) v dQ\right|.$$

Using the integration by parts, the Ostrogradsky-Gauss theorem and taking into account the boundary conditions, we have

$$\int_Q u_{tt} v dQ = \int_Q (u_t v)_t dQ - \int_Q u_t v_t dQ = -\int_Q u_t v_t dQ.$$

Applying the integral Cauchy inequality, we obtain

$$\left|\int_Q u_{tt} v dQ\right| = \left|\int_Q u_t v_t dQ\right| \leq \left(\int_Q v_t^2 dQ\right)^{1/2} \left(\int_Q u_t^2 dQ\right)^{1/2} \leq \|v\|_{W^+_{bd^+}} \|u\|_{W^+_{bd}}.$$

Next,

$$\int_Q \sum_{i,j=1}^{n} \left(A_{ij} u_{x_j}\right)_{x_i} v dQ = \int_Q \sum_{i,j=1}^{n} \left(A_{ij} u_{x_j} v\right)_{x_i} dQ - \int_Q \sum_{i,j=1}^{n} A_{ij} u_{x_j} v_{x_i} dQ.$$

Passing to the integral on the surface in the first term in the right-hand side and taking into account the boundary conditions, we obtain

$$\int_Q \sum_{i,j=1}^{n} \left(A_{ij} u_{x_j} v\right)_{x_i} dQ = \int_\Gamma \sum_{i,j=1}^{n} A_{ij} u_{x_j} v n_{x_i} d\Gamma = 0.$$

Here n_{x_i} is the ith component of the vector of the outward normal to the boundary Γ. Applying to the expression $\int_Q \sum_{i,j=1}^{n} A_{ij} u_{x_j} v_{x_i} dQ$ the Cauchy inequality, we have

$$\left|\int_Q \sum_{i,j=1}^{n} A_{ij} u_{x_j} v_{x_i} dQ\right| \le \left(\int_Q \sum_{i,j=1}^{n} A_{ij} v_{x_i} v_{x_j} dQ\right)^{1/2} \left(\int_Q \sum_{i,j=1}^{n} A_{ij} u_{x_i} u_{x_j} dQ\right)^{1/2},$$

whence

$$\left|\int_Q \sum_{i,j=1}^{n} \left(A_{ij} u_{x_j}\right)_{x_i} v dQ\right| \le \|v\|_{W^+_{bd^+}} \|u\|_{W^+_{bd}}.$$

Then, we obtain with the help of integration by parts and the Cauchy inequality that

$$\left|\int_Q vDu_{yy} dQ\right| \le \left(\int_Q Dv_y^2 dQ\right)^{1/2} \left(\int_Q Du_y^2 dQ\right)^{1/2} \le \|v\|_{W^+_{bd^+}} \|u\|_{W^+_{bd}}.$$

Using obtained inequality, we find from (3) that the lemma holds true for smooth functions $u \in W^+_{bd}$. Approaching the limit, we prove that the lemma holds true for any $u \in W^+_{bd}$.

Lemma 2. *For any functions $v \in W^+_{bd^+}$ the following relation holds true*

$$\|L^* v\|_{W^-_{bd}} \le c\|v\|_{W^+_{bd^+}}$$

where L^* is the operator of the adjoint problem.

Proof. Let $v(t,x,y)$ is a smooth function from $W^+_{bd^+}$. By the definition of the negative norm, taking into account the smoothness of the function $v(t,x,y)$, we can write

$$\|L^* v\|_{W^-_{bd}} = \sup_{\substack{u \ne 0 \\ u \in W^+_{bd}}} \frac{\left|(u, L^* v)_{L_2(Q)}\right|}{\|u\|_{W^+_{bd}}}.$$

Let us consider the numerator in the right-hand side of the equality

$$\left|(u, L^* v)_{L_2(Q)}\right| = \left|\int_Q u\left(v_{tt} + \sum_{i,j=1}^{n}\left(A_{ij}v_{x_j}\right)_{x_i} - Dv_{yy}\right)dQ\right|. \quad (4)$$

Using the integration by parts, the Ostrogradsky-Gauss theorem and taking into account the boundary conditions, we have

$$\left|\int_Q uv_{tt}dQ\right| = \left|\int_Q (uv_t)_t dQ - \int_Q u_t v_t dQ\right| \le \|u\|_{W^+_{bd}} \|u\|_{W^+_{bd^+}}.$$

Integrating by parts the last term in (4), we have

$$\int_Q u\left(\sum_{i,j=1}^{n}\left(A_{ij}v_{x_j}\right)_{x_i} - Dv_{yy}\right)dQ = \int_Q \sum_{i,j=1}^{n}\left(uA_{ij}v_{x_j}\right)_{x_i} dQ -$$

$$- \int_Q \sum_{i,j=1}^{n} A_{ij}u_{x_i}v_{x_j}dQ - \int_Q \left(uDv_y\right)_y dQ + \int_Q u_y Dv_y dQ.$$

By the Ostrogradsky-Gauss formula, let us pass to the integral on the boundary Γ:

$$\int_Q \sum_{i,j=1}^{n}\left(uA_{ij}v_{x_j}\right)_{x_i} dQ - \int_Q \left(uDv_y\right)_y dQ = \int_\Gamma u\left(\sum_{i,j=1}^{n} A_{ij}v_{x_j}n_{x_i} - Dv_y n_y\right)d\Gamma.$$

By the condition $u|_{y=0} = 0$. For the normal vector we can write

$$n_{x_i} = \frac{v_{x_i}}{\sqrt{\sum_{k=1}^{n} v_{x_k}^2 + v_y^2}}, \quad n_y = \frac{v_y}{\sqrt{\sum_{k=1}^{n} v_{x_k}^2 + v_y^2}}.$$

Then,

$$\int_Q \sum_{i,j=1}^{n}\left(uA_{ij}v_{x_j}\right)_{x_i} dQ - \int_Q \left(uDv_y\right)_y dQ =$$

$$= \int_\Gamma \frac{u}{\sqrt{\sum_{k=1}^{n} v_{x_k}^2 + v_y^2}}\left(\sum_{i,j=1}^{n} A_{ij}v_{x_j}v_{x_j} - Dv_y^2\right)d\Gamma = 0,$$

because the expression in the brackets coincides with the equation of the characteristics.

To estimate the expression $\sum_{i,j=1}^{n} A_{ij} u_{x_i} v_{x_j}$, we use the Cauchy inequality. Then, using the integral the Cauchy inequality, we shall prove the lemma for smooth functions $v \in W^{+}_{bd^{+}}$. The final step consists in the passing to the limit.

Lemma 3. *For any function* $u \in W^{+}_{bd}$ *the following inequality holds true*

$$\|Lu\|_{W^{-}_{bd^{+}}} \geq c\|u\|_{L_2(Q)}.$$

Proof. Let us prove that the lemma holds true for smooth functions $u(t,x,y)$ satisfying the conditions(2). For such functions $u(t,x,y)$ the following relation is valid

$$(Lu,v)_{L_2(Q)} \geq c\|v\|_{W^{+}_{bd^{+}}}, \tag{5}$$

where

$$v(t,x,y) = \int_{\varphi=0,y\geq 0}^{y} \eta^{-1} u(t,x,\eta) d\eta$$

in the domain Q.

Indeed, by the definition of the function $v(t,x,y)$, it belongs to $W^{+}_{bd^{+}}$, $u = yv_y$. Consider the functional

$$(Lu,v)_{L_2(Q)} = \left(u_{tt} + \sum_{i,j=1}^{n} \left(A_{ij} u_{x_j}\right)_{x_i} - Du_{yy}, v \right)_{L_2(Q)}. \tag{6}$$

Using the integration by parts, the relation between $u(t,x,y)$ and $v(t,x,y)$ and the Ostrogradsky-Gauss formula, we can write

$$\int_Q u_{tt} v dQ = \int_Q (u_t v)_t dQ - \int_Q v_t y v_{yt} dQ = -\int_Q v_t y v_{yt} dQ =$$

$$= -\frac{1}{2}\int_Q \left(yv_t^2\right)_y dQ + \frac{1}{2}\int_Q v_t^2 dQ = \frac{1}{2}\int_Q v_t^2 dQ. \tag{7}$$

In a similar way we investigate the second term in the right-hand side of (6):

$$\left(\sum_{i,j=1}^n \left(A_{ij}u_{x_j}\right)_{x_i}, v\right)_{L_2(Q)} = \int_Q \sum_{i,j=1}^n \left(A_{ij}u_{x_j}v\right)_{x_i} dQ -$$

$$-\int_Q \sum_{i,j=1}^n v_{x_i} A_{ij}u_{x_j} dQ = -\int_Q \sum_{i,j=1}^n v_{x_i} A_{ij}u_{x_j} dQ = \tag{8}$$

$$= -\frac{1}{2}\int_Q \sum_{i,j=1}^n \left(A_{ij}yv_{x_i}v_{x_j}\right)_y dQ + \frac{1}{2}\int_Q \sum_{i,j=1}^n A_{ij}v_{x_i}v_{x_j} dQ.$$

Next, for the third term in the right-hand side of (6) we obtain

$$-\int_Q vDu_{yy}dQ = -\int_Q \left(vDu_y\right)_y dQ + \int_Q v_y Dyv_{yy} dQ + \tag{9}$$

$$+\int_Q Dv_y^2 dQ = \frac{1}{2}\int_Q \left(yDv_y^2\right)_y dQ + \frac{1}{2}\int_Q Dv_y^2 dQ.$$

Add the equalities (7), (8) and (9):

$$(Lu,v)_{L_2(Q)} = \frac{1}{2}\int_Q \left(v_t^2 + \sum_{i,j=1}^n A_{ij}v_{x_i}v_{x_j} + Dv_y^2 \right) dQ -$$

$$-\frac{1}{2}\int_Q \left(\sum_{i,j=1}^n A_{ij}yv_{x_i}v_{x_j} - yDv_y^2 \right)_y dQ. \tag{10}$$

The second term in the right-hand side of (10) equals to zero, since the integrand expression coincides with the equation of the characteristics on the characteristics $\varphi = 0$. Hence, the relation (5) is valid. Using the Schwarz inequality, we obtain

$$c\|v\|^2_{W^+_{bd^+}} \le (Lu,v)_{L_2(Q)} \le \|v\|_{W^+_{bd^+}} \|Lu\|_{W^-_{bd^+}}.$$

Reducing both parts of the inequality by $\|v\|_{W^+_{bd^+}}$ and using the relation between $u(t,x,y)$ and $v(t,x,y)$ we prove that the lemma holds true for smooth functions $u \in W^+_{bd}$. Passing to the limit, we prove that the lemma holds true for any function $u \in W^+_{bd}$.

Lemma 4. *For any function* $v \in W^+_{bd^+}$ *the following inequality holds true*

$$\|L^*v\|_{W^-_{bd}} \geq c\|v\|_{L_2(Q)}.$$

Proof. Let $v(t,x,y)$ be a smooth function from $W^+_{bd^+}$. Let us prove that for such functions the following relation holds true:

$$(u, L^*v)_{L_2(Q)} \geq \|u\|^2_{W^+_{bd}}, \tag{11}$$

where

$$u(t,x,y) = -\int_0^y \eta v(t,x,\eta)d\eta$$

in the domain Q. By the definition of the function $u(t,x,y)$, it belongs to the space W^+_{bd}, $v = -y^{-1}u_y$. Consider the functional

$$(u, L^*v)_{L_2(Q)} = \left(u, v_{tt} + \sum_{i,j=1}^{n}\left(A_{ij}v_{x_j}\right)_{x_i} - Du_{yy}\right)_{L_2(Q)} \tag{12}$$

Integrating by parts and applying the Ostrogradsky-Gauss formula we obtain

$$\int_Q uv_{tt}dQ = \int_Q (uv_t)_t dQ + \int_Q u_t y^{-1}u_{ty}dQ =$$

$$= \frac{1}{2}\int_Q \left(y^{-1}u_t^2\right)_y dQ + \frac{1}{2}\int_Q y^{-2}u_t^2 dQ \geq \frac{1}{2}\int_Q y^{-2}u_t^2 dQ. \tag{13}$$

In the similar way consider the second term in the right-hand side of (12):

$$\left(u, \sum_{i,j=1}^{n}\left(A_{ij}v_{x_j}\right)_{x_i}\right)_{L_2(Q)} = \int_Q \sum_{i,j=1}^{n}\left(uA_{ij}v_{x_j}\right)_{x_i} dQ -$$

$$-\int_Q \sum_{i,j=1}^{n} u_{x_i} A_{ij} v_{x_j} dQ = \int_\Gamma \sum_{i,j=1}^{n} uA_{ij} v_{x_j} n_{x_i} d\Gamma +$$

$$+\int_Q \sum_{i,j=1}^{n} u_{x_i} A_{ij} y^{-1} u_{yx_j} dQ = \int_\Gamma \sum_{i,j=1}^{n} uA_{ij} v_{x_j} n_{x_i} d\Gamma + \tag{14}$$

$$+\frac{1}{2}\int_Q \sum_{i,j=1}^{n}\left(A_{ij} y^{-1} u_{x_i} u_{x_j}\right)_y dQ + \frac{1}{2}\int_Q \sum_{i,j=1}^{n} y^{-2} A_{ij} u_{x_i} u_{x_j} dQ.$$

Consider the last term in (12):

$$-\left(u, Dv_{yy}\right) = -\int_Q \left(uDv_y\right)_y dQ - \int_Q u_y D y^{-1} u_{yy} dQ +$$

$$+\int_Q Dy^{-2} u_y^2 dQ = \int_\Gamma uDv_y n_y d\Gamma -$$

$$-\frac{1}{2}\int_Q \left(Dy^{-1}u_y^2\right)_y dQ + \frac{1}{2}\int_Q Dy^{-2}u_y^2 dQ. \tag{15}$$

Add the expressions (14) and (15):

$$\left(u, \sum_{i,j=1}^{n}\left(A_{ij}v_{x_j}\right)_{x_i} - Dv_{yy}\right)_{L_2(Q)} =$$

$$= \frac{1}{2}\int_Q y^{-2}\left(\sum_{i,j=1}^{n} A_{ij} u_{x_i} u_{x_j} + Du_y^2\right) dQ +$$

$$+\int_\Gamma \frac{u}{\sqrt{\sum_{k=1}^{n} v_{x_k}^2 + v_y^2}}\left(\sum_{i,j=1}^{n} A_{ij} v_{x_i} v_{x_j} - Dv_y^2\right) d\Gamma +$$

$$+\frac{1}{2}\int_{\Gamma}\frac{y^{-1}n_y}{\sqrt{\sum_{k=1}^{n}v_{x_k}^2+v_y^2}}\left(\sum_{i,j=1}^{n}A_{ij}u_{x_i}u_{x_j}-Du_y^2\right)d\Gamma, \qquad (16)$$

since,

$$n_{x_i}=\frac{v_{x_i}}{\sqrt{\sum_{k=1}^{n}v_{x_k}^2+v_y^2}}, \quad n_y=\frac{v_y}{\sqrt{\sum_{k=1}^{n}v_{x_k}^2+v_y^2}}.$$

The last two expressions in the right-hand side of (16) equal to zero, since it equals to zero on $y=0$ and contain the expressions, which are equal to zero on the characteristics.

Adding the inequalities (13) and (16) and applying the Schwarz inequality, we obtain

$$c\left(\int_{Q}y^{-2}Du_y^2dQ\right)^{1/2}\|u\|_{W_{bd}^{+}}\le\left(u,L^{*}v\right)_{L_2(Q)}\le\|u\|_{W_{bd}^{+}}\|L^{*}v\|_{W_{bd}^{-}}.$$

Reducing both parts of the obtained inequality by $\|u\|_{W_{bd}^{+}}$ and taking into account the relation between $u(t,x,y)$ and $v(t,x,y)$, we obtain that the lemma holds true for smooth functions $v\in W_{bd^{+}}^{+}$. Passing to the limit, we prove the lemma for any function $v\in W_{bd^{+}}^{+}$.

The proven lemmas imply Theorem 1.

Using these lemmas and the results of Section 1.1, we have

Theorem 1. *For any function $f\in L_2(Q)$ there exists a unique solution of the problem (1), (2) in the sense of Definition 1.1.*

Theorem 2. *For any functional $f\in W_{bd^{+}}^{-}$ there exists a unique solution of the problem (1), (2) in the sense of Definition 1.4.*

2. GALERKIN METHOD FOR DISTRIBUTED SYSTEM

In this section we study the Galerkin method of approximate solving distributed system. Let the state function satisfies the boundary value problem from Section 1,

$$Lu \equiv \frac{\partial^2 u}{\partial t^2} + Bu = f,$$

where $f \in L_2(Q)$.

Consider an approximate solution in the form

$$u_n(t,x) = \sum_{i=1}^{n} g_i(t)\omega_i(x,y), \tag{1}$$

where $\omega_i(x)$ is a twice continuously differentiable function, that satisfies boundary condition of the problem, the set $\{a\varphi_i\omega_{iy}\}_{i=1}^{\infty}$ is total in $W^+_{bd^+}$, and the function $g_i(t)$ is a solution of the set of differential equations:

$$\sum_{k=1}^{n} g_{ktt}(a\omega_{iy},\omega_k)_{L_2(\Omega)} - \sum_{k=1}^{n} g_k(a\omega_{iy}, B\omega_k)_{L_2(\Omega)} = (a\omega_{iy}, f)_{L_2(\Omega)}, \tag{2}$$

where

$$g_k|_{t=0} = g_k|_{t=T} = 0;\ k = \overline{1,n};\ i = \overline{1,n},$$

$a = a(y) < 0, a_y \geq c_1 > 0, {a_2}/{D} \leq c_2 < \infty$ on the surface $\varphi(x,y) = 0, y \geq 0$.

Relation (2) can be written in the form

$$(a\omega_{iy}, Lu_n)_{L_2(\Omega)} = (a\omega_{iy}, f)_{L_2(\Omega)}. \tag{3}$$

Theorem 1. *For all functions* $f(t,x,y) \in L_2(Q)$ *the approximate sequence* $\{u_n\}_{n=1}^{\infty}$, *converges to a solution*

$u(t,x,y) \in W_{bd}^{+}$ in the sense that there exists a sequence of functions $u_i \in C_{bd}^2(Q)$ that

$$\|u_n - u\|_{L_2(Q)} \xrightarrow[n\to\infty]{} 0, \|Lu_n - f\|_{W_{bd^+}^-} \xrightarrow[n\to\infty]{} 0. \quad (4)$$

P r o o f. Multiplying both right and left hand sides of the equality (3) on the function $g_i(t)$, summing up over i from 1 to n and integrating with respect to t from 0 to T, we have

$$(au_{ny}, Lu_n)_{L_2(Q)} \doteq (au_{ny}, f)_{L_2(Q)}. \quad (5)$$

For all functions $u \in W_{bd}^{+}$ let us prove the following inequality

$$(au_y, Lu)_{L_2(Q)} \geq C\|u\|_{W_{bd}^+}^2. \quad (6)$$

Assume that $u \in C_{bd}^2(Q)$. Consider a functional

$$(au_y, Lu)_{L_2(Q)} = \left(au_y, u_{tt} + \sum_{i,j=1}^{n} \frac{\partial}{\partial x_i} A_{ij} \frac{\partial u}{\partial x_j} - Du_{yy} \right)_{L_2(Q)}$$

Consider the every summand of the right hand side, separately. We have

$$au_y u_{tt} = (au_y u_t)_y - \frac{1}{2}(au_t^2)_y + \frac{1}{2} a_y u_t^2;$$

$$au_y (A_{ij} u_{x_j})_{x_i} = (au_y A_{ij} u_{x_j})_{x_i} - (au_{x_i} A_{ij} u_{x_j})_y +$$

$$+ a_y A_{ij} u_{x_i} u_{x_j} + (au_{x_i} A_{ij} u_y)_{x_j} - au_y (A_{ij} u_{x_i})_{x_j}.$$

Summing up the equality over i and j from 1 to n and taking into account that $A_{ij} = A_{ji}$, we obtain

$$\sum_{i,j=1}^{n} au_y (A_{ij} u_{x_j})_{x_i} = \sum_{i,j=1}^{n} (au_y A_{ij} u_{x_j})_{x_i} - \frac{1}{2} \sum_{i,j=1}^{n} (au_{x_i} A_{ij} u_{x_j})_y +$$

$$+ \frac{1}{2} \sum_{i,j=1}^{n} a_y A_{ij} u_{x_i} u_{x_j}.$$

Next, we have

$$-au_y Du_{yy}=-\frac{1}{2}\left(aDu_y^2\right)_y+\frac{1}{2}a_y Du_y^2.$$

Integrate the expression over Q, and take account that $u(t,x,y)\in C_{bd}^2(Q)$.

$$\int_Q au_y u_{tt}dQ=\int_Q\left(au_y u_t\right)_t dQ-\frac{1}{2}\int_Q\left(au_t^2\right)_y dQ+\frac{1}{2}\int_Q a_y u_t^2 dQ.$$

Passing to the surface integrals, we find:

$$\int_Q\left(au_y u_t\right)_t dQ=\int_\Gamma au_y u_t n_t d\Gamma=0,$$

$$-\frac{1}{2}\int_Q\left(au_t^2\right)_y dQ=-\frac{1}{2}\int_\Gamma au_t^2 n_y d\Gamma\geq 0,$$

In the second expression we use inequality $an_y\leq 0$ on the characteristic surface $\varphi(x,y)=0, y\geq 0$. Whence, we conclude that

$$\int_Q au_y u_{tt}dQ\geq\frac{1}{2}\int_Q au_t^2 dQ.$$

Next,

$$\int_Q\sum_{i,j=1}^n au_y\left(A_{ij}u_{x_j}\right)_{x_i}dQ=\int_Q\sum_{i,j=1}^n\left(au_y A_{ij}u_{x_j}\right)_{x_i}dQ-$$

$$-\frac{1}{2}\int_Q\sum_{i,j=1}^n\left(aA_{ij}u_{x_i}u_{x_j}\right)_y dQ+\frac{1}{2}\int_Q\sum_{i,j=1}^n a_y A_{ij}u_{x_i}u_{x_j}dQ;$$

$$-\int_Q au_y Du_{yy}dQ=-\frac{1}{2}\int_Q\left(aDu_y^2\right)_y dQ+\frac{1}{2}\int_Q a_y Du_y^2 dQ.$$

Consider the following expression

$$\int_Q\sum_{i,j=1}^n\left(aA_{ij}u_y u_{x_j}\right)_{x_i}dQ-\frac{1}{2}\int_Q\sum_{i,j=1}^n\left(aA_{ij}u_{x_i}u_{x_j}\right)_y dQ-\frac{1}{2}\int_Q\left(aDu_y^2\right)_y dQ.$$

Pass to the surface integration

$$\int_\Gamma \left(\sum_{i,j=1}^{n} aA_{ij}u_y u_{x_j} n_{x_i} - \frac{1}{2}\sum_{i,j=1}^{n} aA_{ij}u_{x_i}u_{x_j}n_y - \frac{1}{2}aDu_y^2 n_y \right) d\Gamma.$$

We have

$$\int_\Gamma \left(\sum_{i,j=1}^{n} aA_{ij}u_y u_{x_j} n_{x_i} - \sum_{i,j=1}^{n} aA_{ij}u_{x_i}u_{x_j}n_y + \right. \tag{7}$$
$$\left. + \frac{1}{2}\sum_{i,j=1}^{n} aA_{ij}u_{x_i}u'_{x_j}n_y - \frac{1}{2}aDu_y^2 n_y \right) d\Gamma.$$

Consider the one part of the upper expression

$$\int_\Gamma \left(\sum_{i,j=1}^{n} aA_{ij}u_y u_{x_j} n_{x_i} - \sum_{i,j=1}^{n} aA_{ij}u_{x_i}u_{x_j}n_y \right) d\Gamma. \tag{8}$$

By the data $u|_{y=0} = 0$, and hence $u_{x_j}\Big|_{y=0} = 0$.

On the characteristic surface the co-ordinates of the normal vector satisfy the equality

$$n_{x_i} = \frac{u_{x_i}}{\sqrt{\sum_{k=1}^{n} u_{x_k}^2 + u_y^2}};\ n_y = \frac{u_y}{\sqrt{\sum_{k=1}^{n} u_{x_k}^2 + u_y^2}}.$$

From (8) we have

$$\int_\Gamma \frac{au_y}{\sqrt{\sum_{k=1}^{n} u_{x_k}^2 + u_y^2}} \left(\sum_{i,j=1}^{n} A_{ij}u_{x_j}u_{x_i} - \sum_{i,j=1}^{n} A_{ij}u_{x_i}u_{x_j} \right) d\Gamma = 0.$$

Consider the other summands of (7):

$$\int_\Gamma \left(\frac{1}{2}\sum_{i,j=1}^{n} aA_{ij}u_{x_i}u_{x_j}n_y - \frac{1}{2}aDu_y^2 n_y \right) d\Gamma.$$

It is easy to see that on the characteristic surface the integrated function equals to zero.

By this reasons, the inequality (6) is valid for all $u \in C^2_{bd}(Q)$. Passing to the limit, we prove the inequality for all functions $u(t,x,y) \in W^+_{bd}$.

By the inequality (6) and the Schwarz inequality

$$c_1 \|u_n\|^2_{W^+_{bd}} \le (au_{ny}, Lu_n)_{L_2(Q)} = (au_{ny}, f)_{L_2(Q)} \le$$

$$\le \left(\int_Q a^2 u^2_{ny} dQ\right)^{1/2} \left(\int_Q f^2 dQ\right)^{1/2} \le c_2 \|u_n\|_{W^+_{bd}} \cdot \|f\|_{L_2(Q)},$$

dividing by $\|u_n\|_{W^+_{bd}}$, we have

$$\|u_n\|_{W^+_{bd}} \le c\|f\|_{L_2(Q)}.$$

From the inequality we conclude that there exists a weakly convergent subsequence $\{u_{n_k}\}_{k=1}^{\infty}$. Let the subsequence $\{u_{n_k}\}_{k=1}^{\infty}$ converges weakly to a function $\hat{u} \in W^+_{bd}$. By the Banach theorem there exists a subsequence $\{u_{n_{k_i}}\}_{i=1}^{\infty}$ that the sequence $\hat{u}_y = \frac{1}{\nu}\sum_{i=1}^{\nu} u_{n_{k_i}}$ converges to the same function $\hat{u}$ in norm of the space W^+_{bd}

$$\|\hat{u}_y - \hat{u}\|_{W^+_{bd}} \xrightarrow[\nu\to\infty]{} 0.$$

Multiplying the both right- and left-hand sides of (3) on a function $\varphi_i(t) \in W^+_{bd^+}(0,T)$ and integrating with respect to t from 0 to T, we have

$$\langle \psi_i, Lu_n \rangle = \langle \psi_i, f \rangle, i = \overline{1,n}, \tag{9}$$

where $a\varphi_i \omega_{iy} = \psi_i$.

Consider fundamental sequence $\{\hat{u}_i\}_{i=1}^{\infty}$. By Lemma 1.6.1. we have

$$\|\hat{u}_i - \hat{u}_j\|_{W^+_{bd}} \ge \|L(\hat{u}_i - \hat{u}_j)\|_{W^-_{bd^+}} = \|L\hat{u}_i - L\hat{u}_j\|_{W^-_{bd^+}}.$$

and hence

$$\left\| L\hat{u}_i - L\hat{u}_j \right\|_{W^-_{bd^+}} \xrightarrow[\substack{i\to\infty \\ j\to\infty}]{} 0.$$

Since the sequence $\{L\hat{u}_i\}_{i=1}^{\infty}$ is fundamental in the complete space $W^-_{bd^+}$, there exists a limiting function $\hat{f}$ of the sequence

$$\left\| L\hat{u}_n - \hat{f} \right\|_{W^-_{bd^+}} \xrightarrow[n\to\infty]{} 0.$$

Show that $\hat{f} = f$ in the space $W^-_{bd^+}$. By (9) and the definition of function $\hat{u}_n$, we have the following equality

$$\langle \psi_i, f \rangle = \langle \psi_i, Lu_n \rangle = \langle \psi_i, L\hat{u}_n \rangle. \tag{10}$$

By the Schwarz inequality,

$$\left\langle \psi_i, L\hat{u}_n - \hat{f} \right\rangle \le \left\| \psi_i \right\|_{W^+_{bd^+}} \left\| L\hat{u}_n - \hat{f} \right\|_{W^-_{bd^+}}.$$

The right-hand side of the inequality vanishes as $n \to \infty$. From (10) we have

$$\left\langle \psi_i, \hat{f} \right\rangle = \langle \psi_i, f \rangle, i = 1,2,\ldots. \tag{11}$$

Since the number i in (11) is arbitrary and the set of functions $\{\psi_i\}_{i=1}^{\infty}$ is dense in the space $W^+_{bd^+}$, we obtain $\hat{f} = f$. That is why

$$\left\| \hat{u}_\nu - u \right\|_{W^+_{bd}} \xrightarrow[\nu\to\infty]{} 0, \left\| L\hat{u}_\nu - f \right\|_{W^-_{bd^+}} \xrightarrow[\nu\to\infty]{} 0,$$

and therefore the function $u(t,x,y)$ is a solution of the problem in the sense of Definition 1.1.1. Since the imbedding operator of the space W^+_{bd}, in $L_2(Q)$ is compact, the chosen weakly convergent subsequence $\{u_{n_k}\}_{k=1}^{\infty}$ converges in the space $L_2(Q)$, hence, $u(t,x,y)$ is a solution of the problem in the sense of Theorem 1.

Remark. *There is no necessity to choose a subsequence because it follows from the results of Chapter 1 that there exists a unique solution of the problem.*

3. PULSE OPTIMAL CONTROL OF THE SYSTEMS WITH HYPERBOLIC OPERATOR COEFFICIENT

Consider the problems of impulse optimal control (Section 3.6) for the systems with hyperbolic operator coefficient

$$Lu = \frac{\partial^2 u}{\partial t^2} + Bu = F.$$

The denotation is the same as in Section 6.1.

Consider the initial boundary value problem

$$Lu = f + A(h), \tag{1}$$

$$u\big|_{t=0} = u\big|_{t=T} = 0, u\big|_{y=0} = \frac{\partial u}{\partial y}\bigg|_{y=0} = 0 \tag{2}$$

Complete the table of the template theorems of Section 3.6.

Table 1.

N	Operator	Space N	Space $W^-(Q)$
1.	$L(\cdot)$	$L_2(Q)$	$W^-_{bd^+}$
2.	$L(\cdot)$	W^+_{bd}	$L_2(Q)$

Table 2.

N	Operator $A_i(\cdot)$	Space $W^-(Q)$
1.	$A_1(\cdot)$	$W^-_{bd^+}$
2.	$A_3(\cdot)$	$W^-_{bd^+}$
3.	$A_5(\cdot)$	$W^-_{bd^+}$

Table 3.

N	Operator $L(\cdot)$	Space $W^-(Q)$	Space $W^+(Q)$
1.	$L(\cdot)$	$W^-_{bd^+}$	W^+_{bd}

Table 4.

N	Space $W^-(Q)$	Mapping $A_i(\cdot)$
1.		

Table 5.

N	Operator $L(\cdot)$	Space $W^-(Q)$
1.	$L(\cdot)$	$W^-_{bd^+}$

Table 6.

N	Operator $A_i(\cdot)$	Space $W^-(Q)$
1.	$A_1(\cdot)$	$W^-_{bd^+}$
2.	$A_3(\cdot)$	$W^-_{bd^+}$
3.	$A_5(\cdot)$	$W^-_{bd^+}$

Table 7.

N	Exponent α	Space $W^-(Q)$	Mapping $A_{i,\varepsilon}(\cdot)$
1.	1/2	$W^-_{bd^+}$	$A_{1,\varepsilon}(\cdot)$
2.	1/2	$W^-_{bd^+}$	$A_{3,\varepsilon}(\cdot)$
3.	1/2	$W^-_{bd^+}$	$A_{5,\varepsilon}(\cdot)$

Chapter 9

SOBOLEV SYSTEMS

In the chapter we study systems that don't satisfy the conditions of the Cauchy-Kovalevskaya theorem. These systems were investigated in [147-150].

1. EQUATION OF DYNAMICS OF VISCOUS STRATIFIED FLUID

Many applied problems (such as oceanographic research, oil barging, etc.) are described by the dynamic equation of viscous stratified fluid.

In the papers [151, 152] the equation of dynamics of plane motion of viscous exponential stratified fluid was obtain by Boussinesk approximation:

$$\frac{\partial^2}{\partial t^2}\Delta u - \nu\frac{\partial}{\partial t}\Delta^2 u + w_0^2 u_{x_1x_1} = 0,\ (x_1,x_2)\in R^2,$$

where Δ is the Laplacian. In [152] the solvability of an initial boundary value problem for this equation is studied. There were shown the existence and uniqueness of a generalized solution with the right-hand side from the negative Hilbert space with respect to x variable.

In this section we study the optimal control problems of the generalized dynamic equation of viscous stratified fluid with the right-hand side from the negative Hilbert space with respect to t and x variables.

Consider the system

$$Lu = \frac{\partial^2}{\partial t^2}Au + \frac{\partial}{\partial t}B^2 u + Cu = f, \tag{1}$$

$$u\big|_{t=0} = \frac{\partial u}{\partial t}\Big|_{t=0} = 0, \quad u\big|_{x\in\partial\Omega} = \frac{\partial u}{\partial \mu_B}\Big|_{x\in\partial\Omega} = 0, \tag{2}$$

in the tube $Q=(0,T)\times\Omega$, where $\Omega\subset R^n$ is a regular domain with boundary $\partial\Omega$ $\partial/\partial\vec{\mu}_B$ is the co-normal derivative, $\vec{\mu}_B=\frac{B\vec{n}}{|B\vec{n}|}$, $\vec{n}$ is the normal vector to the surface $\partial\Omega$, $B=\{B_{ij}\}_{i,j=1}^n$ is a matrix of elements $B_{ij}(x)$.

Operators $A(\cdot)$, $B(\cdot)$ and $C(\cdot)$ are defined by the formal expressions:

$$Au=-\sum_{i,j=1}^n\frac{\partial}{\partial x_i}\left(A_{ij}\frac{\partial u}{\partial x_j}\right),$$

$$Bu=-\sum_{i,j=1}^n\frac{\partial}{\partial x_i}\left(B_{ij}\frac{\partial u}{\partial x_j}\right),$$

$$Cu=-\sum_{i,j=1}^n\frac{\partial}{\partial x_i}\left(C_{ij}\frac{\partial u}{\partial x_j}\right)$$

with sufficiently smooth coefficients in a closed domain $\overline{\Omega}$. We assume that the coefficients satisfy the following conditions: $A_{ij}(x)=A_{ji}(x)$, $B_{ij}(x)=B_{ji}(x)$, $C_{ij}(x)=C_{ji}(x)$ for all $1\le i,j\le n$.

Assume also that for all real $\xi_i, i=\overline{1,n}$:

$$\lambda_A\sum_{i=1}^n\xi_i^2\le\sum_{i,j=1}^n A_{ij}\xi_i\xi_j\le\lambda_A^{-1}\sum_{i=1}^n\xi_i^2,$$

$$\lambda_B\sum_{i=1}^n\xi_i^2\le\sum_{i,j=1}^n B_{ij}\xi_i\xi_j\le\lambda_B^{-1}\sum_{i=1}^n\xi_i^2,$$

$$0\le\sum_{i,j=1}^n C_{ij}\xi_i\xi_j\le\lambda_C^{-1}\sum_{i=1}^n\xi_i^2,$$

where $\lambda_A, \lambda_B, \lambda_C$ are positive constants.

Let $L_2(Q)$ be a space of measurable, squared integrable functions, W^{+}_{bd} be a completion of the set of smooth in $\overline{Q}$ functions, that satisfy conditions (2) in the norm

$$\|u\|^2_{W^{+}_{bd}} = \int_Q \left(\sum_{i=1}^{n} u^2_{ttx_i} + \sum_{i,j=1}^{n} u^2_{tx_i x_j} \right) dQ, \tag{3}$$

$W^{+}_{bd^{+}}$ be a completion of the set of smooth in $\overline{Q}$ functions, that satisfy adjoint conditions

$$v\big|_{t=T} = \frac{\partial v}{\partial t}\bigg|_{t=T} = 0, \quad v\big|_{x\in\partial\Omega} = \frac{\partial v}{\partial \bar{\mu}_B}\bigg|_{x\in\partial\Omega} = 0, \tag{4}$$

in the same norm (3).

Let W^{-}_{bd}, $W^{-}_{bd^{+}}$ be corresponding negative spaces.

Prove the following lemmas.

Lemma 1. *For any function* $u \in W^{+}_{bd}$ *the following inequality holds true*

$$\|Lu\|_{W^{-}_{bd^{+}}} \le c\|u\|_{W^{+}_{bd}}.$$

Proof. For smooth functions $u(t,x)$, that satisfy the conditions (2), we study the following expression

$$\|Lu\|_{W^{-}_{bd^{+}}} = \sup_{\substack{v\neq 0\\ v\in W^{+}_{bd^{+}}}} \frac{|\langle Lu, v\rangle|}{\|v\|_{W^{+}_{bd^{+}}}} = \sup_{\substack{v\neq 0\\ v\in W^{+}_{bd^{+}}}} \frac{|(Lu, v)_{L_2(Q)}|}{\|v\|_{W^{+}_{bd^{+}}}} \tag{5}$$

where $\langle\cdot,\cdot\rangle$ is a bilinear form that is defined on $W^{+}_{bd^{+}} \times W^{-}_{bd^{+}}$.

Applying integration by parts and passing to the surface integrals, we have

$$|(Lu, v)_{L_2(Q)}| = \left| \int_Q v\left(\frac{\partial}{\partial t^2} Au + \frac{\partial}{\partial t} B^2 u + Cu \right) dQ \right| \le$$

$$\leq \left| \int_Q \sum_{i,j=1}^n A_{ij} v_{tx_i} u_{tx_j} dQ \right| + \left| \int_Q (Bv)(Bu_t) dQ \right| + \left| \int_Q \sum_{i,j=1}^n C_{ij} v_{x_i} u_{x_j} dQ \right| . \quad (6)$$

Estimate each summand in the right-hand side of the inequality (6). Taking into account the Schwarz inequality, we obtain

$$\left| \int_Q \sum_{i,j=1}^n A_{ij} v_{tx_i} u_{tx_j} dQ \right| \leq \left| \int_Q \left(\sum_{i,j=1}^n A_{ij} v_{tx_i} v_{tx_j} \right)^{1/2} \left(\sum_{i,j=1}^n A_{ij} u_{tx_i} u_{tx_j} \right)^{1/2} dQ \right| \leq$$

$$\leq \left(\int_Q \sum_{i,j=1}^n A_{ij} v_{tx_i} v_{tx_j} dQ \right)^{1/2} \left(\int_Q \sum_{i,j=1}^n A_{ij} u_{tx_i} u_{tx_j} dQ \right)^{1/2} \leq$$

$$\leq c \left(\int_Q \sum_{i=1}^n v_{tx_i}^2 \right)^{1/2} \left(\int_Q \sum_{i=1}^n u_{tx_i}^2 \right)^{1/2} \leq c \|u\|_{W_{bd}^+} \|v\|_{W_{bd^+}^+} .$$

To estimate the second summand of (6), we use the Schwarz inequality and coefficient boundedness of the operator B:

$$\left| \int_Q (Bv)(Bu_t) dQ \right| \leq \|Bv\|_{L_2(Q)} \|Bu_t\|_{L_2(Q)} \leq$$

$$\leq c \left\| \sum_{i,j=1}^n v_{x_i x_j} \right\|_{L_2(Q)} \left\| \sum_{i,j=1}^n u_{tx_i x_j} \right\|_{L_2(Q)} \leq c \|v\|_{W_{bd^+}^+} \|u\|_{W_{bd}^+} .$$

In the same manner, we show that

$$\left| \int_Q \sum_{i,j=1}^n C_{ij} v_{x_i} u_{x_j} dQ \right| \leq c \|v\|_{W_{bd^+}^+} \|u\|_{W_{bd}^+} .$$

Taking into account proven inequalities and (5), we obtain the desired inequality for smooth functions $u(t,x) \in C_{bd}^\infty(Q)$. Passing to the limit, we have the inequality for all functions $u \in W_{bd}^+$.

In the same manner we prove the following assertion

Lemma 2. *For any function* $v \in W^{+}_{bd^{+}}$ *the following inequality holds true*

$$\left\|L^{*}v\right\|_{W^{-}_{bd}} \le c\left\|v\right\|_{W^{+}_{bd^{+}}},$$

where $L^{*}(\cdot)$ *is the adjoint operator*

$$L^{*}v = \frac{\partial^{2}}{\partial t^{2}}Au - \frac{\partial}{\partial t}B^{2}u + Cu.$$

Lemma 3. *For any function* $u \in W^{+}_{bd}$ *the following inequality holds true*

$$\left\|Lu\right\|_{W^{-}_{bd}} \ge c\left\|u\right\|_{L_{2}(Q)}.$$

Proof. Show the desired inequality for smooth functions $u(t,x)$ that satisfy the conditions (2). Consider the auxiliary operator

$$v(t,x) = \int_{T}^{t} \varphi(\xi) \int_{T}^{\xi} \psi(\eta) u(\eta, x) d\eta d\xi,$$

where

$$\varphi(t) = \exp\left(-2\lambda_{A}^{-1}\lambda_{C}^{-1}(t+2)\exp(T-t)\right),$$

$$\psi(t) = \exp\left(-2\lambda_{A}^{-1}\lambda_{C}^{-1}(t+2)\exp(T-t) + T - t\right).$$

For the functions $u(t,x), v(t,x)$ prove the inequality

$$(v, Lu)_{L_{2}(Q)} \ge c\left\|v\right\|^{2}_{W^{+}_{bd^{+}}}.$$

Since the function $u(t,x)$ satisfies the conditions (2), then the function $v(t,x)$ satisfies the conditions (4). In addition the functions $u(t,x), v(t,x)$ satisfy the following relation

$$u(t,x) = \exp(t-T)v_{tt}(t,x) - 2\lambda_{A}^{-1}\lambda_{C}^{-1}(t+1)v_{t}(t,x).$$

Estimate the expression

$$(v, Lu)_{L_{2}(Q)} = \left(u, L^{*}v\right)_{L_{2}(Q)} =$$

$$= \left(\exp(t-T)v_{tt} - 2\lambda_{A}^{-1}\lambda_{C}^{-1}(t+1)v_{t}, L^{*}v\right)_{L_{2}(Q)}.$$

Employing the integration by parts and passing to the surface integrals, we obtain

$$(v, Lu)_{L_2(Q)} =$$

$$= \left(\exp(t-T)v_{tt} - 2\lambda_A^{-1}\lambda_C^{-1}(t+1)v_t, Av_{tt} - B^2 v_t + Cv\right)_{L_2(Q)} =$$

$$= \left(\exp(t-T)v_{tt}, Av_{tt}\right)_{L_2(Q)} +$$

$$+\left(\left(\frac{1}{2}\exp(t-T) + 2\lambda_A^{-1}\lambda_C^{-1}(t+1)\right)Bv_t, Bv_t\right)_{L_2(Q)} +$$

$$+ I_1 + I_2 + I_3 + I_4 + I_5 + I_6,$$

where

$$I_1 = -\left(\exp(t-T)v_t, Cv_t\right)_{L_2(Q)} + \lambda_A^{-1}\lambda_C^{-1}\left(v_t, Av_t\right)_{L_2(Q)} \geq$$

$$\geq -\lambda_C^{-1}\int_Q \left(\sum_{i=1}^n v_{tx_i}^2\right) dQ + \lambda_C^{-1}\int_Q \left(\sum_{i=1}^n v_{tx_i}^2\right) dQ = 0,$$

$$I_2 = \frac{1}{2}\left(\left(\exp(t-T) + 2\lambda_A^{-1}\lambda_C^{-1}\right)v, Cv\right)_{L_2(Q)} \geq 0,$$

$$I_3 = \frac{1}{2}\exp(-T)\left\|Bv_t\big|_{t=0}\right\|_{L_2(\Omega)}^2 \geq 0,$$

$$I_4 = -\frac{1}{2}\exp(-T)\left(v_t, Cv_t\right)_{L_2(\Omega)}\Big|_{t=0} + \lambda_A^{-1}\lambda_C^{-1}\left(v_t, Av_t\right)_{L_2(\Omega)}\Big|_{t=0} \geq$$

$$\geq -\frac{1}{2}\lambda_C^{-1}\exp(-T)\int_\Omega \left(\sum_{i=1}^n v_{tx_i}^2\right)\Bigg|_{t=0} d\Omega + \lambda_C^{-1}\int_\Omega \left(\sum_{i=1}^n v_{tx_i}^2\right)\Bigg|_{t=0} d\Omega \geq 0,$$

$$I_5 = \lambda_A^{-1}\lambda_C^{-1}\left(v, Cv\right)_{L_2(\Omega)}\Big|_{t=0} \geq 0,$$

$$I_6 = \frac{1}{2}\lambda_C^{-1}\exp(-T)\left((v_t - v), C(v_t - v)\right)_{L_2(\Omega)}\Bigg|_{t=0} \geq 0.$$

Thus,

$$(v, Lu)_{L_2(Q)} \geq c\left[\left(v_{tt}, Av_{tt}\right)_{L_2(Q)} + \left\|Bv_t\right\|_{L_2}^2\right]. \tag{7}$$

Using inequality of coercivity [74] for uniformly elliptic operator B, it is not difficult to prove, that

$$\|Bv\|_{L_2(Q)} \geq c\left(\int_Q \left(\sum_{i,j=1}^{n} v_{x_i x_j}^2\right) dQ\right)^{1/2}.$$

From the estimate of the norm of the elliptic operator B, we have

$$\left|(v, Lu)_{L_2(Q)}\right| \geq c\int_Q \left(\sum_{i=1}^{n} v_{ttx_i}^2 + \sum_{i,j=1}^{n} v_{tx_i x_j}^2\right) dQ = c\|v\|_{W_{bd^+}^+}^2.$$

Applying to the left-hand side the Schwarz inequality, we find

$$\left|(v, Lu)_{L_2(Q)}\right| \leq \|v\|_{W_{bd^+}^+} \|Lu\|_{W_{bd^+}^-},$$

then

$$\|Lu\|_{W_{bd^+}^-} \geq c\|v\|_{W_{bd^+}^+}.$$

From the relation between $u(t,x)$ and $v(t,x)$, we obtain

$$\|u\|_{L_2(Q)} = \left\|\exp(t-T)v_{tt} - 2\lambda_A^{-1}\lambda_C^{-1}v_t\right\|_{L_2(Q)} \leq c\|v_{tt}\|_{L_2(Q)} \leq c\|v\|_{W_{bd^+}^+}.$$

Thus, for all smooth functions $u(t,x)$, that satisfy the conditions (2), the inequality in Lemma 3 is proved.

To prove the inequality for all functions $u \in \dot{W}_{bd}^+$, we employ passing to the limit.

Lemma 4. *For any function* $v \in W_{bd^+}^+$ *the following inequality holds true*

$$\|L^* v\|_{W_{bd}^-} \geq c\|v\|_{L_2(Q)}.$$

Proof of Lemma 4 is much analogous the previous proof. The auxiliary integral operator is of the form

$$u(t,x) = \int_0^t \varphi_1(\xi)\int_0^t \psi_1(\eta)v(\eta,x)d\eta d\xi.$$

where

$$\varphi_1(t) = \exp\left(-2\lambda_A^{-1}\lambda_C^{-1}(T-t+2)\exp(t)\right),$$

$$\psi_1(t) = \varphi_1^{-1}(t)\exp(t).$$

Theorem 1. *For any function* $f \in L_2(Q)$ *there exists a unique solution of the equation (1), (2) in the sense of Definition 1.1.1.*

Theorem 2. *For any function* $f \in W^{-}_{bd^+}$ *there exists a unique solution of the equation (1), (2) in the sense of Definition 1.1.4.*

2. ONE SOBOLEV PROBLEM

Consider a function $u(t,x)$ that is a solution of the following differential equation:

$$Lu = A\left(\frac{\partial^2 u}{\partial t^2}\right) + B(u) = F \tag{1}$$

with boundary conditions

$$u\big|_{t=0} = \frac{\partial u}{\partial t}\Big|_{t=0} = 0, \quad u\big|_{x\in\partial\Omega} = 0, \tag{2}$$

where A, B are differential operators:

$$Au = \sum_{i,j=1}^{n} \frac{\partial}{\partial x_i}\left(A_{ij}\frac{\partial u}{\partial x_j}\right)_{x_i}, \quad Bu = \sum_{i,j=1}^{n} \frac{\partial}{\partial x_i}\left(B_{ij}\frac{\partial u}{\partial x_j}\right), \tag{3}$$

where $A_{ij} = A_{ji}(x)$, $B_{ij} = B_{ji}(x)$, $\{A_{ij}\}_{i,j=1}^{n}$, $\{B_{ij}\}_{i,j=1}^{n}$ are continuously differentiable functions in the domain $\overline{\Omega}$.

The differential operators (3) satisfy the following conditions:

$$\sum_{i,j=1}^{n} A_{ij}(x)\xi_i\xi_j \ge \alpha\sum_{i=1}^{n}\xi_i^2, \quad \sum_{i,j=1}^{n} B_{ij}(x)\xi_i\xi_j \ge 0. \tag{4}$$

We study the system (1), (2) in a tube $Q = (0,T)\times\Omega$, $\Omega \subset R^n$.

Let $W_{bd}^{+}(Q)$ be a completion of the set of smooth in $\overline{Q}$ functions, which satisfy the following conditions

$$u\big|_{t=0} = \frac{\partial u}{\partial t}\Big|_{t=0} = 0, \quad u\big|_{x\in\partial\Omega} = 0$$

in the norm

$$\|u\|^2_{W^+_{bd}} = \int_Q \left(\sum_{i,j=1}^{n} A_{ij} u'_{x_i t} u_{x_j t} + \sum_{i,j=1}^{n} B_{ij} u_{x_i} u_{x_j} \right) dQ, \tag{5}$$

$W^{+}_{bd^{+}}$ be a completion of the set of smooth in $\overline{Q}$ functions, which satisfy the adjoint conditions

$$v\big|_{t=T} = v_t\big|_{t=T} = 0, \quad v\big|_{x\in\partial\Omega} = 0, \tag{6}$$

in the same norm (5), W^{-}_{bd} , $W^{-}_{bd^{+}}$ be corresponding negative spaces.

Lemma 1. *For any function* $u(t,x) \in W^{+}_{bd}$ *the following inequality holds true*

$$\|Lu\|_{W^{-}_{bd^{+}}} \le c\|u\|_{W^{+}_{bd}}.$$

Proof. First prove the inequality for smooth functions $u(t,x)$ which satisfy the conditions (2).

By the negative norm definition,

$$\|Lu\|_{W^{-}_{bd^{+}}} = \sup_{\substack{v\neq 0 \\ v\in W^{+}_{bd^{+}}}} \frac{|\langle v, Lu\rangle|}{\|v\|_{W^{+}_{bd^{+}}}} = \sup_{\substack{v\neq 0 \\ v\in W^{+}_{bd^{+}}}} \frac{\left|(v, Lu)_{L_2(Q)}\right|}{\|v\|_{W^{+}_{bd^{+}}}}, \tag{7}$$

Applying integration by parts and passing to the surface integrals, we have

$$\int_Q v \frac{\partial^2}{\partial t^2}\left(\sum_{i,j=1}^{n} \frac{\partial}{\partial x_i} A_{ij} \frac{\partial u}{\partial x_j} \right) dQ = \int_Q \frac{\partial}{\partial t}\left(v \frac{\partial}{\partial t} \sum_{i,j=1}^{n} \frac{\partial}{\partial x_i} A_{ij} \frac{\partial u}{\partial x_j} \right) dQ -$$

$$-\int_Q \sum_{i,j=1}^{n} \frac{\partial}{\partial x_i}\left(\frac{\partial v}{\partial t} A_{ij} \frac{\partial^2 u}{\partial x_j \partial t} \right) dQ + \int_Q \sum_{i,j=1}^{n} A_{ij} v_{x_i t} u_{x_j t} dQ =$$

$$= \int_Q \sum_{i,j=1}^{n} A_{ij} v_{x_i t} u_{x_j t} dQ$$

Employing the Schwarz inequality, we obtain

$$\int_Q v \frac{\partial^2}{\partial t^2}\left(\sum_{i,j=1}^{n} \frac{\partial}{\partial x_i} A_{ij} \frac{\partial u}{\partial x_j} \right) dQ \le c \|v\|_{W^+_{bd^+}} \|u\|_{W^+_{bd}} .$$

In the same manner estimate the other summands of L:

$$\int_Q v \sum_{i,j=1}^{n} \frac{\partial}{\partial x_i} B_{ij} \frac{\partial u}{\partial x_j} dQ = \int_Q \sum_{i,j=1}^{n} \frac{\partial}{\partial x_i} \left(v B_{ij} \frac{\partial u}{\partial x_j} \right) dQ -$$

$$- \int_Q \sum_{i,j=1}^{n} B_{ij} v_{x_i} u_{x_j} dQ \le c \|v\|_{W^+_{bd^+}} \|u\|_{W^+_{bd}} .$$

Substituting these inequalities into (7) and dividing by $\|v\|_{W^+_{bd^+}}$, we obtain the desired inequality for smooth functions $u(t,x)$, which satisfy the conditions (2). Continuously extending the operator L to the space W^+_{bd}, we have the inequality for all functions.

Lemma 2. *For any function* $v(t,x) \in W^+_{bd^+}$ *the following inequality holds true*

$$\|L^* v\|_{W^-_{bd}} \le c \|v\|_{W^+_{bd^+}} .$$

Lemma 2 is proved in much the same manner as Lemma 1.

From the proved inequalities it follows that the operators L and $L^*(\cdot)$ can be extended to the continuous operators mapping W^+_{bd} and $W^+_{bd^+}$ into $W^-_{bd^+}$ and W^-_{bd}, respectively.

Lemma 3. *For any function* $u(t,x) \in W^+_{bd}$ *the following inequality holds true*

$$\|Lu\|_{W^-_{bd^+}} \ge c \|u\|_{H^+_{bd}},$$

where H^+_{bd} is a completion of the set of smooth in $\overline{Q}$ functions, which satisfy the conditions (2), in the norm

$$\|u\|^2_{H^+_{bd}} = \int_Q \sum_{i,j=1}^{n} A_{ij} u_{x_i} u_{x_j} dQ .$$

P r o o f. First prove the lemma for smooth functions $u(t,x)$ that satisfy conditions (2).

Consider the auxiliary functions $v(t,x)$ of the form:

$$v(t,x) = \int_T^t \left[tg\left(\pi\tau/2T\right)+1\right]^{-1} u(\tau,x)d\tau .$$

It is obvious that $v \in W^+_{bd^+}$. Prove the following inequality:

$$(v, Lu)_{L_2(Q)} \geq c\|v\|^2_{W^+_{bd^+}} . \tag{8}$$

Employing the integration by parts and taking into account relation between $u(t,x)$ and $v(t,x)$, we obtain

$$\int_Q v \sum_{i,j=1}^{n} \frac{\partial^2}{\partial t^2} \frac{\partial}{\partial x_i} A_{ij} \frac{\partial u}{\partial x_j} dQ =$$

$$= \int_Q \left(v \sum_{i,j=1}^{n} \frac{\partial}{\partial t} \frac{\partial}{\partial x_i} A_{ij} \frac{\partial u}{\partial x_j} \right)_t dQ - \int_Q \frac{\partial}{\partial x_i} \left(v_t \sum_{i,j=1}^{n} A_{ij} u_{tx_j} \right) dQ + \tag{9}$$

$$+ \frac{\pi}{2T} \int_Q \sum_{i,j=1}^{n} v_{tx_i} A_{ij} \cos^{-2} \frac{\pi t}{2T} v_{tx_j} dQ +$$

$$+ \int_Q \left(tg \frac{\pi t}{2T} + 1 \right) \sum_{i,j=1}^{n} A_{ij} v_{tx_i} v_{ttx_j} dQ.$$

Passing in the formula (9) to the surface integration and taking into account boundary conditions, we have

$$\int_Q v \sum_{i,j=1}^{n} \frac{\partial^2}{\partial t^2} \frac{\partial}{\partial x_i} A_{ij} \frac{\partial u}{\partial x_j} dQ \geq c_1 \int_Q \cos^{-2} \frac{\pi t}{2T} \sum_{i,j=1}^{n} A_{ij} v_{tx_i} v_{tx_j} dQ . \tag{10}$$

In the same way consider the other summands

$$\int_Q v\sum_{i,j=1}^{n}\frac{\partial}{\partial x_i}B_{ij}\frac{\partial u}{\partial x_j}dQ \geq c_2\int_Q \cos^{-2}\frac{\pi t}{2T}\sum_{i,j=1}^{n}B_{ij}v_{x_i}v_{x_j}dQ. \quad (11)$$

Adding (10) and (11), we obtain (8). Applying the Schwarz inequality to left-hand side of (8), we have desired inequality for smooth functions $u \in W_{bd}^{+}$. Passing to the limit, we obtain the inequality for all functions $u \in W_{bd}^{+'}$.

Lemma 4. *For any function* $v(t,x) \in W_{bd^+}^{+}$ *the following inequality holds true*

$$\|L^* v\|_{W_{bd}^{-}} \geq c\|v\|_{H_{bd^+}^{+}},$$

where $H_{bd^+}^{+}$ *is a completion of the set of smooth in* $\overline{Q}$ *functions, which satisfy the conditions (6), in the norm*

$$\|v\|_{H_{bd^+}^{+}}^2 \doteq \int_Q \sum_{i,j=1}^{n} A_{ij}v_{x_i}v_{x_j}dQ.$$

The operators L^* and L are defined by the same differential expression but $D(L^*) \neq D(L)$. Lemma 4 is proved in much the same way as previous one. The auxiliary function is in the form

$$u(t,x) = -\int_0^t \left[ctg\left(\pi\tau/2T\right)+1\right]^{-1} v(\tau,x)d\tau.$$

Theorem 1. *For any function* $f \in H_{bd^+}^{-}$ *there exists a unique solution of the equation (1), (2) in the sense of Definition 1.1.1.*

Theorem 2. *For any function* $f \in W_{bd^+}^{-}$ *there exists a unique solution of the equation (1), (2) in the sense of Definition 1.1.4.*

3. PULSE OPTIMAL CONTROL OF THE DYNAMIC EQUATION OF VISCOUS STRATIFIED FLUID

Complete the templates of the theorems of Section 3.6

Table 1.

N	Operator	Space N	Space $W^-(Q)$
1.	$L(\cdot)$	$L_2(Q)$	$W^-_{bd^+}$
2.	$L(\cdot)$	W^+_{bd}	$L_2(Q)$

Table 2.

N	Operator $A_i(\cdot)$	Space $W^-(Q)$
1.	$A_1(\cdot)$	$W^-_{bd^+}$
2.	$A_2(\cdot)$, $k=1$	$W^-_{bd^+}$
3.	$A_3(\cdot)$	$W^-_{bd^+}$
4.	$A_4(\cdot)$	$W^-_{bd^+}$
5.	$A_5(\cdot)$	$W^-_{bd^+}$

Table 3.

N	Operator $L(\cdot)$	Space $W^-(Q)$	Space $W^+(Q)$
1.	$L(\cdot)$	$W^-_{bd^+}$	$W^+_{bd^+}$

Table 4.

N	Space $W^{-}(Q)$	Map $A_i(\cdot)$
1.	$W^{-}_{bd^{+}}$	$A_1(\cdot)$
2..	$W^{-}_{bd^{+}}$	$A_3(\cdot)$
3.	$W^{-}_{bd^{+}}$	$A_4(\cdot)$
4.	$W^{-}_{bd^{+}}$	$A_5(\cdot)$

Table 5.

N	Operator $L(\cdot)$	Space $W^{-}(Q)$
1.	$L(\cdot)$	$W^{-}_{bd^{+}}$

Table 6.

N	Operator $A_i(\cdot)$	Space $W^{-}(Q)$
1.	$A_1(\cdot)$	$W^{-}_{bd^{+}}$
2.	$A_2(\cdot),\, k=1$	$W^{-}_{bd^{+}}$
3.	$A_3(\cdot)$	$W^{-}_{bd^{+}}$
4.	$A_4(\cdot)$	$W^{-}_{bd^{+}}$
5.	$A_5(\cdot)$	$W^{-}_{bd^{+}}$

Table 7.

N	Exponent α	Space $W^{-}(Q)$	Map $A_{i,\varepsilon}(\cdot)$
1.	1/2	$W^{-}_{bd^{+}}$	$A_{1,\varepsilon}(\cdot)$
2.	1/2	$W^{-}_{bd^{+}}$	$A_{2,\varepsilon}(\cdot)$

Table 7 (continuation)

N	Exponent α	Space $W^{-}(Q)$	Map $A_{i,\varepsilon}(\cdot)$
3.	$1/2$	$W^{-}_{bd^{+}}$	$A_{3,\varepsilon}(\cdot)$
4.	$1/2$	$W^{-}_{bd^{+}}$	$A_{4,\varepsilon}(\cdot)$
5.	$1/2$	$W^{-}_{bd^{+}}$	$A_{5,\varepsilon}(\cdot)$

4. PULSE OPTIMAL CONTROL OF ONE SOBOLEV SYSTEM

Using the templates of Section 3.6, complete the tables.

Table 1.

N	Operator	Space N	Space $W^{-}(Q)$
1.	$L(\cdot)$	H^{+}_{bd}	$W^{-}_{bd^{+}}$
2.	$L(\cdot)$	W^{+}_{bd}	$H^{-}_{bd^{+}}$

Table 2.

N	Operator $A_i(\cdot)$	Space $W^{-}(Q)$
1.	$A_1(\cdot)$	$W^{-}_{bd^{+}}$
2.	$A_3(\cdot)$	$W^{-}_{bd^{+}}, H^{-}_{bd^{+}}$
3.	$A_4(\cdot)$	$W^{-}_{bd^{+}}$
4.	$A_5(\cdot)$	$W^{-}_{bd^{+}}, H^{-}_{bd^{+}}$

Table 3.

N	Operator $L(\cdot)$	Space $W^{-}(Q)$	Space $W^{+}(Q)$
1.	$L(\cdot)$	$W^{-}_{bd^{+}}$	$W^{+}_{bd^{+}}$

Table 4 is empty.

Table 5.

N	Operator $L(\cdot)$	Space $W^-(Q)$
1.	$L(\cdot)$	$W^-_{bd^+}$

Table 6.

N	Operator $A_i(\cdot)$	Space $W^-(Q)$
1.	$A_1(\cdot)$	$W^-_{bd^+}$
2.	$A_3(\cdot)$	$W^-_{bd^+}, H^-_{bd^+}$
3.	$A_4(\cdot)$	$W^-_{bd^+}$
4.	$A_5(\cdot)$	$W^-_{bd^+}, H^-_{bd^+}$

Table 7.

N	Exponent α	Space $W^-(Q)$	Map $A_{i,\varepsilon}(\cdot)$
1.	1/2	$W^-_{bd^+}$	$A_{1,\varepsilon}(\cdot)$
2.	1/2	$W^-_{bd^+}$	$A_{2,\varepsilon}(\cdot)$
3.	1/2	$W^-_{bd^+}, H^-_{bd^+}$	$A_{3,\varepsilon}(\cdot)$
4.	1/2	$W^-_{bd^+}$	$A_{4,\varepsilon}(\cdot)$
5.	1/2	$W^-_{bd^+}, H^-_{bd^+}$	$A_{5,\varepsilon}(\cdot)$

Chapter 10

CONTROLLABILITY OF LINEAR SYSTEMS WITH GENERALIZED CONTROL

1. TRAJECTORY CONTROLLABILITY

In investigations of various dynamic systems the problem of its controllability is one the most important.

The problem of the controllability was studied in [6] for the linear ordinary systems, which allow the generalized controls. It was pointed out there that the introduction of such controls does not extend the conditions of the complete controllability. In the case of distributed systems the situation is much more difficult. It was shown in the paper [14] that the solvability of the problem of the controllability for distributed systems with a point control significantly depends on the fact whether the point of the control application may be approximated by the Diophant approximations well enough. Various conditions of the controllability of lumped and distributed systems were obtained, for examples, in [30-32, 84, 86, 153-157, 177].

Let $\|u\|_W = \|\ell f\|_{W_{bd}^{+}}$ be a solution of the problem (1.1.1.12). By controllability of the system (1.1.1.12) we shall mean the possibility to reach any state $u(t,x)$ as a result of admissible controls $h \in U_{ad}$.. To investigate the controllability, it is necessary to study the properties of the operator L. This problem is solved with the help of the apparatus of equipped Hilbert spaces and inequalities in negative norms [63, 122, 142, 143, 158-162, 177], and also with the help of correspondent inequalities in positive norms, which follow from them.

Definition 1. *The system is controllable in a Banach space W by the set of admissible controls U_{ad} if the set $\{u(t,x;h) | h \in U_{ad}\}$ covers the space W, i.e. $\forall z(t,x) \in W\ \exists h_z \in U_{ad}:\ u(t,x;h_z) = z(t,x)$ in W.*

Definition 2. *The system is* ε*-controllable in the Banach space* W *by the set of admissible controls* U_{ad} *if the set* $\{u(t,x;h) | h \in U_{ad}\}$ *is dense in* W, *i.e.* $\forall z \in L_2(Q)\, \forall \varepsilon > 0\, \exists h_{z,\varepsilon} \in U_{ad}$ *such that*

$$\left\| u(h_{z,\varepsilon}) - z \right\|_{W} < \varepsilon . \tag{1}$$

Let the following inequalities in the negative norms still hold true for $u(t,x) \in W_{bd}^{+l}, v(t,x) \in W_{bd^+}^{+l}$:

$$\begin{aligned} \|u\|_{H_{bd}^+} &\le C_1 \|Lu\|_{W_{bd^+}^{-l}} \le C_2 \|u\|_{W_{bd}^{+l}},\ u \in W_{bd}^{+l}, \\ \|v\|_{H_{bd}^+} &\le C_1 \left\|L^* u\right\|_{W_{bd}^{-l}} \le C_2 \|u\|_{W_{bd^+}^{+l}},\ v \in W_{bd^+}^{+l}. \end{aligned} \tag{2}$$

The inequalities in the negative norms and the results of Section 1 imply the following theorems.

Theorem 1. *If operator* $A(\cdot)$ *maps* $U_{ad} \subset H$ *surjectively into the whole space* $W_{bd^+}^{-l}$ *then the system (1.1.12) is controllable in the space* W_{bd}^{+l}.

Proof. Let $u^*(t,x)$ be an arbitrary element belonging to W_{bd}^{+l}. Due to the inequalities (2), we obtain that $Lu^*(t,x) \in W_{bd^+}^{-l}$. As far as the operator $A(\cdot)$ surjectively maps $U_{ad} \subset H$ into $W_{bd^+}^{-l}$ there exists such element $h^* \in U_{ad} \subset H$ that $A(h^*) = Lu^*(t,x) - f$ in $W_{bd^+}^{-l}$. Hence, by Lemma 1.3, we obtain that

$$\left\| u(t,x;^*) - u^*(t,x) \right\|_{H_{bd}^+} \le c \left\| A(h^*) + f - Lu^*(t,x) \right\|_{W_{bd^+}^{-l}} = 0,$$

where $u(t,x;h^*)$ is a solution of the problem (1.1.12) with the right-hand side $A(h^*) + f$ in the sense of Definition 1.1.4. For this reason, $u(t,x;h^*) - u^*(t,x) = 0$ in H_{bd}^+, and hence, in W_{bd}^{+l} also. Taking into account that $u^*(t,x) \in W^{+l}_{bd^+}$, we conclude that $u(t,x;h^*) \in W^{+l}_{bd^+}$. It follows from Lemma 1.4 that $u(t,x;h^*)$ is a solution of problem (1.1.12) in the sense of Definition 1.1 also. Thus, the system is controllable in W^{+l}_{bd}.

Theorem 2. *If operator $A(\cdot)$ maps $U_{ad} \subset H$ in $W^{-l}_{bd^+}$ so that the set $A(U_{ad})$ is dense in $W^{-l}_{bd^+}$, then system (1.1) is ε-controllable in H^+_{bd}.*

Proof. Let $u^*(t,x)$ is an arbitrary element belonging to H^+_{bd}. Since W^{+l}_{bd} is densely imbedded in H^+_{bd}, there exists a sequence $u^*_k(t,x) \in W^{+l}_{bd^+}$ such that $\|u^*_k - u^*\|_{H^+_{bd^+}} \xrightarrow[k\to\infty]{} 0$. Consider $Lu^*_k(t,x) \in W^{-l}_{bd^+}$ with fixed k. As far as $A(U_{ad})$ is dense in $W^{-l}_{bd^+}$, there exists a sequence of controls $h^*_{i,k}$ such that $\|f(h^*_{i,k}) - Lu^*_k\|_{W^{-l}_{bd^+}} \xrightarrow[i\to\infty]{} 0$. Choose such $i = i(k) \in N$ that $\|f(h^*_{i(k),k}) - Lu^*_k\|_{W^{-l}_{bd^+}} \le \frac{1}{k}$. Let $u_{i(k),k} = u(t,x;h^*_{i(k),k})$ be a solution of the problem (1.1.12) with the right-hand side $A(h^*_{i(k),k}) + f$ in the sense of Definition 1.1.4. Granting Lemma 1.3 we obtain

$$\|u_{i(k),k} - u^*\|_{H^+_{bd}} \le \|u_{i(k),k} - u^*_k\|_{H^+_{bd}} + \|u^*_k - u^*\|_{H^+_{bd}} \le$$
$$\le c\|f(h^*_{i(k),k}) - Lu^*_k\|_{W^{-l}_{bd^+}} + \|u^*_k - u^*\|_{H^+_{bd}} \le$$
$$\le \frac{c}{k} + \|u^*_k - u^*\|_{H^+_{bd}} \xrightarrow[k\to\infty]{} 0.$$

In other words, the system is ε-controllable in H^+_{bd}.

It is well known [78], that the linear span of the functions $M=\{g(t)\cdot\omega(x)\}$, $g(t)\cup C_0^\infty(0,T)$, $\omega(x)\in C_0^\infty(\Omega)$ is dense in the set $L_2(Q)$,, and hence in $W^{-1}_{2,bd^+}\otimes L_2(Q)$. also. On the other hand, it is easy to prove that the sequence

$$\sum_{i=1}^{n} g(t_i)\delta(t-t_i)\omega(x)(t_{x+1}-t_i)$$

converges to the function $g'(t)\omega(x)$, as the diameter of the decomposition $\tau=\{t_1,t_2,\ldots t_{n+1}\}$ of the interval $(0,T)$ tends to zero.

Thus, the set

$$\left\{\sum_{i=1}^{n}\delta(t-t_i)\varphi_i(x)\right\}$$

is dense in $W^{-1}_{2,bd^+}\otimes L_2(\Omega)$, and hence, the previous theorems are valid in the case of pulse controllability.

2. TRAJECTORY-FINAL CONTROLLABILITY OF SOME LINEAR SYSTEMS

The problems of controllability of linear systems have one essential difference from the classic problems for systems with concentrated or distributed parameters. In this problems it is necessary to provide the required state of the system with the help of controls during the whole time interval of the system functioning (trajectory controllability) rather than to transfer the system into a desired state in a finite time interval (final controllability).

It is naturally to consider the problem of the final controllability in a finite time interval in the case of linear distributed systems with generalized controls. The singularity of the right-hand side makes this consideration difficult enough, since the function $u(T,x)$, $x\in\Omega$ may does not belong to $L_2(\Omega)$, but it may be an element of some negative space.

However, in the case of hyperbolic and pseudo-hyperbolic systems it is possible to obtain a positive results concerning the final controllability in a finite time interval. Moreover, it is possible to provide the desired state of the system with an arbitrary given precision (in an integral metrics). Thus, we may' say about the trajectory-final controllability.

Let us consider in detail the system described by the Newton pseudo-hyperbolic initial boundary value problem.

Let Ω be a bounded domain in R^n with a regular bound $\partial\Omega$. The state $u(t,x)$ of the considered system is a solution of the following problem

$$Lu = \frac{\partial^2 u}{\partial t^2} - \frac{\partial}{\partial t}\Delta u - \Delta u = f(t,x;h) \quad \text{in } Q = (0,T] \times \Omega, \tag{1}$$

$$u\big|_{t=0} = \frac{\partial u}{\partial t}\Big|_{t=0} = 0, \; x \in \Omega, \tag{2}$$

$$\frac{\partial}{\partial t}\frac{\partial u}{\partial n} + \frac{\partial u}{\partial n} + u = 0 \text{ on } \Gamma = (0,T] \times \partial\Omega, \tag{3}$$

where the parameter h is a control from the set of admissible controls $U_{ad} \subset V$, V is a space of controls.

Remark. *The following results will be valid also if we shall replace the Laplacian by symmetric elliptic operators of the second order with smooth coefficients and the conditions (3) will be replaced by the corresponding Dirichlet or Neumann conditions.*

Next, we shall investigate the generalized solvability of the problem (1)-(3), basing on which we can prove the trajectory-final ε-controllability of the system.

2.1. Generalized solutions and ε-controllability

Denote by W_{bd}^{+} a completion of $D(L)$ (the set of smooth in $\overline{Q}$ functions satisfying the conditions (2), (3)) in the norm

$$\|u\|_{W_{bd}^{+}} = \left(\int_Q u_t^2 + \sum_{i=1}^{n} u_{x_i t}^2 \, dQ \right)^{1/2}.$$

$W_{bd^+}^{+}$ is a completion of $D(L^+)$ in the same norm. Here $L^+ v = \frac{\partial^2 v}{\partial t^2} + \Delta \frac{\partial v}{\partial t} - \Delta v$, and $D(L^+)$ is the set of smooth in $\overline{Q}$ functions satisfying the adjoint conditions

$$v|_{t=T} = \frac{\partial v}{\partial t}\Big|_{t=T} = 0, \; x \in \Omega,$$

$$-\frac{\partial}{\partial n}\frac{\partial v}{\partial t} + \frac{\partial v}{\partial n} + v = 0 \text{ on } \Gamma = (0,T] \times \partial\Omega.$$

Let us construct on the space $L_2(Q)$ the negative spaces W_{bd}^{-}, $W_{bd^+}^{-}$ using the positive spaces introduced above

Introduce the pair of the spaces H, H_+ as completions of $D(L)$, $D(L^+)$ in the following norms

$$|u|_H = \left(\int_Q u^2 + \sum_{i=1}^{n} u_{x_i}^2 \, dQ + \int_\Omega u^2|_{t=T} d\Omega \right)^{1/2},$$

$$|v|_{H_+} = \left(\int_Q v^2 + \sum_{i=1}^{n} v_{x_i}^2 \, dQ + \int_\Omega v^2|_{t=0} d\Omega \right)^{1/2}$$

The spaces H, H_+ are isometric to the direct sum of the Hilbert spaces $(L_2(0,T) \otimes W_2^1(\Omega)) \oplus L_2(\Omega)$ and there are no imbeddings $H \subseteq L_2(Q)$, $H_+ \subseteq L_2(Q)$ [74].

Lemma 1. *For any function* $u \in D(L)$, $v \in D(L^+)$ *the following inequalities holds true*

$$|u|_H \le c_1 \|Lu\|_{W^-_{bd^+}} \le c_2 \|u\|_{W^+_{bd}}, \tag{4}$$

$$|v|_{H_+} \le c_1 \|L^+ v\|_{W^-_{bd}} \le c_2 \|v\|_{W^+_{bd^+}}. \tag{5}$$

Proof. Let us prove the inequality (4) ((5) can be proved in the similar way). The right-hand side of the inequality (4) is proved by applying of the integration by part, the Ostrogradsky-Gauss formula and the integral Cauchy inequality to $(Lu,v)_{L_2(Q)}$. Here v is an arbitrary smooth in $\overline{Q}$ function satisfying the adjoint boundary conditions.

To prove the left-hand side of (5), consider the expression

$$(Lu, Iu)_{L_2(Q)} = \int_Q Lu\left(-\int_T^t e^{-\tau} u(\tau,x)\, d\tau\right) dQ,$$

where $u \in D(L)$ is an arbitrary smooth in $\overline{Q}$ function satisfying the boundary conditions (3), (4). Note, that $Iu = -\int_T^t e^{-\tau} u(\tau,x) d\tau \in W^+_{bd^+}$. Indeed, the initial condition $Iu|_{t=T} = 0$ holds true as a result of the construction of the auxiliary operator. The others boundary conditions disappear after completion of $D(L^+)$ in the norm $\|\cdot\|_{W^+_{bd^+}}$. Expressing u in terms of Iu, we obtain $u = -e^t \dfrac{\partial Iu}{\partial t}$.

We have

$$(Lu, Iu)_{L_2(Q)} = J_1 + J_2,$$

where

$$J_1 = \int_Q u_{tt} Iu dQ, \quad J_2 = -\int_Q \left(\frac{\partial}{\partial t}\Delta u + \Delta u\right) Iu dQ.$$

Consider every term J_1, J_2, separately.

J_1. Using the integration by parts and taking into account the conditions

$$u\big|_{t=0} = \frac{\partial u}{\partial t}\Big|_{t=0} = Iu\big|_{t=T} = 0$$

and the fact, that $\dfrac{\partial Iu}{\partial t} = -e^{-t}u$, we have

$$J_1 = \int_Q u_{tt} Iu\, dQ = \int_Q (u_t Iu)_t dQ - \int_Q u_t (Iu)_t dQ =$$

$$= \int_Q e^{-t} uu_t dQ = \frac{1}{2}\int_Q e^{-t}\left(u^2\right)_t dQ = \frac{1}{2}\int_Q \left(e^{-t}u^2\right)_t dQ +$$

$$+\frac{1}{2}\int_Q e^{-t}u^2 dQ = \frac{1}{2}\int_\Omega e^{-t}u^2\Big|_{t=0}^{t=T} d\Omega + \frac{1}{2}\int_Q e^{t}(Iu)_t (Iu)_t dQ \geq$$

$$\geq \frac{1}{2}\|(Iu)_t\|^2_{L_2(Q)} + \frac{1}{2}e^{-T}\int_\Omega u^2\big|_{t=T} d\Omega.$$

J_2. Using the Ostrogradsky-Gauss formula and taking into account the boundary condition $\dfrac{\partial}{\partial t}\dfrac{\partial u}{\partial n} + \dfrac{\partial u}{\partial n} + u\big|_\Gamma = 0$, we have

$$J_2 = -\int_Q \left(\frac{\partial}{\partial t}\Delta u + \Delta u\right) Iu\, dQ = -\int_\Gamma \left(\frac{\partial}{\partial t}\frac{\partial u}{\partial n} + \frac{\partial u}{\partial n}\right) Iu\, d\Gamma +$$

$$+\int_Q \sum_{i=1}^n u_{x_i t} Iu_{x_i}\, dQ + +\int_Q \sum_{i=1}^n u_{x_i} Iu_{x_i}\, dQ = \int_\Gamma uIu\, d\Gamma +$$

$$+\int_Q \sum_{i=1}^n u_{x_i t} Iu_{x_i}\, dQ + \int_Q \sum_{i=1}^n u_{x_i} Iu_{x_i}\, dQ.$$

Consider every term separately.

a) $\displaystyle\int_\Gamma uIu\, d\Gamma = -\int_\Gamma e^t Iu(Iu)_t d\Gamma = -\frac{1}{2}\int_\Gamma e^t (IuIu)_t d\Gamma =$

$$=-\frac{1}{2}\int_{\Gamma}\left(e^{t}IuIu\right)_{t}d\Gamma+\frac{1}{2}\int_{\Gamma}e^{t}IuIu\,d\Gamma=$$

$$=-\frac{1}{2}\int_{\partial\Omega}e^{t}IuIu\Big|_{t=0}^{t=T}d\partial\Omega+\frac{1}{2}\int_{\Gamma}e^{t}IuIu\,d\Gamma\geq 0\,.$$

b) To estimate the second integral, we must use the integration by parts and the conditions $u\big|_{t=0}=Iu\big|_{t=T}=0$. We obtain

$$\int_{Q}\sum_{i=1}^{n}u_{x_{i}t}Iu_{x_{i}}\,dQ=\int_{Q}\sum_{i=1}^{n}\left(u_{x_{i}}Iu_{x_{i}}\right)_{t}dQ-\int_{Q}\sum_{i=1}^{n}u_{x_{i}}(Iu)_{x_{i}t}\,dQ=$$

$$=-\int_{Q}\sum_{i=1}^{n}u_{x_{i}}(Iu)_{x_{i}t}\,dQ.$$

Granting that $u_{x_{i}}=-e^{t}\left(Iu\right)_{x_{i}t}$, we can write

$$-\int_{Q}\sum_{i=1}^{n}u_{x_{i}}(Iu)_{x_{i}t}dQ=\int_{Q}e^{t}\sum_{i=1}^{n}(Iu)_{x_{i}t}(Iu)_{x_{i}t}\,dQ\geq$$

$$\geq\int_{Q}\sum_{i=1}^{n}(Iu)_{x_{i}t}(Iu)_{x_{i}t}\,dQ.$$

c) Let us show, that the third integral in nonnegative. To do this, we must apply the integration by parts.

$$\int_{Q}\sum_{i=1}^{n}u_{x_{i}}(Iu)_{x_{i}}\,dQ=-\int_{Q}\sum_{i=1}^{n}e^{t}(Iu)_{x_{i}t}(Iu)_{x_{i}}\,dQ=$$

$$=-\frac{1}{2}\int_{Q}\sum_{i=1}^{n}\left(e^{t}(Iu)_{x_{i}}(Iu)_{x_{i}}\right)_{t}dQ+\frac{1}{2}\int_{Q}e^{t}\sum_{i=1}^{n}(Iu)_{x_{i}}(Iu)_{x_{i}}\,dQ=$$

$$=-\frac{1}{2}\int_{\Omega}\sum_{i=1}^{n}e^{t}(Iu)_{x_{i}}(Iu)_{x_{i}}\Big|_{t=0}^{t=T}d\Omega+\frac{1}{2}\int_{Q}e^{t}\sum_{i=1}^{n}(Iu)_{x_{i}}(Iu)_{x_{i}}\,dQ\geq 0\,.$$

Collecting together all the obtained inequalities and taking into account, that $\left(Iu\right)_{t}=-e^{-t}u$, we have

$$\|Lu\|_{W^-_{bd^+}}\left(\int_Q e^{-2t}\left(u^2+\sum_{i=1}^n u_{x_i}^2\right)dQ\right)^{1/2} \geq (Lu, Iu)_{L_2(Q)} \geq$$

$$\geq c\int_Q e^{-2t}\left(u^2+\sum_{i=1}^n u_{x_i}^2\right)dQ + c\left\|u|_{t=T}\right\|^2_{L_2(\Omega)},\ c>0.$$

Applying to the right-hand side the inequality $\frac{a^2+b^2}{2}\geq\sqrt{ab}$ and reducing by $\int_Q e^{-2t}\left(u^2+\sum_{i=1}^n u_{x_i}^2\right)dQ$, we have $2c\left\|u|_{t=T}\right\|_{L_2(\Omega)}\leq\|Lu\|_{W^-_{bd^+}}$. Granting that

$$\|Lu\|_{W^-_{bd^+}}\geq c\left(\int_Q e^{-2t}\left(u^2+\sum_{i=1}^n u_{x_i}^2\right)dQ\right)^{1/2},$$

we have that (4) is valid. The lemma is proved.

Basing on the right-hand sides of the inequalities (4), (5), we can extend with respect to continuity the operator L (L^+, respectively) onto the whole space W^+_{bd} ($W^+_{bd^+}$, respectively) and consider that it continuously maps into $W^-_{bd^+}$ (W^-_{bd}, respectively). We save the same denotations for the extended operators. The inequalities (4), (5) are valid also for the extended operators for any $u\in W^+_{bd}$, $v\in W^+_{bd^+}$.

Using inequalities obtained above in the similar way, it is possible to prove the unique dense solvability of the operator equations

$$Lu=f,\ f\in W^-_{bd^+},$$

$$L^+v=g,\ g\in W^-_{bd}.$$

Let H^-_{bd}, $H^-_{bd^+}$ be the negative spaces constructed on $L_2(Q)$ and the corresponding positive spaces H^+_{bd}, $H^+_{bd^+}$, which, in turn, are completions of $D(L)$, $D(L^+)$ in the norm $\|u\| = \left(\int_Q u^2 + \sum_{i=1}^{n} u_{x_i}^2 \, dQ \right)^{1/2}$. The imbeddings $H_{bd}^- \subset W_{bd}^-$, $H_{bd^+}^- \subset W_{bd^+}^-$ are valid and dense. The following theorems hold true.

Theorem 1. *For any element* $f \in H_{bd^+}^-$ *there exists a unique solution* $u \in W_{bd}^+$ *of the equatioh* $Lu = f$.

Theorem 2. *For any element* $g \in H_{bd}^-$ *there exists a unique solution* $v \in W_{bd^+}^+$ *of the equation* $L^+ v = g$.

Let us define a generalized solution of the problem (1), (2), (3) as a function $u \in H$ for which there exists a sequence of smooth functions $u_m \in D(L)$, $m = 1,2,\ldots$, such that $|u_m - u|_H \xrightarrow[m \to \infty]{} 0$, $\|Lu_m - f\|_{W_{bd^+}^-} \xrightarrow[m \to \infty]{} 0$.

It should been noted, that the similar conception of a generalized solution of the operator equation with closed linear operator in a Banach space was proposed in the paper [165].

Theorem 3. *For any element* $f \in W_{bd^+}^-$ *there exists a unique generalized solution* $u \in H$ *of the problem (1)-(3).*

Proof. Let $f \in W_{bd^+}^-$. By virtue of the density $H_{bd^+}^-$ in $W_{bd^+}^-$ there exists such sequence $f_m \in H_{bd^+}^-$, that $\|f - f_m\|_{W_{bd^+}^-} \xrightarrow[m \to \infty]{} 0$.

By Theorem 1 for any function $f_m \in H_{bd^+}^-$ there exists a unique solution $\bar{u}_m \in W_{bd}^+$ of the operator equation $Lu = f_p$. Using the left-hand side of the inequality (4) and the imbedding $W_{bd}^+ \subset H$, we have

$$|\bar{u}_{m_1} - \bar{u}_{m_2}|_H \le c_1 \|L\bar{u}_{m_1} - L\bar{u}_{m_2}\|_{W_{bd^+}^-} = c_1 \|f_{m_1} - f_{m_2}\|_{W_{bd^+}^-} \xrightarrow[m_1, m_2 \to \infty]{} 0 .$$

Thus, by virtue of the completeness of the space H there exists such function $u \in H$ that $|u-\overline{u}_m|_H \xrightarrow[m\to\infty]{} 0$.

Let us show that $u \in H$ is desired generalized solution. From the density $D(L)$ in W^+_{bd} it follows that for any natural number m there exists a sequence of smooth functions $u_{m,i} \in D(L)$, $i=1,2,\dots$, such that $\|\overline{u}_m - u_{m,i}\|_{W^+_{bd}} \xrightarrow[i\to\infty]{} 0$. Choose $i(m) \in N$ from the condition $\|\overline{u}_m - u_{m,i(m)}\|_{W^+_{bd}} < 1/m$ and put $u_m' = u_{m,i(m)}$. Then

$$|u-u_m|_H \le |u-\overline{u}_m|_H + |\overline{u}_m - u_m|_H \le |u-\overline{u}_m|_H + c\|\overline{u}_m - u_m\|_{W^+_{bd}} \le$$

$$\le |u-\overline{u}_m|_H + \frac{c}{m} \xrightarrow[m\to\infty]{} 0$$

On the other hand,

$$\|Lu_m - f\|_{W^-_{bd^+}} \le \|Lu_m - L\overline{u}_m\|_{W^-_{bd^+}} + \|f_m - f\|_{W^-_{bd^+}} \le c\|u_m - \overline{u}_m\|_{W^+_{bd}} +$$

$$+\|f_m - f\|_{W^-_{bd^+}} \le \frac{c}{m} + \|f_m - f\|_{W^-_{bd^+}} \xrightarrow[m\to\infty]{} 0.$$

The uniqueness is proved in the standard way by contradiction. The theorem is proved.

Definition 1. *The system (1)-(3) is* ε*-controllable in the space* H *by the set of admissible controls* U_{ad}, *if the set* $\{u(h) | h \in U_{ad}\}$ *is dense in* H.

Theorem 4. *If the set* $f(U_{ad})$ *is dense in* $W^-_{bd^+}$ *then the system (1)-(3) is* ε*-controllable in the space* H.

Proof. Let an arbitrary function $z \in H$ be given. Consider the space W, which is a completion of the set $D(L)$ in the norm $|u|_W = \|Lu\|_{W^-_{bd^+}}$. It is possible to prove that the imbeddings $W^+_{bd} \subset W \subset H$ are valid and dense, and in addition, the following inequalities hold true

$$|u|_H \le c_1 |u|_W \le c_2 \|u\|_{W^+_{bd^+}},\ \forall\, u \in W^+_{bd},\ |y|_H \le c_1 |y|_W,\ \forall\, y \in W .$$

Moreover, it follows from the proof of Theorem 3 that generalized solutions belong to $H \cap W$ and $|u|_W = \|f\|_{W^-_{bd^+}}$.

As far as W is dense in H, there exists such sequence $z_i \in W$ that $|z - z_i|_H \xrightarrow[i\to\infty]{} 0$. By virtue of the density of $f(U_{ad})$ in $W^-_{bd^+}$ for any natural number i there exists such sequence of the admissible controls $h^i_m \in U_{ad}$ that $\|Lz_i - f(h^i_m)\|_{W^-_{bd^+}} < 1/m$. Then, we have

$$|u(h^i_m) - z|_H \le |u(h^i_m) - z_i|_H + |z_i - z|_H \le c_1 |u(h^i_m) - z_i|_W +$$

$$+ |z_i - z|_H \le c_1/m + |z_i - z|_H \xrightarrow[i,m\to\infty]{} 0 .$$

The theorem is proved.

In this case the imbedding $W^{-1}_{2,bd^+}(0,T) \otimes L_2(\Omega) \subset W^-_{bd^+}$ is valid and dense, so we can conclude that the system (1)-(3) is trajectory-final controllable in the class of pulse controls.

3. PULSE CONTROLLABILITY OF PARABOLIC SYSTEMS

Next, we shall consider some other problems of controllability on the examples of parabolic systems with impulse impact. Urgency of such problems is stipulated by both the arising of new technologies and the simplicity of the control on the basis of spatially distributed impulse of given class with simply regulative control. For the system with continuous and discrete control similar problems were posed and investigated in [161, 162].

Let the functioning of the system in a tube $Q=(0,T)\times\Omega$, $\Omega\subset R^m$ with a regular boundary is described by the equation

$$Lu=\frac{\partial u}{\partial t}+Bu=\sum_{i=1}^{N}\delta(t-t_i)\otimes\sum_{k=1}^{p}c_{ik}\varphi_k(x), \tag{1}$$

$$u\big|_{t=0}=0,\ u\big|_{x\in\partial\Omega}=0, \tag{2}$$

$$Bu=-\sum_{i,j=1}^{m}\frac{\partial}{\partial x_i}\left(b_{ij}(x)\frac{\partial u}{\partial x_j}\right)+b(x)u, \tag{3}$$

where B is uniformly elliptic operator in $\overline{\Omega}$, $b_{ij}(x)\in C^1(\overline{\Omega})$, $b_{ij}=b_{ji}$, $b(x)\in C(\overline{\Omega})$, $b(x)\geq 0$. The control is made with the help of coefficients $c=\{c_{ik}\}_{i=\overline{1,N};k=\overline{1,p}}$, $c\in R^{Np}$, t_i are moments of impulse impacts $t_{i+1}>t_i$, $0\leq t_i\leq T$, $\varphi_k(x)\in L_2(\Omega)$. In consequence of the presence of Dirac δ-function with respect to the time variable in the right-hand side of the equation (1) the solution of the problem (1), (2) can be represented in the following form: $u(t,x)=\sum_{j=1}^{N}u_j$, where u_j is the solution of the problem (1), (2) with the right-hand side $\delta(t-t_j)f_j$, $f_j=\sum_{k=1}^{p}c_{jk}\varphi_k$, $j=\overline{1,N}$.

Obviously, $u_j(t,x)\equiv 0$ when $t<t_j$ and satisfies the identity

$$(u_j,L^*v)_{L_2(Q)}=(u_j,L^*v)_{L_2([t_j,T];L_2(\Omega))}=(v(t_j,x),f_j(x))_{L_2(\Omega)}. \tag{4}$$

It follows from the relation (4), that in $Q_j=[t_j,T]\times\Omega$ the function u_j is a solution of the problem

$$Lu_j=0,\ u_j\big|_{t=t_j}=f_j(x),\quad u_j\big|_{x\in\partial\Omega}=0, \tag{5}$$

which, as well-known, belongs to the space $C([t_j, T], L_2(\Omega))$. Hence, function u_j is continuous on the segment $[0, T]$ everywhere, except the point $t = t_j (j = \overline{1, N})$, and moreover, it is continuous from the right.

The solution of problem (5) we may represent in the form [166]

$$u_j(t,x) = \begin{cases} 0, 0 \le t < t_j, \\ T(t - t_j) f_j(x), t \ge t_j, \end{cases} \tag{6}$$

where $T(t)$ is a semi-group generated by the operator B.

Under the above mentioned restrictions the operator B generate an orthonormal basis in $L_2(\Omega)$ which consists of the eigenfunctions

$$B\omega_{jk} = \lambda_j \omega_{jk},\ \omega_{jk} \in \overset{0}{W}{}_2^1,\ \lambda_j > 0,\ \lambda_{j+1} > \lambda_j,\ \lambda_j \xrightarrow[j \to \infty]{} \infty,$$

and, moreover, only finite number of the eigenfunctions $\omega_{jk} (k = \overline{1, m_j})$ may correspond to each eigenvalue λ_j.

For any $f \in L_2(\Omega)$

$$T(t) f = \sum_{j=1}^{\infty} e^{-\lambda_j t} \sum_{k=1}^{m_j} f_{jk} \omega_{jk},\ f_{jk} = (f, \omega_{jk})_{L_2(\Omega)}. \tag{7}$$

The solution the of problem (1), (2) at the moment $t = t_i (i = \overline{1, N})$ we may represent in the form

$$u(t_i, x) = \sum_{j=1}^{i} T(t_i - t_j) \sum_{k=1}^{p} c_{jk} \varphi_k(x) =$$

$$= \sum_{j=1}^{i} \sum_{n=1}^{\infty} e^{-\lambda_n (t_i - t_j)} \sum_{k=1}^{p} c_{jk} \sum_{s=1}^{m_n} \varphi_{kns} \omega_{ns}(x), \tag{8}$$

$$\varphi_{kns} = (\varphi_k, \omega_{ns})_{L_2(\Omega)}.$$

Definition 1. *The system is impulse controllable if the set* $\{u(t_i,x;c), c\in R^{Np}, i=\overline{1,N}, N=1,2,...\}$ *is dense in* $L_2(\Omega)$ *where* $0\le t_1<...<t_j<...,\ t_j \xrightarrow[j\to\infty]{} +\infty$.

Definition 2. *The system is impulse controllable at* N *steps in* $G\subset L_2(\Omega)$, *if the set* $\{u(t_i,x;c), c\in R^{Np}, i=\overline{1,N},\}$, *where* $0\le t_1<...<t_N\le T$, *is dense in* G

Theorem 1. *In order that the system (9), (10) be controllable in the sense of Definition 1, it is necessary and sufficient that for any* $n=1,2,...$, *the following conditions is satisfied*

$$rank\{\varphi_{kns}\}_{k=\overline{1,p};s=\overline{1,m_n}} = m_n \le p. \tag{9}$$

Proof. *Sufficiency.* On the basis of the criterion of the density in $L_2(\Omega)$ the system (1), (2) is impulse controllable if for $z(x)\in L_2(\Omega)$ it follows from the equality

$$(u(t_i,x;c),z)_{L_2(\Omega)}=0,\quad c\in R^{Np}, i=\overline{1,N}, N=1,2,... \tag{10}$$

that $z=0$. Granting the representation (8), in particular, we have that

$$\sum_{n=1}^{\infty} e^{-\lambda_n(t_i-t_j)}\sum_{s=1}^{m_n}\varphi_{kns}z_{ns}=0,\ k=\overline{1,p},\ i=1,2,... \tag{11}$$

This implies that

$$\sum_{s=1}^{m_1}\varphi_{k1s}z_{1s}+\sum_{n=2}^{\infty}e^{-(\lambda_n-\lambda_1)(t_i-t_1)}\sum_{s=1}^{m_n}\varphi_{kns}z_{ns}=0. \tag{12}$$

Under $i\to\infty$ (hence, $t_i\to\infty$) the second term in the left-hand side in (12) tends to zero. Hence,

$$\sum_{s=1}^{m_1}\varphi_{k1s}z_{1s}=0,\ k=\overline{1,p}.$$

In the similar way, we can show that for any $n=1,2,...$ we have

$$\sum_{s=1}^{m_n} \varphi_{kns} z_{ns} = 0, \; k = \overline{1, p}. \tag{13}$$

If condition (9) is satisfied then from (13) we obtain that $z_{ns} = 0$, $n = 1,2,\ldots$, $s = \overline{1, m_n}$. Hence, $z = 0$.

Necessarity. Let the system (1), (2) is controllable in the sense of Definition 1 but for some n the condition (9) is not valid. Then there exists a nonzero element $z \in L_2(\Omega)$ such that the relation (11), and hence (10), hold true, that contradicts to Definition 2.

Theorem 2. *The system (1), (2) is not controllable in $L_2(\Omega)$ in the sense of Definition 2.*

This result follows from the fact that under finite number of impulses N the system of linear algebraic equations (11) $(k = \overline{1, p}, i = \overline{1, N})$ with respect to the variables z_{ns}, $s = \overline{1, m_n}$, $n = 1,2,\ldots$, has infinite number of nonzero solutions. Thus, the system is not controllable.

Denote by H_N a subspace generated by the eigenfunctions of the operator B corresponding to the first eigenvalues, and let u_N be an approximate solution of the problem (1), (2) obtained by the Galerkin method, where the same functions were taken as a basic ones. The, as well-known [142], $u_N(t,x) \xrightarrow[N \to \infty]{} u(t,x)$ strongly in $L_2(\Omega)$. The functions $u_N(t,x)$ at the points t_i, $i = \overline{1, N}$ have the form

$$u_N(t_i, x) = \sum_{j=1}^{i} \sum_{n=1}^{N} e^{-\lambda_n (t_i - t_j)} \sum_{k=1}^{p} c_{jk} \sum_{s=1}^{m_n} \varphi_{kns} \omega_{ns}. \tag{14}$$

Theorem 3. *In order that the system (1), (2) be controllable in $H_N \subset L_2(\Omega)$ in the sense of Definition 2 $(\Delta t_i = t_{i+1} - t_i = const)$ it is necessary and sufficient that the condition (9) holds true for any $n = \overline{1, N}$.*

P r o o f. *Sufficiency.* Let $z_N(t,x)$ is an arbitrary element from H_N. It is obvious, that

$$z_N = \sum_{n=1}^{N}\sum_{m=1}^{m_n} z_{nm}\omega_{nm}(x),\ z_{nm} = (z,\omega_{nm})_{L_2(\Omega)}.$$

The system (1), (2) will be controllable in the sense of Definition 2 if it follows from the equalities

$$(u_N(t_i,x;c), z_M)_{L_2(\Omega)} = 0,\ c\in R^{Np},\ i=\overline{1,N} \tag{15}$$

that $z_N = 0$. Granting (14), it follows from (15) that, in particular,

$$\sum_{n=1}^{N} e^{-\lambda_n(t_i-t_1)}\sum_{m=1}^{m_n}\varphi_{knm}z_{nm} = 0,\quad k=\overline{1,p};\ i=\overline{1,N}. \tag{16}$$

Relations (16) form the system of linear equations

$$\Phi\bar{z} = 0, \tag{17}$$

where $\bar{z} = \left(z_{11},\dots,z_{1m_1},z_{21},\dots,z_{2m_2},\dots,z_{M1},\dots,z_{Mm_M}\right)^T$, and matrix Φ has block form

$$\Phi = \begin{pmatrix} \{\varphi_{k1m}\}_{k=\overline{1,p}}^{m=\overline{1,m_1}} & \dots & \{\varphi_{kNm}\}_{k=\overline{1,p}}^{m=\overline{1,m_N}} \\ e^{-\lambda_1(t_2-t_1)}\{\varphi_{k1m}\}_{k=\overline{1,p}}^{m=\overline{1,m_1}} & \dots & e^{-\lambda_N(t_2-t_1)}\{\varphi_{kNm}\}_{k=\overline{1,p}}^{m=\overline{1,m_N}} \\ \dots & \dots & \dots \\ e^{-\lambda_1(t_N-t_1)}\{\varphi_{k1m}\}_{k=\overline{1,p}}^{m=\overline{1,m_1}} & \dots & e^{-\lambda_N(t_N-t_1)}\{\varphi_{kNm}\}_{k=\overline{1,p}}^{m=\overline{1,m_N}} \end{pmatrix}.$$

Introducing the notation $F_i = \{\varphi_{kim}\}_{k=\overline{1,p}}^{m=\overline{1,m_i}}$, $i=\overline{1,N}$, we have

$$\Phi = \begin{pmatrix} F_1 & \dots & F_N \\ e^{-\lambda_1(t_2-t_1)}F_1 & \dots & e^{-\lambda_N(t_2-t_1)}F_N \\ \dots & \dots & \dots \\ e^{-\lambda_1(t_N-t_1)}F_1 & \dots & e^{-\lambda_N(t_N-t_1)}F_N \end{pmatrix}.$$

Solving the system (17) by the Gauss method, we transfer the matrix to the following form:

$$\Phi' = \begin{pmatrix} F_1 & F_2 & \dots & F_N \\ 0 & \alpha_{22} F_2 & \dots & \alpha_{2N} F_N \\ \dots & \dots & \dots & \dots \\ 0 & 0 & \dots & \alpha_{NN} F_N \end{pmatrix}, \tag{18}$$

where $\alpha_{ij}\left(1=\overline{2,N}, j=\overline{2,N}, i \le j\right)$ are some constants and $\alpha_{ii}\left(i=\overline{2,N}\right)$ do not equal to zero.

Indeed, to prove this fact we may just show that the determinant

$$F = \begin{vmatrix} 1 & \dots & 1 \\ e^{-\lambda_1(t_2-t_1)} & \dots & e^{-\lambda_N(t_2-t_1)} \\ \dots & \dots & \dots \\ e^{-\lambda_1(t_N-t_1)} & \dots & e^{-\lambda_N(t_N-t_1)} \end{vmatrix}$$

does not equal to zero. It is easy to see that the determinant F is a Vandermond determinant with respect to $\left(e^{-\lambda_1 \Delta t}, e^{-\lambda_2 \Delta t}, \dots, e^{-\lambda_N \Delta t}\right)$, which does not equal to zero as far as $\lambda_i \neq \lambda_j$ when $i \neq j$.

Granting (9) and (18) we can conclude that the rank of the matrix Φ', and hence Φ, is equal to $\sum_{n=1}^{N} m_n$ and coincide with the dimension of the vector $\vec{z}$. It implies that the system (17) can have only zero solution.

Necessarity. Suppose that for some n the condition (9) is not valid. Hence, the rank of the matrix Φ of the system of linear algebraic equations (17) is less than the number of variables. It means that there exist the nonzero solutions of the systems. Thus, there exists $z_N \neq 0$, $z_N \in H_N$ such that (15) is valid, and hence the system (1), (2) is not controllable.

4. SUBDOMAIN CONTROLLABILITY OF PSEUDO-PARABOLIC SYSTEMS

Consider a pseudo-parabolic system with controls concentrated in subdomains of the space domain.

Remark. *The result represented below holds true also for the case of point control in a one-dimensional space domain.* .

For hyperbolic equations one of the ways of the investigation of the point controllability on the basis of the concept of “exact controllability” described in the J.-L.Lions' work [31, 32].

Let the state of the system is a solution of the initial boundary value problem

$$Lu = u_t + \varepsilon(Au)_t + Au = \sum_{j=1}^{N} \chi_{\omega_j} v_j(x) h_j(t)$$

$$\text{in } Q \doteq (0,T)\times\Omega, \tag{1}$$

$$u|_{t=0} = 0, \quad u|_{\Gamma} = 0, \tag{2}$$

where χ_{ω_j} is the indicator of the subset (subdomain) $\omega_j \subset \Omega$ with positive Lebesque measure, $v_j \in L_2(\omega_j)$ is a given "elementary" intensivities. The control of the system is carried out with the help of the vector-function $h(t) = (h_1(t), h_2(t), \ldots, h_N(t)) \in (L_2(0,T))^N$.

The differential expression A has the form

$$Au = -\sum_{i,j=1}^{m} \left(a_{ij}(x) u_{x_j}\right)_{x_i} + a(x)u,$$

$$a_{ij}(x) \in C^1(\overline{\Omega}), \; a_{ij} = a_{ji}, \; a(x) \in C(\overline{\Omega}), \; a(x) \geq 0.$$

It is uniformly elliptic in $\overline{\Omega}$, ε is a positive constant.

By virtue of the restrictions introduced above the differential expression A yields in the space $L_2(\Omega)$ a symmetric and positive definite operator with a discrete spectrum, namely: there exists a countable set of the eigenvalues which can be arranged in ascending order:

$$0<\lambda_1<\lambda_2<\ldots<\lambda_j<\ldots,\ \lambda_j \xrightarrow[j\to\infty]{} +\infty,$$

and a complete orthonormal in $L_2(\Omega)$ system of the corresponding eigenfunctions $u_{jk}(\bar{x})\in H_0^1(\Omega)$, and only finite number of linearly independent eigenfunctions $u_{jk}(x)$, $k=\overline{1,m_j}$ can be assigned to every eigenvalue λ_j.

The generalized solution the problem (1), (2) belongs to the space $C([0,T],H_0^1(\Omega))$, which is the space of continuous on $[0,T]$ functions taking on values in the Sobolev space $H_0^1(\Omega)$. It follows from the imbedding of the space W^+ into the space $C([0,T],H_0^1(\Omega))$. With the help of the Fourier method the solution of the problem (1), (2) can be represented in the following form

$$u(t,x)=\sum_{j=1}^{N}\sum_{n=1}^{\infty}\int_0^t e^{-\frac{\lambda_n}{1+\varepsilon\lambda_n}(t-\tau)} h_j(\tau)\,d\tau\sum_{s=1}^{m_n}\left(v_j,u_{ns}\right)_{L_2(\omega_j)}u_{ns}(x). \quad (3)$$

It is requested to transfer the system to the desired state before the time moment T with the help of the controls $h(t)$. The controllability of the system (1), (2) we shall mean in the sense of the following definitions.

Definition 1. *The system is a subdomain controllable one if the set* $\{u(T,x;h) |\ h\in (L_2(0,T))^N\}$ *is dense in* $L_2(\Omega)$.

Theorem 1. *The system (1), (2) is a subdomain controllable one in the space* $L_2(\Omega)$ *if and only if for an arbitrary natural number* n *the following condition is satisfied*

$$rank\Pi_n = m_n \le N, \quad (4)$$

where

$$\Pi_n = \begin{pmatrix} (v_1, u_{n1})_{L_2(\omega_1)} & \cdots & (v_1, u_{nm_n})_{L_2(\omega_1)} \\ \cdots & \cdots & \cdots \\ (v_N, u_{n1})_{L_2(\omega_N)} & \cdots & (v_N, u_{nm_n})_{L_2(\omega_N)} \end{pmatrix}.$$

P r o o f. At first, let us prove the sufficiency. Let us show that for any $z(x) \in L_2(\Omega)$ the condition

$$(u(T,\cdot;h), z)_{L_2(\Omega)} = 0 \ \ \forall\, h \in (L_2(0,T))^N$$

implies that $z(x) = 0$. Granting the representation of the solution (3), we have

$$\sum_{j=1}^{N} \sum_{n=1}^{\infty} \int_0^T e^{-\frac{\lambda_n}{1+\varepsilon\lambda_n}(T-t)} h_j(t)\,dt \sum_{s=1}^{m_n} (v_j, u_{ns})_{L_2(\omega_j)} (z, u_{ns})_{L_2(\Omega)} = 0, \tag{5}$$

$$h_j \in L_2(0,T).$$

Whence,

$$\int_0^T \left(\sum_{n=1}^{\infty} e^{-\frac{\lambda_n}{1+\varepsilon\lambda_n}(T-t)} \sum_{s=1}^{m_n} (v_j, u_{ns})_{L_2(\omega_j)} (z, u_{ns})_{L_2(\Omega)} \right) h_j(t)\,dt = 0, \tag{6}$$

$$h_j \in L_2(0,T),\ j = \overline{1\ N}.$$

It follows from (6) that

$$\sum_{n=1}^{\infty} e^{-\frac{\lambda_n}{1+\varepsilon\lambda_n}(T-t)} \sum_{s=1}^{m_n} (v_j, u_{ns})_{L_2(\omega_j)} (z, u_{ns})_{L_2(\Omega)} = 0, \ \ j = \overline{1,N}. \tag{7}$$

The function $F(\tau) = \sum_{n=1}^{\infty} e^{-\frac{\lambda_n}{1+\varepsilon\lambda_n}\tau} \sum_{s=1}^{m_n} (v_j, u_{ns})_{L_2(\omega_j)} (z, u_{ns})_{L_2(\Omega)}$ is analytic in the right-hand complex half-plane $\mathrm{Re}\ \tau > 0$ and it is equal to zero in the interval $(0,T)$. Thus, $F(\tau) = 0$ on $\mathrm{Re}\ \tau > 0$. Reasoning as in the proof of Theorem 3.1, we obtain that

$$\sum_{s=1}^{m_n} (v_j, u_{ns})_{L_2(\omega_j)} (z, u_{ns})_{L_2(\Omega)} = 0, \ \ j = \overline{1,N}, \ \ n = 1,2,\ldots. \tag{8}$$

Denote

$$\vec{z}_n = \begin{pmatrix} (z, u_{n1})_{L_2(\Omega)} \\ (z, u_{n2})_{L_2(\Omega)} \\ \vdots \\ (z, u_{nm_n})_{L_2(\Omega)} \end{pmatrix}.$$

We can rewrite (8) in the form $\Pi_n \vec{z}_n = 0$, $n = 1,2,\ldots$. Since (4) is valid, $\vec{z}_n = 0$, $n = 1,2,\ldots$. Thus, $z(x) = 0$.

Let the system (1), (2) is controllable in the sense of Definition 1 but for some natural number k the condition (4) does not satisfied. The there exists such vector $\vec{z}_k \in R^{m_k}$ that

$$\vec{z}_k \neq 0, \; \Pi_k \vec{z}_k = 0$$

Let the function $z(x) \in L_2(\Omega)$ be such that

$$(z, u_{ls})_{L_2(\Omega)} = 0, \; s = \overline{1, m_l}, \; l \neq k \text{ and } (z, u_{ks})_{L_2(\Omega)} = z_{ks},$$
$$s = \overline{1, m_k}.$$

We have that for any n

$$\sum_{s=1}^{m_n} (v_j, u_{ns})_{L_2(\omega_j)} (z, u_{ns})_{L_2(\Omega)} = 0, \; j = \overline{1, N}.$$

Whence, we obtain

$$(u(T, \cdot; h), z)_{L_2(\Omega)} = 0 \quad \forall\, h \in (L_2(0,T))^N,$$

that contradicts to Definition 1. The theorem is proved.

The following example is sufficiently instructive. The system

$$Lu = u_t - u_{xxt} - u_{xx} = \chi_{(\alpha,\beta)} h(t) \text{ â } Q = (0,T) \times (0,1),$$
$$u|_{t=0} = u|_{x=0} = u|_{x=1} = 0,$$

is controllable if $\alpha - \beta$ is irrational number and is not controllable if $\alpha - \beta$ is rational number $([\alpha, \beta] \subseteq [0,1])$.

Remark. *For the case when* $\Omega = (a,b) \subset R$ *Theorem 1 can be generalized for the systems with point controls*

$$Lu = u_t + \varepsilon(Au)_t + Au = \sum_{j=1}^{N} c_j \delta(x - x_j) \otimes h_j(t) \quad \text{in } Q = (0,T)\times(a,b), \tag{9}$$

$$u|_{t=0} = 0, \quad u|_{\Gamma} = 0, \tag{10}$$

in the following way: the system '(9), (10) is controllable if and only if for an arbitrary natural n *the following condition holds true*

$$rank \Pi_n = 1,$$

where

$$\Pi_n = \begin{pmatrix} c_1 u_n(x_1) \\ c_2 u_n(x_2) \\ \vdots \\ c_N u_n(x_N) \end{pmatrix}$$

Here $u_n(x)$ *is an eigenfunction of the operator* A *corresponding to the eigenvalue* λ_n. *The condition* $rank \Pi_n = 1$ *is equivalent to the condition that not all components of the vector* Π_n *equal to zero.*

Chapter 11

PERSPECTIVES

1. GENERALIZED SOLVABILITY OF LINEAR SYSTEMS

1.1. Basic definitions and facts

Concept of a generalized solution of an operator equation

$$Lu = f \tag{1}$$

with an arbitrary linear closed operator L was proposed in [165]. Recall the basic principles of this approach.

Let E, F be Banach spaces and L be a closed linear injective operator mapping E into F with everywhere dense in E domain of definition $D(L)$ and with dense in F range of values $R(L)$. In a parallel way (1), we shall investigate the adjoint equation

$$L^{*}\varphi = l \tag{2}$$

where $L^{*}: F^{*} \to E^{*}$ is the adjoint operator.

Suppose also that the set $D(L^{*})$ is total in F^{*}, and the set $R(L^{*})$ is total in E^{*} in duality (F, F^{*}). The condition of totality of $R(L^{*})$ may be replaced by one of the following assumptions:

A) the kernel of L^{*} consists of zero-vector $Ker(L^{*}) = \{0\}$ and $R(L^{*})$ is weakly dense in E^{*} (this condition independently implies the injectivity of L and the density of $R(L)$ in F);

B) the operator L is continuous $(D(L) = E)$;

C) the space E is reflexive (if in addition F is reflexive then the set $D(L^{*})$ is total in F^{*} also).

Under the assumptions A) and C) the totality $R(L^{*})$ follows from the well-known results, and in the case B) it follows from the formula

$$R\left(L^*\right)^\circ \cap D(L) = Ker(L), \tag{3}$$

where $M^\circ \subset E$ is the polar of the set $M \subset E^*$ in duality $\left(E, E^*\right)$. Let us prove that (3) holds true for an arbitrary closed linear operator

$$R\left(L^*\right)^\circ \cap D(L) = \left\{u \in E : u \in D(L), l(u) = 0 \ \forall l \in R\left(L^*\right)\right\} =$$
$$= \left\{u \in E : u \in D(L), \varphi(Lu) = 0 \ \forall \varphi \in D\left(L^*\right)\right\}$$

as far as $D\left(L^*\right)$ is a total linear subspace then

$$R\left(L^*\right)^\circ \cap D(L) = \left\{u \in E : u \in D(L), \, Lu = 0\right\} = Ker(L).$$

Thus, the formula (3) is proved. It follows from (3) that

$$\left(R\left(L^*\right)^\circ \cap D(L)\right)^\circ = \left(Ker(L)\right)^\circ$$

If L is a continuous injective operator then $D(L) = E$, $Ker(L) = \{0\}$. Therefore,

$$\left(R\left(L^*\right)\right)^{\circ\circ} = \left(Ker(L)\right)^\circ = E^*.$$

Thus, the bipolar of the set $R\left(L^*\right)$, i.e. the weak closure of $R\left(L^*\right)$, coincides with E^*, that implies the totality of $R\left(L^*\right)$ in E^*

Denote by $\widetilde{E}_L$ the completion of the set E in topology $\sigma\left(E, R\left(L^*\right)\right)$; as far as the sets E and $R\left(L^*\right)$ are in duality then $\widetilde{E}_L$ is a separable locally convex linear topological space. Consider an arbitrary linear functional $\varphi \in D\left(L^*\right)$. The equation (1) implies that

$$\varphi(Lu) = \varphi(f), l(u) = L^*\varphi(u) = \varphi(f).$$

By the Banach theorem about weakly continuous linear functional [78], the functional $l = L^*\varphi$ allows the unique extension by continuity on the whole space $\widetilde{E}_L$.

Definition 1. *The generalized solution of the operator equation (1) is an element* $u \in \widetilde{E}_L$ *obeying the relation*

$$l(u) = L^*\varphi(u) = \varphi(f) \qquad \forall \varphi \in D(L^*). \tag{4}$$

Definition 2. *The generalized solution of the operator equation (1) is such element* $u \in \tilde{E}_L$ *that there exists a sequence* $u_n \in E$ *(which we shall call as almost solution) obeying the relations*

$$u_n \to u \text{ in topology } \tilde{E}_L, \|Lu_n - f\|_F \to 0, \text{ as } n \to \infty \tag{5}$$

The concept of a generalized solution arises when the right-hand side of the operator (1), i.e. the element $f \in F$, does not belong to the range of values $R(L)$ of the operator L (in this case there are no any classic solutions). But if $f \in R(L)$ than the generalized solution becomes a classic one. In [165] the following theorem was proved.

Theorem 1. *For any element* $f \in F$ *there exists a unique element* $u \in \tilde{E}_L$ *which is a generalized solution of the equation (1) in the sense of Definitions 1,2.*

Let us introduce once more definition of a generalized solution.

Let $\overline{E}_L$ is a completion of a linear set $D(L)$ with the norm $\|u\|_{\overline{E}_L} = \|Lu\|_F$. The equality $\|u\|_{\overline{E}_L} = \|Lu\|_F$ allows extending L by continuity from the set $D(L)$ of the space E onto the whole space $\overline{E}_L$. The extended operator we shall denote by $\overline{L}$. The operator $\overline{L}: \overline{E}_L \to F$ is linear and continuous.

Definition 3. *The solution of the equation (1) is such element* $u \in \tilde{E}_L$ *that the equality (1) holds true for the extended operator* $\overline{L}$.

It is easy to see that the extended operator $\overline{L}$ determine an isometric isomorphism between the spaces $\overline{E}_L$ and F.

Theorem 2. *For any element* $f \in F$ *there exists a unique generalized solution of the equation (1) in the sense of Definition 3.*

As well-known, the case of closed linear operator L reduced easy to the case of a linear continuous operator L. Indeed, introducing the norm of the graphics on $D(L)$

$$\|u\|_{D(L)} = \|u\|_E + \|Lu\|_F,$$

relatively to which the linear set $D(L)$ is a Banach space, we obtain that the operator $L_1 : D(L) \to F$ is linear and continuous $(L_1 u = Lu, u \in D(L))$. Taking into consideration the mentioned above, we shall further investigate only the case of continuous linear operator $L\,(D(L) = E)$.

At first, we shall determine the relations between the solvability in the sense of Definitions 1,2 and 3.

Theorem 3. *The space $\overline{E}_L$ is àlgebraically and topologically imbedded into the space $\widetilde{E}_L$.*

Proof. Let some net $\{u_\alpha\}_{\alpha\in A}, u_\alpha \in E$ converges to 0 in the topology of the space $\overline{E}_L$. Then $Lu_\alpha \to 0$ in the space F, and hence, $\varphi(Lu_\alpha) \to 0$ for any $\varphi \in F^*$. Thus, $l(u_\alpha) \to 0\ \forall l \in R(L^*)$. Whence, we have that the topology $\widetilde{E}_L$ is weaker than the topology $\overline{E}_L$. It still remains to prove that if $u_\alpha \to u$ in the topology of the space $\overline{E}_L$ and $u_\alpha \to 0$ in the topology $\widetilde{E}_L$ then $u = 0$ (the condition π [165]). It follows from the condition $u_\alpha \to u$ in $\overline{E}_L$ that $l(u_\alpha) = L^*\varphi(u_\alpha) \to \varphi(\overline{L}u)\ \forall l \in R(L^*)$, and $u_\alpha \to 0$ implies that $l(u_\alpha) \to 0\ \forall l \in R(L^*)$, whence we have that $\varphi(\overline{L}u) = 0\ \forall \varphi \in D(L^*)$. By the totality of the set $D(L^*)$ and the injectivity of the operator $\overline{L}$ we have that $u = 0$. Thus, $\overline{E}_L \subset \widetilde{E}_L$.

Theorem 4. *Definitions 1,2 and 3 are equivalent.*

P r o o f . Let us prove that the solution $u \in \overline{E}_L$ in the sense of Definition 3 is a solution in the sense of Definition 1 also (inverse statement is obvious). Indeed, in $\overline{E}_L$ there exists a solution $u^* \in \overline{E}_L$ of the same equation $Lu = f$. It is clear that $L^*\varphi(u) = \varphi(f) = \varphi(\overline{L}u^*)$. Whence we have that $u = Ou^*$ is an operator of imbedding of the set $\overline{E}_L$ into $\tilde{E}_L$.

Often constructive description of the spaces $\overline{E}_L$ and $\tilde{E}_L$ is sophisticated and in general case still be an unsolved problem. Therefore it is necessary to find out an algebraic and topological dense imbedding of the space $\overline{E}_L$ or $\tilde{E}_L$ into some other well-known Banach or locally convex linear topological space H. In this section we shall describe such spaces H. Besides, we shall investigate the properties of the generalized solutions in these spaces H.

1.2 A priori inequalities

Taking into account the mentioned above, suppose that the space $\overline{E}_L$ is algebraically and topologically imbedded into the Banach space H. By the condition of the topological imbedding $c_1\|u\|_H \leq \|u\|_{\overline{E}_L}$. Whence, we conclude that

$$c_1\|u\|_H \leq \|Lu\|_F \leq c_2\|u\|_E \quad \forall u \in E \tag{6}$$

where c_1, c_2 is positive constants.

The estimations of such kind are usual in applications and they are called a priori inequalities. Except for (6), it is easy to see that the following a priori inequalities hold true

$$c_1\|u\|_{\overline{E}_L} \leq \|Lu\|_F \leq c_2\|u\|_E \quad \forall u \in E,$$

$$c_1\|u\|_H \leq \|\overline{L}u\|_F \leq c_2\|u\|_{\overline{E}_L} \quad \forall u \in \overline{E}_L.$$

Note that the inequalities (6) themselves do not guarantee the existence of a topological imbedding $\overline{E}_L \subset H$, but it only compare the topologies induced by the norms $\|\cdot\|_{\overline{E}_L}$ and $\|\cdot\|_H$ in the set E.

Hereinafter, we shall prove that the estimations (6) can serve as a basis of the theory of the generalized solvability of the operator equation (1).

1.3. Generalized solution of operator equation in Banach space

It follows from the results of the previous subsection that the inequalities (6) are necessary conditions for the construction of the theory of generalized solvability of the operator equation in the Banach space H. We shall prove that the inequalities (6) are also sufficient conditions of the solvability of the equation (1) in some generalized sense. Note that proposed scheme cover various approaches to the construction of generalized solutions of differential equations (see [167], for example).

Thus, we shall suppose that the linear operator $L(D(L)) = E, \overline{R(L)} = F$ satisfy the inequalities (6) where $u \in E, c_1, c_2 > 0$, H is a completion of the space E with the norm $\|u\|_H$. It is clear that the right-hand side of the inequalities (6) implies the continuity of the operator L, and the left-hand side implies the injectivity. Besides, by virtue of the density of the imbedding $E \subset H$ the set H^* is total in E^* and, hence, the spaces E and H^* are in duality.

Consider the operator $\tilde{L} : H \to F\left(D(\tilde{L}) = E\right)$ defined in the following way: $\tilde{L}u = Lu, u \in E$.

Lemma 1. *The operator equation* $L^*\varphi = l$ *is solvable in the subset* H^* *of the space* E^*.

Proof. The left-hand side of the inequalities (6) implies the correct solvability of the operator $\widetilde{L}$. Consider also the adjoint operator $\widetilde{L}^*: F^* \to H^*, D(\widetilde{L}^*) \subset D(L^*) = F^*$. It is clear that if $\varphi \in D(\widetilde{L}^*)$ then $\widetilde{L}^*\varphi|_E = L^*\varphi$, where $\widetilde{L}^*\varphi|_E$ is a contraction of the functional $\widetilde{L}^*\varphi \in H^*$ from the set H onto E. As well known, the correct solvability of the operator $\widetilde{L}'$ implies the everywhere solvability of the operator $\widetilde{L}^*$ [165], whence, taking into account the mentioned above, we obtain the solvability of the operator L^* on the set H^* (considered as a subspace of E) of the right-hand sides in (2).

Remark 1. *If the operator* L *satisfies the inequalities (6) then* $H^* \subset R(L^*) \subset E^*$.

Definition 4. *A generalized solution of the equation (1) with the right-hand side* $f \in F$ *is such element* $u \in H$ *that the equality*

$$\widetilde{L}^*\varphi(u) = \varphi(f) \tag{7}$$

holds true for any $\varphi \in D(\widetilde{L}^*)$.

It is obviously that the equality (7) is equivalent to

$$L^*\varphi(u) = \varphi(f) \qquad \forall\varphi \in F^*, L^*\varphi \in H^*.$$

Theorem 5. *For any right-hand side* $f \in F$ *there exists a unique solution* $u \in H$ *of the equation (1) in the sense of Definition 4.*

Proof. Choose a sequence $f_p \in R(L)$ such that $f_p \to f$ in the space F. If $u_p \in E$ is a solution of the equation $Lu = f_p$ then granting (6) and the fact that the sequence $\{f_p\}$ is fundamental, we have

$$\left\|u_{p_1} - u_{p_2}\right\|_H \le c_1^{-1}\left\|Lu_{p_1} - Lu_{p_2}\right\|_F = c_1^{-1}\left\|f_{p_1} - f_{p_2}\right\|_F \to 0$$

as $p_1, p_2 \to \infty$.

Thus, there exists such $u^* \in H$ that $u_p \to u^*$ in the space H. Further, we have

$$L^*\varphi(u_p) = \varphi(Lu_p) = \varphi(f_p), \varphi \in F^*.$$

Passing in the last equality to the limit as $p \to \infty$, we obtain

$$L^*\varphi(u^*) = \varphi(f) \qquad \varphi \in F^*, L^*\varphi \in H^*.$$

Thus, u^* is a solution of (1) in the sense of Definition 4.

As long as $H^* \subset R(L^*)$, the equality

$$l(u^*) = L^*\varphi(u^*) = 0 \qquad\qquad \forall\varphi \in F^*, L^*\varphi \in H^*$$

implies that $u^* = 0$, and hence, the solution is unique.

Definition 5. *A generalized solution of the equation (1) with the right-hand side $f \in F$ is such element $u \in H$ that there exists a sequence $u_i \in E$ satisfying the conditions*

$$\|u_i - u\|_H \to 0, \qquad \|Lu_i - f\|_F \to 0 \text{ as } i \to \infty$$

Theorem 6. *Definitions 4 and 5 are equivalent.*

Proof. Let u be a solution of the equation $Lu = f$ in the sense of Definition 4, i.e. $L^*\varphi(u) = \varphi(f)$. Reasoning similarly to Theorem 5, we conclude that $u = u^*$, and hence, $\|u_p - u\|_H \to 0$. On the other hand, $\|Lu_p - f\|_F = \|f_p - f\|_F \to 0$. Thus, u is a solution of the equation (1) in the sense of Definition 5.

Let us prove the inverse assertion. Let u be a solution of the equation (1) in the sense of Definition 5. Then,

$$L^*\varphi(u) = L^*\varphi(u_i) + L^*\varphi(u - u_i) = \varphi(Lu_i) + L^*\varphi(u - u_i) =$$
$$= \varphi(Lu_i - f) + \varphi(f) + L^*\varphi(u - u_i)$$

for any $\varphi \in F^*, L^*\varphi \in H^*$.

Estimate the first and third summands in the right-hand side

$$|\varphi(Lu_i - f)| \le \|\varphi\|_{F^*} \cdot \|Lu_i - f\|_F \to 0,$$
$$|L^*\varphi(u - u_i)| \le \|L^*\varphi\|_{H^*} \cdot \|u - u_i\|_H \to 0 \quad \text{as} \quad i \to \infty.$$

Whence we have that

$$L^*\varphi(u) = \varphi(f), \qquad \varphi \in F^*, L^*\varphi \in H^*,$$

i.e. u is a solution of the equation (1) in the sense of Definition 4.

Remark 2. *It is obviously that the solutions in the sense 4 and 5 coincide with a classic solution* $u \in E$ *if* $f \in R(L)$. *It is also easy to testify that in this case the classic solution is a generalized one. In the case when the generalized solution* u *belongs to* $D(L)$ *it is a classic one.*

Theorem 7. *If the space* $\overline{E}_L$ *is continuously imbedded into* H *then Definitions 3, 4, and 5 are equivalent.*

Proof. Let us prove that Definition 3 is equivalent to definition 5. Let $u \in H$ be a solution of (1) by Definition 5. As it was indicated above, for any right-hand side $f \in F$ there exists a unique solution $u^* \in \overline{E}_L$ in the sense of Definition 3. Considering u^* as an element of the space H (by virtue of the imbedding $\overline{E}_L \subset H$), we can prove that $u^* = u$. Indeed,

$$\|u - u^*\|_H \le \|u - u_i\|_H + \|u_i - u^*\|_H \le \|u - u_i\|_H + c_1^{-1}\|u_i - u^*\|_{\overline{E}_L} =$$
$$= \|u - u_i\|_H + c_1^{-1}\|\overline{L}u_i - \overline{L}u^*\|_F =$$
$$= \|u_i - u^*\|_H + c_1^{-1}\|Lu_i - f\|_F \to 0 \quad as \quad i \to \infty$$

where $u_i \in E$ is a sequence defining the solution $u \in H$.

Thus, u is a solution of (1) in the sense of Definition 3.

Let now, conversely, u be a solution of (1) by Definition 3. Choose an arbitrary sequence $u_i \in E$ such that $\|u_i - u\|_{\overline{E}_L} \to 0$.

Then by the continuity of the imbedding $\overline{E}_L \subset H$ we have

$$\|u-u_i\|_H \to 0, \|u_i-u\|_{\bar{E}_L} = \|\bar{L}u_i - \bar{L}u\|_F = \|Lu_i - f\|_F \to 0$$
$$\text{as } i \to 0.$$

That is what had to be proved.

Remark 3. *The theorem implies that there exists a constant* $c>0$ *such that the following inequality holds true:*

$$\|u\|_H \le c\|f\|_F \qquad \forall f \in F, (c>0),$$

where u *is a solution of (1) with the right-hand side* f *in the sense of Definition 3, 4, and 5.*

Remark 4. *The imbedding* $\bar{E}_L \subset H$ *follows from the density of* $D(\tilde{L}^*)$ *in the space* F^* *in the weak topology* $\sigma(F^*, F)$ *or from the fact that the operator* $\tilde{L}$ *is closable..*

Usually from the inequalities (6) for the operator L it is possible to prove the similar inequalities for the adjoint operator

$$c_1\|\varphi\|_G \le \|L^*\varphi\|_{E^*} \le c_2\|\varphi\|_{F^*} \ \forall u \in F^* \tag{8}$$

where G is a completion of the set F^* with some norm. Consider this situation for reflexive Banach spaces E, F. In this case $L^{**} = L$ and similarly to Lemma 1 we obtain that the operator equation (1) is solvable in $G^* \subset F$. Besides, we have the theory of solvability of the adjoint equation (2) (analogous of the Theorems 5, 6, and 7).

Theorem 8. *There exists a constant* $c>0$ *such that for any* $f \in G^* \subset F$ *and any* $l \in H^* \subset E^*$ *the following inequalities hold true*

$$\|u\|_E \le c\|f\|_{G^*} \tag{9}$$
$$\|\varphi\|_{F^*} \le c\|l\|_{H^*} \tag{10}$$

where $u \in E, \varphi \in F^*$ *are solutions of the equations* $Lu = f$ *and* $L^*u = l$.

Proof. Let us prove the inequality (9) (the inequality (10) is proved in the similar way). As far as the equation (2) is solvable (in the sense of the analogous of Definitions 4 and 5 for the adjoint operator) for any $l \in E^*$ then for all $u \in E : Lu \in G^*$ the following equality holds true (the second adjoint space is identified with the original space)

$$Lu(\varphi) = u(l)\,, \tag{11}$$

where $\varphi \in G$ is a solution of (2) with the right-hand side $l \in E^*$. Whence, we have

$$|u(l)| \le \|Lu\|_{G^*} \cdot \|\varphi\|_G$$

or

$$\left|\left(\frac{u}{\|Lu\|_{G^*}}\right)(l)\right| \le \|\varphi\|_G.$$

Thus, the set of functionals

$$\left\{\left(\frac{u}{\|Lu\|_{G^*}}\right) : u \in E, Lu \in G^*\right\} \subset E^{**} = E$$

is bounded in every point $l \in E^*$, and hence, by the Banach-Steinhaus theorem it is bounded by the norm of the space $E^{**} = E$, hence, the inequality (9) is proved.

Theorem 9. *There exists a constant $c > 0$ such that for any $f \in F$ and any $l \in E^*$ the following inequalities hold true*

$$\|u\|_H \le c\|f\|_F \tag{12}$$

$$\|\varphi\|_G \le c\|l\|_{E^*} \tag{13}$$

where $u \in H, \varphi \in G$ are solutions of the equations $Lu = f$ and $L^* u = l$ in the sense of Definitions 4 and 5.

Proof. Granting the inequality (11), we have

$$|\varphi(Lu)| = |Lu(\varphi)| \le \|u\|_E \cdot \|l\|_{E^*}$$

Applying to the right-hand side the inequality (9), we obtain

$$|\varphi(Lu)| \le c\|Lu\|_{G^*} \cdot \|l\|_{E^*}$$

or

$$\|\varphi\|_G = \|\varphi\|_{G^{**}} = \sup_{Lu \in G^*} \frac{|\varphi(Lu)|}{\|Lu\|_{G^*}} \le c\|l\|_{E^*}.$$

Taking into account that $G^* \subset R(L)$, we obtain the inequality (13). The inequality (12) is proved in the similar way.

Remark 5. *As far as* $\|f\|_F = \|\overline{L}u\|_F = \|u\|_{\overline{E}_L}$ *and the inequality (12) holds true for any* $f \in F$ *it may be seemed that (12) provides the existence of imbedding* $\overline{E}_L \subset H$. *However, this is not true. In the space* $\overline{E}_L$ *really exists an element* u^* *such that* $\|u^*\|_{\overline{E}_L} = \|f\|_F$. *But if the imbedding* $\overline{E}_L \subset H$ *is not proved, we cannot to compare* $u \in H$ *as a solution of (1) in the sense of Definitions 4 and 5 and* $u^* \in \overline{E}_L$ *as a solution of (1) in the sense of Definition 3.*

1.4. Generalized solutions in locally convex linear topological spaces

Let us introduse several more definitions of a generalized solution. As before, we suppose that L is a linear continuous operator. Select in the space E^* some total linear set $M \subset R(L^*) \subset E^*$. Let $\tilde{M}$ is a completion of the set E by topology $\sigma(E, M)$. By the Banach theorem about weakly continuous linear functional [78] the functional $l = L^*\varphi \in M$ allows a unique extension by continuity on the whole space $\tilde{M}$.

Definition 6. A generalized solution of the equation (1) is an element $u \in \tilde{M}$ obeying the relation

$$L^*\varphi(u) = \varphi(f) \qquad \forall \varphi \in F^*, L^*u \in M.$$

Definition 7. A generalized solution of the equation (1) is an element $u \in \tilde{M}$ for which there exists a sequence $u_i \in E$ such that

$$u_i \to u \text{ in topology } \tilde{M}, \; \|Lu_i - f\|_F \to 0 \text{ as } i \to \infty.$$

In the case when $M = R(L^*)$ Definitions 6 and 7 coincide with Definitions 1 and 2 of a weak solution and an almost solution.

As in [165], we can prove the following

Theorem 10. *Definitions 6 and 7 are equivalent and for any element $f \in F$ there exists a unique generalized solution $u \in \tilde{M}$ of the equation (1)in the sense of Definitions 6 and 7.*

Remark 6. *Definition 7 implies that if $u \in \tilde{M}$ is a generalized solution of (1) then the point (u, f) is an point of accumulation of the graphics $\Gamma(L)$ of the operator L, i.e. of the set*

$$\{(u, f) : Lu = f, u \in E \subset F\}$$

in the topology $\tilde{M} \times F$.

Remark 7. Moreover, the point (u, f) belong to the sequential closure of the graphics $\Gamma(L)$ in the topology $\tilde{M} \times F$, and hence, $u \in \tilde{M}_s$, $\tilde{M}_s$ is a sequential closure of the set E in the topology $\tilde{M}$.

Taking into consideration Remark 7, we could investigate the concept of generalized solution in sequentially complete spaces $\tilde{M}_s$.

In addition, it is easy to prove that a classic solution of (1) is a generalized one in the sense of Definitions 6 and 7. If $f \in R(L)$ or the generalized solution belongs to $D(L)$ then the generalized solution becomes classic.

As long as $M \subset R(L^*)$ then there exists such set $M_F \in F^*$ that $M = L^*(M_F)$.

Theorem 11. *In order that there exists an algebraic and topological imbedding of the space $\overline{E}_L$ into the space $\tilde{M}$ it is necessary and sufficient that M_F be total linear subset of the space F^*. In this case Definitions 6, 7 and 3 are equivalent.*

Proof. Similarly to Theorem 3 we can prove that on the set E the topology induced by the norm $\|\cdot\|_{\overline{E}_L}$ stronger than the topology $\sigma(E, M)$. In the similar way we consider the condition π. We have that $\varphi(\overline{L}u) = 0 \ \forall \varphi \in M_F$. By virtue of the fact that the operator $\overline{L}$ determines an isometric isomorphism between the spaces $\overline{E}_L$ and F, we conclude that the condition $u = 0$ is equivalent to the totality of the set M_F. The equivalence if the definitions is proved similarly to Theorem 4.

Remark 8. *Theorem 11 implies that the space $\tilde{M}$ can be introduced (under the assumptions of the theorem) as a completion of the set $\overline{E}_L$ by the topology $\sigma(E, M)$.*

Remark 9. *Note that the condition of the totality of M_F has the principal character and not always holds true. It is easy to give an example of a linear continuous injective operator mapping not total sets into total.*

Indeed, the operator $A: l_2 \to l_2$ acting on the vector

$$x = (\xi_1, \xi_2, \ldots)$$

according to the rule

$$Ax = \left(\frac{\xi_1}{2^0}, \frac{\xi_2}{2^1}, \frac{\xi_3}{2^2}, \ldots \right)$$

is obviously injective, linear and continuous (even completely continuous) operator that maps the vector system

$$g_1 = (1, 2, 0, 0, 0, \ldots),$$

$$g_2 = (0, 2, 4, 0, 0, \ldots),$$

$$g_3 = (0,0,4,8,0,\dots),$$
$$\dots$$

into the system

$$e_1 = (1,1,0,0,0,\dots),$$
$$e_2 = (0,1,1,0,0,\dots),$$
$$e_3 = (0,0,1,1,0,\dots),$$
$$,\dots$$

The vector system $\{g_i\}$ is orthogonal to the vector

$$x^* = \left(1,-\frac{1}{2},\frac{1}{4},-\frac{1}{8},\dots\right)$$

and, hence, is not total, on the other hand the totality of the system $\{e_i\}$ in the space l_2 is obvious.

Besides, if we suppose that the operator L obeys the a priori inequalities (6), this implies the following

Theorem 12. *The Bdnach space H is algebraically and topologically imbedded into the space $\tilde{H}^*$. Definitions 6 and 7 for the set H^* and Definitions 4 and 5 , respectively, are equivalent.*

Proof. It is clear that the topology induced by the norm $\|\cdot\|_H$ stronger than the topology $\sigma(E, H^*)$. Let us test the condition π. Let the net $\{u_\alpha\}_{\alpha\in A}$ converges to u in the topology of the space H and $\{u_\alpha\}_{\alpha\in A} \to 0$ in the topology $\sigma(E, H^*)$. Then by the Hahn-Banach theorem the norm $\|\cdot\|_H$ may be represented in the form

$$\|u_\alpha - u\|_H = \sup_{l\in H^*} \frac{|l(u_\alpha - u)|}{\|l\|_{H^*}} \xrightarrow[A]{} 0$$

therefore, $l(u_\alpha) \to l(u)\ \forall l \in H^*$. On the other hand, $l(u_\alpha) \to 0\ \forall l \in H^*$. Thus, $l(u) = 0\ \forall l \in H^*$, and hence, $u = 0$.

Remark 10. *The concept of a generalized solution investigated in [167] is equivalent to the concept considered in this subsection.*

1.5 Connection between generalized solutions in Banach spaces and in locally convex spaces

Note that the constructions of generalized solutions in the sense of Definitions 4 and 5 and constructions of generalized solutions in the sense of Definition 6 and 7 are analogous in some sense.

On the basis of the total set $M \subset R(L^*) \subset E^*$ it is possible to construct the Banach space $\overline{M}$ giving a generalized solution in the sense of Definitions 4 and 5 in a way analogous to that used in the construction of a separable locally convex linear topological space $\widetilde{M}$ on the basis of the total set. Namely, let $\overline{M}$ is a completion of the set E by the norm

$$\|u\|_{\overline{M}} = \sup_{L^*\varphi \in M} \frac{|L^*\varphi(u)|}{\|\varphi\|_{F^*}}. \tag{14}$$

The norm (14) can be rewritten in the form

$$\|u\|_{\overline{M}} = \sup_{\varphi \in M_F} \frac{|\varphi(Lu)|}{\|\varphi\|_{F^*}}. \tag{15}$$

The spaces induced by the norms (15) were investigated in detail in the case of total sets M_F.

On the other hand if $M = R(L^*)$ then the norm $\|u\|_{\overline{M}}$ coincides with the norm of the space $\overline{E}_L$, and hence, in this case $\overline{M} = \overline{E}_L$ (an analogy of the space $\widetilde{E}_L$). Indeed,

$$\|u\|_{\overline{M}} = \sup_{L^*\varphi \in M} \frac{|L^*\varphi(u)|}{\|\varphi\|_{F^*}} = \sup_{\varphi \in F^*} \frac{|\varphi(Lu)|}{\|\varphi\|_{F^*}} = \|Lu\|_F$$

The last equality holds true, since by the Hahn-Banach theorem for any element $Lu \in F$ there exists such functional $\varphi \in F^*$ with the unit norm that $\varphi(Lu) = \|Lu\|_F$.

Lemma 2. *If* $M \subset R(L^*)$ *and* M_F *is total subset of the space* F^* *then* $\overline{E}_L$ *algebraically and topologically imbedded into the space* $\overline{M}$.

Proof. Totality of the set M_F and injectivity of the operator L implies the totality of the set M. In is easy to see that the norm $\|\cdot\|_{\overline{E}_L}$ stronger than the norm $\|\cdot\|_{\overline{M}}$. It still remains to testify the condition π. Let the sequence $u_n \in E$ converges to $u \in \overline{E}_L$ in the norm $\|u\|_{\overline{E}_L}$ and also u_n converges to zero in the norm $\|u\|_{\overline{M}}$, then on the one hand $\varphi(Lu_n) \to \varphi(\overline{L}u)$, but on the other hand $\varphi(Lu_n) \to 0$ for any $\varphi \in M_F$. Whence, $\varphi(\overline{L}u) = 0 \ \forall \varphi \in M_F$. By virtue of the totality of M_F and injectivity of $\overline{L}$ we have $u = 0$.

Remark 11. *The theorem implies that under the assumption of the totality of* M_F *the space* $\overline{M}$ *can be constructed by completing the set* $\overline{E}_L$ *with norm (14).*

Thus, on the basis of any total set $M \subset R(L^*)$ it is possible to obtain a priori estimation (under the assumption of the totality of M_F we can prove this estimation even with the imbedding $\overline{E}_L \subset \overline{M}$)

$$c_1\|u\|_{\overline{M}} \le \|Lu\|_F \le c_2\|u\|_E \quad \forall u \in E,$$

which implies the assertions about solvability of the equation (1) (in the sense of Definitions 3, 4, and 5) similar to Theorems 5, 6 and 7.

It easy to see that the topology induced by the norm $\|\cdot\|_{\overline{M}}$ on the set E is stronger than the topology of the space $\widetilde{M}$.

Thus, developing the indicated idea about connection between generalized solutions in Banach and in locally convex spaces, we could represent the results of Subsection 1.3 in the style of Subsection 1.4 basing on the total linear subset of the set $R(L^*)$, and vice versa, we could represent the results of Subsection 1.4 in the style of Subsection 1.3 using the analogies of the a priori inequalities, i.e. assuming that the separable locally cônvex topology defined on the set E is weaker than the norm $\|\cdot\|_{\overline{E}_L}$.

Finally, note that as far as many aspects of the methods used above have a topological character then the equation (1) we can investigate using the same approach in a locally convex topological spaces E and F (which can be, possibly, general topological spaces). In this case the role of $\overline{E}_L$ is played by the completion of E by the topology induced by the systems of semi-norms

$$p_{\alpha,\overline{E}_L}(u) = p_{\alpha,F}(Lu), \qquad \alpha \in \Re .$$

where $\{p_{\alpha,F}\}_{\alpha\in\Re}$ is a semi-norm system giving the topology of the space F, and instead of the estimations (6) we obtain the chain of the dense continuous imbeddings

$$E \subset \overline{E}_L \subset H \qquad \forall u \in E$$

where H is a completion of the set E in some locally convex topology which is weaker than the norm of the space $\overline{E}_L$.

2. PARAMETRIZATION OF SINGULAR CONTROL IN GENERAL CASE

Consider the general problem of singular optimal control of systems with distributed parameters. Let L be a linear partial differential operator (for example, pseudo-parabolic) defined in the space $L_2(Q)$

($Q=(0,T)\times\Omega\subset R^{n+1}$ is a regular tube) and having the domain of definition $D(L)$ consisting of sufficiently smooth in $\overline{Q}$ functions obeying some uniform boundary conditions (bd). The formally adjoint operator and its domain of definition we shall denote by L^+ and $D(L^+)$, respectively, where $D(L^+)$ is a set of smooth in $\overline{Q}$ functions satisfying the uniform adjoint boundary conditions (bd^+).

Define on the linear manifolds $D(L)$, $D(L^+)$ the positive norms $\|\cdot\|_{W^+}$, $\|\cdot\|_{H^+}$, $\|\cdot\|_{W^+_+}$, $\|\cdot\|_{H^+_+}$, induced by the corresponding inner products for which the following relations hold true

$$\|u\|_{L_2(Q)}\le c_1\|u\|_{H^+}\le c_2\|u\|_{W^+}\quad (u\in D(L)),$$

$$\|v\|_{L_2(Q)}\le c_3\|v\|_{H^+_+}\le c_4\|v\|_{W^+_+}\quad (v\in D(L^+)),\ c_i>0,\ i=\overline{1,4},$$

and the topological condition π [165] for the pairs of the neighbour norms in the inequalities is valid. By completing $D(L)$, $D(L^+)$ we obtain the chain of the positive Hilbert spaces

$$W^+\subset H^+\subset L_2(Q),\ W^+_+\subset H^+_+\subset L_2(Q),$$

and this imbeddings are dense and the imbedding operators are continuous. Adding to the chains spaces which are negative with respect to $L_2(Q)$ and are constructed by the positive norm introduced above, we obtain

$$W^+\subset H^+\subset L_2(Q)\subset H^-\subset W^-,$$

$$W^+_+\subset H^+_+\subset L_2(Q)\subset H^-_+\subset W^-_+.$$

Suppose that for the operators L, L^+ the following a priori inequalities hold true

$$c_1\|u\|_{H^+}\le\|Lu\|_{W^-_+}\le c_2\|u\|_{W^+}\quad (u\in D(L)),$$

$$c_1\|v\|_{H^+_+}\le\left\|L^+v\right\|_{W^-}\le c_2\|v\|_{W^+_+}\quad (v\in D(L^+)).$$

Then the operator L (L^{+}) can be considered as an operator continuously mapping the whole space W^{+} (W_{+}^{+}) into W_{+}^{-} (W^{-}). For the operator equation $Lu = f$ the following results are valid:

1) for any element $f \in H_{+}^{-}$ there exists a unique function $u \in W^{+}$ such that $Lu = f$ and the estimation $\|u\|_{W^{+}} \leq C\|f\|_{H_{+}^{-}}$ is valid;

2) for any element $f \in W_{+}^{-}$ there exists a unique function $u \in H^{+}$ such that $\langle u, L^{+}v\rangle_{H} = \langle f, v\rangle_{W_{+}}$ $\forall v \in W_{+}^{+}$: $L^{+}v \in H^{-}$ (where $\langle\cdot,\cdot\rangle_{H}$, $\langle\cdot,\cdot\rangle_{W_{+}}$ are bilinear forms constructed with the help of extension of the inner product $(\cdot,\cdot)_{L_2(Q)}$ onto $H^{+} \times H^{-}$ and onto $W_{+}^{-} \times W_{+}^{+}$, respectively) by continuity and the following estimation holds true $\|u\|_{H^{+}} \leq C\|f\|_{W_{+}^{-}}$.

Similar results are valid also for the equation with the adjoint operator L^{+}.

Let the system state $u(h) \in H^{+}$ is a solution of the equation

$$Lu(h) = f(h),\ h \in U_{ad}, \tag{1}$$

in the sense 2), where $f(\cdot): V \to W_{+}^{-}$ is a mapping of the reflexive Banach space of controls V into the negative space W_{+}^{-}, U_{ad} is bounded, closed, and convex subset of V. On the solutions of the equation (1) some functional $\Phi(\cdot): H^{+} \to R$ is defined. The optimization problem consists in finding out such controls $h \in U_{ad}$ on which the performance functional $J(h) = \Phi(u(h))$ attains its minimal values. The existence of the optimal controls is provided by the following assumptions:

a) $\Phi(\cdot): H^{+} \to R$ is a weakly lower semicontinuous functional;

b) $f(\cdot): V \to W_{+}^{-}$ is a weakly continuous mapping.

The natural and generally accepted idea of approximate solving infinite-dimensional optimization problems consists in reducing this problem to finite-dimensional extremal problems. Consider the general principles allowing to approximate a solution of a singular optimization problem by a solution of some finite-dimensional problem. This approach we shall call parametrization of control.

Suppose that V is a separable reflexive Banach space. Then there exists a sequence of finite-dimensional subspaces (V_s) of the space V satisfying the condition of the boundary density: $\lim_{s\to\infty} \inf_{h'\in V_s} \|h'-h\|_V = 0 \quad (h \in V)$.

Next, let (U_{ad}^s) be a sequence of closed and convex subsets V_s, which are bounded uniformly with respect to s in the norm of V and approximate the set of the admissible controls U_{ad} in the following sense:

$$\forall h \in U_{ad} \ \exists h^s \in U_{ad}^s : h^s \xrightarrow[s\to\infty]{} h \text{ strongly in } V, \tag{2}$$

$$\forall h \in V, \ \ h^s \in U_{ad}^s : h^s \xrightarrow[s\to\infty]{} h \text{ weakly in } V \Rightarrow h \in U_{ad}. \tag{3}$$

The issue of the nature of the spaces, which provide the best approximation, is the most important and fine in the numerical analysis.

The problem of oprimization of the system (1) we replace by the following problem:

$$J(h) = \Phi(u(h)) \to \inf_{h\in U_{ad}^s} \tag{4}$$

$$Lu(h) = f(h),\ h \in U_{ad}^s \subset V_s \subset V. \tag{5}$$

Theorem 1. *Under the conditions a), b), (2), (3) and also*

1) $\Phi(\cdot): H^+ \to R$ *is a strongly upper semicontinuous functional;*

2) $f(\cdot): V \to W_+^-$ *is a strongly continuous mapping;*

for an arbitrary $s \in \{1,2,3,\ldots\}$ *the optimal control problem (4), (5) has a solution* $h_*^s \in U_{ad}^s$ *and*

$$\overline{\lim_{s\to\infty}} J(h_*^s) \le \inf_{h\in U_{ad}} J(h).$$

It is possible to extract from the sequence (h_*^s) *such subsequence* $(h_*^{s_m})$ *that* $h_*^{s_m} \xrightarrow[m\to\infty]{} h_*$ *weakly in* V, *where* $h_* \in U_{ad}$ *is an optimal control of the system (1).*

P r o o f. Since the conditions a) and b) hold true and U_{ad}^s are closed and bounded subsets of the finite-dimensional subspaces V_s, the problem of optimal control of the system (1) and the problem (4) and (5) are solvable. Let $h_* \in U_{ad}$ be some optimal control of the system (1), and (h_*^s): $h_*^s \in U_{ad}^s$ is a sequence of solutions of the problems (4) and (5). Then by the condition (2) there exists a sequence of controls (h^s): $h^s \in U_{ad}^s$ such that $\|h^s - h_*\|_V \xrightarrow[s\to\infty]{} 0$. By virtue of the strong continuity of the mapping $f(\cdot): V \to W_+^-$, we have

$$\|f(h^s) - f(h_*)\|_{W_+^-} \xrightarrow[s\to\infty]{} 0. \tag{6}$$

The a priori inequality $\|u(h^s) - u(h_*)\|_{H^+} \le C\|f(h^s) - f(h_*)\|_{W_+^-}$, where $u(h^s) \in H^+$, $u(h_*) \in H^+$ are generalized solutions of the state equation of the system (1) corresponding to the controls h^s and h_*, and (6) imply that

$$\|u(h^s) - u(h_*)\|_{H^+} \xrightarrow[s\to\infty]{} 0.$$

Since $\Phi(\cdot): H^+ \to R$ is a strongly upper semicontinuous functional,

$$\overline{\lim_{s\to\infty}} J(h_*^s) \le \overline{\lim_{s\to\infty}} J(h^s) = \overline{\lim_{s\to\infty}} \Phi(u(h^s)) \le \Phi(u(h_*)) = \inf_{h\in U_{ad}} J(h). \tag{7}$$

Consider the sequence (h_*^s): $h_*^s \in U_{ad}^s$. U_{ad}^s is bounded uniformly with respect to s in the reflexive Banach space V. Thus, by the Eberlein-Shmuljan theorem there exists such subsequence $(h_*^{s_m})$ that

$$h_*^{s_m} \xrightarrow[m\to\infty]{} h_* \text{ weakly in } V.$$

The condition (3) implies that $h_* \in U_{ad}$. The mapping $f(\cdot): V \to W_+^-$ is a weakly continuous (assumption b)), hence

$$f\left(h_*^{s_m}\right) \xrightarrow[m\to\infty]{} f(h_*) \text{ weakly in } W_+^-.$$

The sequence $\left(f\left(h_*^{s_m}\right)\right)$ is bounded in the norm W_+^-. The estimation $\left\|u\left(h_*^{s_m}\right)\right\|_{H^+} \le C\left\|f\left(h_*^{s_m}\right)\right\|_{W_+^-}$ implies that $\left(u\left(h_*^{s_m}\right)\right)$ is a sequence bounded in H^+. Extract from $\left(u\left(h_*^{s_m}\right)\right)$ a weakly convergent subsequence $\left(u\left(h_*^{s_{m_k}}\right)\right)$

$$u\left(h_*^{s_{m_k}}\right) \xrightarrow[k\to\infty]{} u_* \text{ weakly in } H^+.$$

Let us prove that $u_* = u(h_*)$. Indeed, we have

$$\left\langle u\left(h_*^{s_{m_k}}\right), L^+ v\right\rangle_H = \left\langle f\left(h_*^{s_{m_k}}\right), v\right\rangle_{W_+} \quad \forall v \in W_+^+ : L^+ v \in H^-. \tag{8}$$

Passing to the limit in (8) as $k \to \infty$, we obtain

$$\left\langle u_*, L^+ v\right\rangle_H = \left\langle f(h_*), v\right\rangle_{W_+} \quad \forall v \in W_+^+ : L^+ v \in H^-$$

Thus, $u_* = u(h_*) \in H^+$ is a generalized solution of the system state equation (1) corresponding to the control $h_* \in U_{ad}$. Since this solution is unique, the sequence $\left(u\left(h_*^{s_m}\right)\right)$ weakly converges to $u(h_*) \in H^+$.

Let us prove that $h_* \in U_{ad}$ is the optimal control of the system (1). The weak lower semi-continuity of the functional $\Phi(\cdot): H^+ \to R$ and (7) imply that

$$J(h_*) \le \underline{\lim}_{m\to\infty} J\left(h_*^{s_m}\right) = \underline{\lim}_{m\to\infty} \Phi\left(u\left(h_*^{s_m}\right)\right) \le$$
$$\le \overline{\lim}_{m\to\infty} \Phi\left(u\left(h_*^{s_m}\right)\right) = \overline{\lim}_{m\to\infty} J\left(h_*^{s_m}\right) \le \inf_{h\in U_{ad}} J(h),$$

thus, $J(h_*) = \inf_{h\in U_{ad}} J(h)$, $h_* \in U_{ad}$ is the optimal control of the system (1). The theorem is proved.

Remark. *If the functional $\Phi(\cdot)$ is strongly continuous then* $\lim_{s\to\infty} J(h_*^s) = \inf_{h\in U_{ad}} J(h)$. *If the optimal control of the system (1)* $h_* \in U_{ad}$ *is unique then* $h_*^s \xrightarrow[s\to\infty]{} h_*$ *weakly in* V.

3. DIFFERENTIAL PROPERTIES OF PERFORMANCE CRITERION IN GENERAL CASE

To solve numerically the problems of simulation and optimization of systems the gradient descent methods are usually used. To use these methods correctly we must at first to investigate the smoothness of the performance functional and to prove that the necessary conditions of optimality hold true.

The purpose of this section is to study the issue of the smoothness of the performance functional for the singular optimization problem in general case [169-172].

Consider the optimization problem for the system (2.1)

$$J(h) = \Phi(u(h)) \to \inf_{h\in U_{ad}}, \quad Lu(h) = f(h), \; h \in U_{ad}.$$

Theorem 1. *Let the system state* $u(h) \in H^+$ *is defined as a solution of the equation (2.1) with the right-hand side* $f(h) \in W_+^-$. *If*

1) *there exists a Fréchet derivative* $f_h(\cdot)$ *of the mapping* $f(\cdot): V \to W_+^-$ *at the point* $h \in V$;

2) *there exists a Fréchet derivative* $\Phi_u(u(h)) \in H^-$ *of the functional* $\Phi(\cdot): H^+ \to R$ *at the point* $u = u(h) \in H^+$,

then the performance functional $J(\cdot): V \to R$ *is differentiable by Fréchet at the point* $h \in V$ *and the derivative is defined by the expression*

$$J_h(\cdot) = \langle f_h(\cdot), v_h \rangle_{W_+},$$

where $v_h \in W_+^+$ *is a solution of the adjoint problem* $L^+ v_h = \Phi_u(u(h))$.

P r o o f. It is clear that $\langle f_h(\cdot), v_h \rangle_{W_+}$ is meaningful, since the adjoint problem $L^+ v_h = \Phi_u(u(h))$ has a unique generalized solution $v_h \in W_+^+$, and $f_h(\cdot)$ take on its values in W_+^-. Since $f_h(\cdot)$ is a linear mapping, $\langle f_h(\cdot), v_h \rangle_{W_+}$ is a linear functional in V, whose continuity follows from the continuity of $f_h(\cdot)$. Indeed, for all $\overline{h} \in V$:

$$\left|\langle f_h(\overline{h}), v_h \rangle_{W_+}\right| \le \left\|f_h(\overline{h})\right\|_{W_+^-} \left\|v_h\right\|_{W_+^+} \le \left\|f_h\right\| \left\|v_h\right\|_{W_+^+} \left\|\overline{h}\right\|_V .$$

Let $\Delta h \in V$ is a increment of a control. Consider

$$\left|J(h+\Delta h) - J(h) - \langle f_h(\Delta h), v_h \rangle_{W_+}\right| =$$

$$= \left|\Phi(u(h+\Delta h)) - \Phi(u(h)) - \langle f_h(\Delta h), v_h \rangle_{W_+}\right|.$$

Denote $\Delta u(h) = u(h+\Delta h) - u(h) \in H^+$. We have

$$\left|J(h+\Delta h) - J(h) - \langle f_h(\Delta h), v_h \rangle_{W_+}\right| \le$$

$$\le \left|\Phi(u(h)+\Delta u) - \Phi(u(h)) - \langle \Delta u(h), \Phi_u(u(h)) \rangle_H\right| +$$

$$+ \left|\langle \Delta u(h), \Phi_u(u(h)) \rangle_H - \langle f_h(\Delta h), v_h \rangle_{W_+}\right|.$$

In is clear that $\Delta u(h) \in H^+$ is a unique solution of the problem

$$\langle \Delta u(h), L^+ v \rangle_H = \langle f(h+\Delta h) - f(h), v \rangle_{W_+} \quad \forall v \in W_+^+ : L^+ v \in H^-$$

Let the function v be a solution of the adjoint problem $L^+ v_h = \Phi_h(u(h))$. We can write that

$$\langle \Delta u(h), \Phi_u(u(h)) \rangle_H = \langle f(h+\Delta h) - f(h), v \rangle_{W_+}$$

and

$$\left|J(h+\Delta h) - J(h) - \langle f_h(\Delta h), v_h \rangle_{W_+}\right| \le$$

$$\leq \left|\Phi(u(h)+\Delta u(h))-\Phi(u(h))-\left\langle \Delta u(h),\Phi_u(u(h))\right\rangle_H\right|+ \quad (1)$$
$$+\left|\left\langle f(h+\Delta h)-f(h)-f_h(\Delta h),v_h\right\rangle_{W_+}\right|.$$

Applying the Schwarz inequality to the second summand in the right-hand side of (1), we have

$$\left|\left\langle f(h+\Delta h)-f(h)-f_h(\Delta h),v_h\right\rangle_{W_+}\right|\leq$$
$$\leq \left\|f(h+\Delta h)-f(h)-f_h(\Delta h)\right\|_{W_+^-}\left\|v_h\right\|_{W_+^+}.$$

For $\Delta u(h)\in H^+$ the a priori estimation is valid

$$\left\|\Delta u(h)\right\|_{H^+}\leq C\left\|f(h+\Delta h)-f(h)\right\|_{W_+^-}\leq$$
$$\leq C\left(\left\|f(h+\Delta h)-f(h)-f_h(\Delta h)\right\|_{W_+^-}+\left\|f_h(\Delta h)\right\|_{W_+^-}\right)\leq$$
$$\leq C\left(\left\|f(h+\Delta h)-f(h)-f_h(\Delta h)\right\|_{W_+^-}+\left\|f_h\right\|\left\|\Delta h\right\|_V\right).$$

The assumptions of the theorem imply that for any $\varepsilon>0$

1) $\exists\,\delta_1>0:\left\|\Delta h\right\|_V<\delta_1\Rightarrow$

$$\left\|f(h+\Delta h)-f(h)-f_h'(\Delta h)\right\|_{W_+^-}\leq\frac{\varepsilon}{2}\left(1+\left\|v_h\right\|_{W_+^+}\right)^{-1}\left\|\Delta h\right\|_V;$$

2) $\exists\,\delta_2>0:\left\|\Delta u\right\|_{H^+}<\delta_2\Rightarrow$

$$\left|\Phi(u(h)+\Delta u)-\Phi(u(h))-\left\langle \Delta u,\Phi_u(u(h))\right\rangle_H\right|\leq$$
$$\leq\frac{\varepsilon}{2C\left(\frac{\varepsilon}{2}+\left\|f_h\right\|\right)}\left\|\Delta u\right\|_{H^+}.$$

Choose $\delta>0$ such that $\delta<\min\left\{\delta_1,\dfrac{\delta_2}{C\left(\frac{\varepsilon}{2}+\left\|f_h\right\|\right)}\right\}$. Then the condition $\left\|\Delta h\right\|_V<\delta$ implies that

$$\left|J(h+\Delta h)-J(h)-\left\langle f_h(\Delta h),v_h\right\rangle_{W_+}\right|<\varepsilon\left\|\Delta h\right\|_V.$$

Thus, the theorem is proved..

Theorem 2. *Let*

1) *the mapping* $f(\cdot): V \to W_+^-$ *has a Fréchet derivative in the neighbour of the point* $\bar{h} \in V$ *,which is continuous at the point-*$\bar{h}$ *;*
2) *the functional* $\Phi(\cdot): H_r^+ \to R$ *has a Fréchet derivative in the neighbour of the point* $\bar{u} = u(\bar{h}) \in H^+$ *,which is continuous at the point* $\bar{u}$ *.*

The derivative $J_h(\cdot)$ *of the performance functional is continuous at the point* $\bar{h} \in V$

Proof. Let h is an arbitrary point in the vicinity of the point $\bar{h} \in V$. Consider for all $h' \in V$ the following difference

$$|J_h(h') - J_{\bar{h}}(h')| = \left|\langle f_h(h'), v_h\rangle_{W_+} - \langle f_{\bar{h}}(h'), v_{\bar{h}}\rangle_{W_+}\right| =$$

$$= \left|\langle f_h(h'), v_h - v_{\bar{h}}\rangle_{W_+} - \langle (f_{\bar{h}} - f_h)(h'), v_{\bar{h}}\rangle_{W_+}\right| \le$$

$$\le \left|\langle f_h(h'), v_h - v_{\bar{h}}\rangle_{W_+}\right| + \left|\langle (f_h - f_{\bar{h}})(h'), v_{\bar{h}}\rangle_{W_+}\right|,$$

where $v_{\bar{h}} \in W_+^+$, $v_h \in W_+^+$ is solutions of the adjoint problems:

$$L^+ v_{\bar{h}} = \Phi_u(u(\bar{h})),\ L^+ v_h = \Phi_u(u(h)).$$

Using the Schwarz inequality and the continuity of the linear operators $f_h(\cdot)$, $f_{\bar{h}}(\cdot)$:

$$|J_h(h') - J_{\bar{h}}(h')| \le \left(\|f_h\| \|v_h - v_{\bar{h}}\|_{W_+^+} + \|f_h - f_{\bar{h}}\| \|v_{\bar{h}}\|_{W_+^+}\right)\|h'\|_V .$$

Whence,

$$\|J_h - J_{\bar{h}}\| \le \|f_h\| \|v_h - v_{\bar{h}}\|_{W_+^+} + \|f_h - f_{\bar{h}}\| \|v_{\bar{h}}\|_{W_+^+}$$

The derivative of the mapping $f(\cdot)$ is continuous at the point $\bar{h} \in V$, since there exists a neighbour $O_{\delta_0}(\bar{h})$ ($O_\delta(\bar{h}) = \{h : \|h - \bar{h}\|_V < \delta\}$), in which $\|f_h\| \le C_1 < +\infty$. Thus,

$$\|J_h - J_{\bar{h}}\| \le C_1 \|v_h - v_{\bar{h}}\|_{W_+^+} + \|f_h - f_{\bar{h}}\| \|v_{\bar{h}}\|_{W_+^+}.$$

The following a priori estimation is valid

$$\|v_h - v_{\bar{h}}\|_{W_+^+} \le C_2 \|\Phi_u(u(h)) - \Phi_u(u(\bar{h}))\|_{H^-},$$

therefore

$$\|J_h - J_{\bar{h}}\| \le C_3 \|\Phi_u(u(h)) - \Phi_u(u(\bar{h}))\|_{H^-} + \|f_h - f_{\bar{h}}\| \|v_{\bar{h}}\|_{W_+^+}. \quad (2)$$

The assumptions of the theorem imply that for any $\varepsilon > 0$

1) $\exists \delta_1 > 0 : \|h - \bar{h}\|_V < \delta_1 \Rightarrow \|f_h - f_{\bar{h}}\| < \frac{\varepsilon}{2}\left(1 + \|v_{\bar{h}}\|_{W_+^+}\right)^{-1};$

2) $\exists \delta_2 > 0 : u \in H^+, \|u - u(\bar{h})\|_{H^+} < \delta_2 \Rightarrow$

$$\Rightarrow \|\Phi_u(u) - \Phi_u(u(\bar{h}))\|_{H^-} < \frac{\varepsilon}{2}.$$

For $u(h) - u(\bar{h}) \in H^+$ the following a priori inequality holds true:

$$\|u(h) - u(\bar{h})\|_{H^+} \le C_4 \|f(h) - f(\bar{h})\|_{W_+^-}.$$

Since the mapping $f(\cdot)$ has a Fréchet derivative at the point $\bar{h} \in V$, it is continuous at the point $\bar{h}$, i.e.

$$\exists \delta_3 > 0 : \|h - \bar{h}\|_V < \delta_3 \Rightarrow \|f(h) - f(\bar{h})\|_{W_+^-} < \frac{\delta_2}{C_4}.$$

Choose such $\delta > 0$ that $\delta < \min\{\delta_0, \delta_1, \delta_3\}$. Then the assumption $\|h - \bar{h}\|_V < \delta$ in (2) implies that $\|J_h - J_{\bar{h}}\| < \varepsilon$. The theorem is proved.

Definition 1. *A mapping $F(\cdot)$ from a linear normalized space X into a linear normalized space Y obeys the Lipschitz condition with index $\alpha \in (0,1]$ in the set $M \subseteq X$, if*

$$\exists C > 0: \ \|F(x') - F(x'')\|_Y \le C\|x' - x''\|_X^\alpha \quad \forall \{x', x''\} \subseteq M .$$

Theorem 3. *Let*

1) *the mapping $f(\cdot): V \mapsto W_+^-$ has a Fréchet derivative obeying the Lipschitz condition with index $\alpha \in (0,1]$ in the bounded convex set $U_{ad} \subset V$;*
2) *the functional $\Phi(\cdot): H^+ \to R$ has a Fréchet derivative obeying the Lipschitz condition with index $\beta \in (0,1]$ in the space H^+.*

Under these assumptions the derivative $J_h(\cdot)$ of the performance functional satisfies the Lipschitz condition with index $\gamma = \min\{\alpha, \beta\}$ in the set U_{ad}.

Proof is similar to the proof of the previous theorem. Let h_1, h_2 be arbitrary points in the set U_{ad}. Consider for all $h' \in V$ the difference

$$\left|J_{h_1}(h') - J_{h_2}(h')\right| = \left|\left\langle f_{h_1}(h'), v_{h_1}\right\rangle_{W_+} - \left\langle f_{h_2}(h'), v_{h_2}\right\rangle_{W_+}\right| =$$

$$= \left|\left\langle f_{h_1}(h'), v_{h_1} - v_{h_2}\right\rangle_{W_+} - \left\langle \left(f_{h_2} - f_{h_1}\right)(h'), v_{h_2}\right\rangle_{W_+}\right| \le$$

$$\le \left|\left\langle f_{h_1}(h'), v_{h_1} - v_{h_2}\right\rangle_{W_+}\right| + \left|\left\langle \left(f_{h_1} - f_{h_2}\right)(h'), v_{h_2}\right\rangle_{W_+}\right|,$$

where $v_{h_1} \in W_+^+$, $v_{h_2} \in W_+^+$ are solutions of the adjoint problems:

$$L^+ v_{h_1} = \Phi_u(u(h_1)), \ L^+ v_{h_2} = \Phi_u(u(h_2)).$$

Using the Schwarz inequality and the continuity of the linear operators $f_{h_1}(\cdot)$, $f_{h_2}(\cdot)$:

$$\left|J_{h_1}(h')-J_{h_2}(h')\right|\le\left(\left\|f_{h_1}\right\|\left\|v_{h_1}-v_{h_2}\right\|_{W_+^+}+\left\|f_{h_1}-f_{h_2}\right\|\left\|v_{h_2}\right\|_{W_+^+}\right)\left\|h'\right\|_V .$$

Whence,

$$\left\|J_{h_1}-J_{h_2}\right\|\le\left\|f_{h_1}\right\|\left\|v_{h_1}-v_{h_2}\right\|_{W_+^+}+\left\|f_{h_1}-f_{h_2}\right\|\left\|v_{h_2}\right\|_{W_+^+} .$$

Let us prove that the derivative $f_h(\cdot)$ is bounded in the set U_{ad}. Let $h^*\in U_{ad}$, and h is an arbitrary point that belongs to U_{ad}.

Then the following inequalities are valid

$$\begin{gathered}\left\|f_h\right\|\le\left\|f_h-f_{h^*}\right\|+\left\|f_{h^*}\right\|\le C_0\left\|h-h^*\right\|_V^\alpha+\left\|f_{h^*}\right\|\le\\ \le C_0 d^\alpha+\left\|f_{h^*}\right\|\le C_1<+\infty\end{gathered},$$

where $d=\operatorname{diam}\left(U_{ad}\right)=\sup\limits_{h'\in U_{ad},h''\in U_{ad}}\left\|h'-h''\right\|$ is the diameter of the set U_{ad}.

Next,

$$\left\|J_{h_1}-J_{h_2}\right\|\le C_2\left(\left\|v_{h_1}-v_{h_2}\right\|_{W_+^+}+\left\|h_1-h_2\right\|_V^\alpha\left\|v_{h_2}\right\|_{W_+^+}\right). \tag{3}$$

The following a priori estimations hold true ($h^*\in U_{ad}$ is some fixed admissible control)

$$\begin{gathered}\left\|v_{h_1}-v_{h_2}\right\|_{W_+^+}\le C_3\left\|\Phi_u(u(h_1))-\Phi_u(u(h_2))\right\|_{H^-}\le\\ \le C_4\left\|u(h_1)-u(h_2)\right\|_{H^+}^\beta\le C_5\left\|f(h_1)-f(h_2)\right\|_{W_+^-}^\beta .\end{gathered}$$

$$\begin{gathered}\left\|v_{h_2}\right\|_{W_+^+}\le C_6\left\|\Phi_u(u(h_2))\right\|_{H^-}\le\\ \le C_6\left\|\Phi_u(u(h_2))-\Phi_u\left(u\left(h^*\right)\right)\right\|_{H^-}+C_6\left\|\Phi_u\left(u\left(h^*\right)\right)\right\|_{H^-}\le\\ \le C_7\left\|u(h_2)-u\left(h^*\right)\right\|_{H^+}^\beta+C_6\left\|\Phi_u\left(u\left(h^*\right)\right)\right\|_{H^-}\le\\ \le C_8\left\|f(h_2)-f\left(h^*\right)\right\|_{W_+^-}^\beta+C_6\left\|\Phi_u\left(u\left(h^*\right)\right)\right\|_{H^-} .\end{gathered}$$

Applying to $\left\|f(h_1)-f(h_2)\right\|_{W_+^-}$, $\left\|f(h_2)-f(h^*)\right\|_{W_+^-}$ the formula of finite increments and taking into account the fact that $f_h(\cdot)$ is bounded in U_{ad}, we have

$$\left\|f(h_1)-f(h_2)\right\|_{W_+^-} \leq \sup_{\theta\in[0,1]}\left\|f_{h_1+\theta(h_2-h_1)}\right\|\left\|h_1-h_2\right\|_V \leq C_1\left\|h_1-h_2\right\|_V,$$

$$\left\|f(h_2)-f(h^*)\right\|_{W_+^-} \leq \sup_{\theta\in[0,1]}\left\|f_{h_2+\theta(h^*-h_2)}\right\|\left\|h_2-h^*\right\|_V \leq C_1 d .$$

Next, we obtain

$$\left\|J_{h_1}-J_{h_2}\right\| \leq C_9\left(\left\|h_1-h_2\right\|_V^{\beta}+\left\|h_1-h_2\right\|_V^{\alpha}\right).$$

Whence,

$$\left\|J_{h_1}-J_{h_2}\right\| \leq C_9\left(\left\|h_1-h_2\right\|_V^{\max\{\alpha,\beta\}-\gamma}+1\right)\left\|h_1-h_2\right\|_V^{\gamma} \leq$$

$$\leq C_9\left(d^{\max\{\alpha,\beta\}-\gamma}+1\right)\left\|h_1-h_2\right\|_V^{\gamma},$$

where $\gamma=\min\{\alpha,\beta\}$. The theorem is proved.

Consider applications of these theorems to the specific singular optimization problems. Consider various performance functionals of the systems (2.1) under the assumptions of Theorem 1 for the mapping f.

1. If $\Phi(u)=(p,u)_{L_2(Q)}$, $p\in L_2(Q)$, then the derivative of the functional $J(\cdot)$ defined by the formula

$$J_h(\cdot)=\langle f_h(\cdot),v\rangle_{W_+}, \tag{4}$$

and the adjoint problem has the form $L^+v=Op$, where O is an operator of imbedding of the space $L_2(Q)$ into H^-.

2. For the quadratic functional $\Phi(u)=\sum_{i=1}^m\alpha_i\left\|u-u_i\right\|_{L_2(Q)}^2$, $u_i\in L_2(Q)$, $\alpha_i>0$ Theorem 1 implies the formula:

$$L^+v=2\sum_{i=1}^m\alpha_i O(u(h)-u_i).$$

3. If $\Phi(u) = \frac{1}{2}\|u - u_{ad}\|_{H^+}^2$, $u_{ad} \in H^+$, then the adjoint problem has the form

$$L^+ v = I^{-1}(u(h) - u_{ad}),$$

where I^{-1} is the inverse operator for the operator I, defining the isometry between the whole space H^- and the whole space H^+. In the case of the pseudo-parabolic system investigated in Chapter 5 the operator I^{-1} is defined on the smooth functions $u \in H^+$ obeying the conditions $u|_{x \in \partial\Omega} = 0$ in the following way:

$$I^{-1} u = -\Delta u + u.$$

4. Let $\Phi(u) = \frac{1}{2}\Psi\left(\|u\|_{H^+}^2\right)$, where $\Psi(\cdot) \in C^1([0, +\infty))$. The derivative of the functional $J(\cdot)$ is defined by the formula (4), and the adjoint problem has the form

$$L^+ v = \Psi'\left(\|u(h)\|_{H^+}^2\right) I^{-1} u(h).$$

4. CONVERGENCE OF GRADIENT METHODS ANALOGIES

The theorems of the previous Section allows us to construct gradient iterative methods for finding out optimal controls of systems. For this purpose at every step we must solve direct and adjoint boundary value problems, which cannot be solved exactly in practice. In addition, a singular right-hand side of the state equation are usually approximated by piecewise constant or piecewise linear functions in order to regularize problem for computer simulation. Computer computations have round-off errors. Hence, it is necessary to investigate convergence and stability of methods under the perturbations of data. Since, perturbations and round-off errors in our class of problems have an additive character, we shall consider only the case when the right-hand side of the system state equation is perturbed and all sub-problems are solved exactly.

Consider the optimal control problem for the system

$$J(h)=\Phi(u(h))\to\inf_{h\in U_{ad}},\ Lu(h)=f(h),$$

where $f(\cdot):V\to W_+^-$, U_{ad} is the set of admissible controls from the Hilbert space of controls V. We suppose that the operator L, the functional $\Phi(\cdot)$ and the mapping $f(\cdot)$ obey the assumptions of Sections 2 and 3, U_{ad} is compact in the strong topology and convex. Although in the problems of oprimization of systems with distributed parameters the strong compactness does not hold as a rule, using parametrization or regularization we can approximate the original problem in a such way, that the strong compactness holds.

Consider the problem with a perturbation in the right-hand side

$$Lu(h)=f^\varepsilon(h),$$

where $\{f^\varepsilon(\cdot):V\to W_+^-\}_{\varepsilon>0}$ is one-parameter set of differentiable by Fréchet mappings, which approximate in some sense the mapping $f(\cdot):V\to W_+^-$. Instead of the Fréchet derivative $J_h(\cdot)=\langle f_h(\cdot),v\rangle_{W_+}$ we have its estimation $J_h^\varepsilon(\cdot)=\left\langle f_h^\varepsilon(\cdot),v\right\rangle_{W_+}$.

Since V is a Hilbert space, and $J_h(\cdot)$, $J_h^\varepsilon(\cdot)$ is a linear continuous functional in V, there exists such elements $l\in V$, $l^\varepsilon\in V$ that $J_h(\Delta h)=(l,\Delta h)_V$, $J_h^\varepsilon(\Delta h)=(l^\varepsilon,\Delta h)_V$, respectively. We shall denote them by $l=J'(h)$, $l^\varepsilon=J'_\varepsilon(h)$.

The methods investigated below have the following structure: the sequence of controls is constructed

$$h^{s+1}=h^s+\rho_s\left(\overline{h}^s-h^s\right),$$

which satisfies the conditions $h^s\in U_{ad}$, and the control $\overline{h}^s$ is a solution of some auxiliary extremal problem. The precision of computation of the Fréchet derivative is selected according to the conditions

$$\left\| f\left(h^s\right) - f^{\varepsilon_s}\left(h^s\right) \right\|_{W_+^-} < \varepsilon'_s, \ \left\| f_{h^s} - f_{h^s}^{\varepsilon_s} \right\| < \varepsilon''_s,$$

where (ε'_s), (ε''_s) are infinitesimal sequences of positive numbers.

To prove the convergence of these methods we shall use the sufficient conditions of the convergence of the algorithms of non-linear and stochastic programming.

4.1 Analogy of Rosen's gradient projection method

Let the sequence of controls (h^s) is generated by the following procedure:

(i) Start from $h^0 \in U_{ad}$. Put $s = 0$.

(ii) For all integer positive s compute

$$\bar{h}^s = \ddot{I}\left(h^s - J'_{\varepsilon_s}\left(h^s\right)\right),$$

$$h^{s+1} = h^s + \rho_s\left(\bar{h}^s - h^s\right),$$

where s is the number of iteration, $h^0 \in U_{ad}$ is the initial approximation, $\ddot{I}\,(\cdot)$ is the operator of projection in the set U_{ad}, ρ_s is the step multiplier selected by the condition

$$\sum_{s=0}^{\infty} \rho_s = +\infty, \ \rho_s \xrightarrow[s\to\infty]{} 0, \ \rho_s \in (0,1).$$

This gradient procedure is an analogy of the well-known Rosen's gradient projection method.

Consider the set U^* of admissible controls, for which the necessary condition of the local minimum of the functional $J(\cdot)$ in the set U_{ad} holds true, i.e.

$$U^* = \left\{h^* \in U_{ad} : \left(J'\left(h^*\right), h - h^*\right)_V \geq 0, \ \forall\, h \in U_{ad}\right\}.$$

The necessary condition of the local minimum can be written in the form $h^* = \ddot{I}\left(h^* - J'(h^*)\right)$, which is convenient for justification of the convergence of the method.

Theorem 1. *Let*

1) $f : V \to W_+^-$ *be a differentiable by Fréchet mapping, whose derivative obeys the Lipschitz condition with index* $\alpha \in (0,1]$;
2) $\Phi : H^+ \to R$ *is a differentiable by Fréchet functional, whose derivative obeys the Lipschitz condition with index* $\beta \in (0,1]$.

If the functional $J(\cdot)$ *takes on at the most countable number of the values in the set* $U^* = \{h^* \in U_{ad} : (J'(h^*), h - h^*)_V \geq 0, \forall\, h \in U_{ad}\}$, *then all limit points (which exist necessarily) of the sequence* (h^s) *belong to the compact connected subset* U^* *and numerical sequence* $(J(h^s))$ *has a limit.*

Proof. Let us test the assumptions of the theorem of sufficient conditions of the iterative algorithms convergence. By construction, all entries of the sequence (h^s) belong to the compact U_{ad}. Consider a sequence (h^{s_k}): $h^{s_k} \xrightarrow[k\to\infty]{} h' \in U^*$. We have

$$\left\|h^{s_k+1} - h^{s_k}\right\|_V = \rho_{s_k}\left\|\bar{h}^{s_k} - h^{s_k}\right\|_V \leq \rho_{s_k}\operatorname{diam}(U_{ad}) \xrightarrow[k\to\infty]{} 0 .$$

Let (h^{s_k}) is a subsequence convergent to the control $h' \notin U^*$. Let us prove that there exists such $\delta_0 > 0$ that for all k and $\delta \in (0, \delta_0]$

$$\tau_k = \min_{s > s_k}\left\{s : \left\|h^s - h^{s_k}\right\|_V > \delta\right\} < +\infty .$$

Suppose the contrary. Let for all $\delta_0 > 0$ there exists such $k_0 = k_0(\delta_0)$ that $\left\|h^s - h^{s_{k_0}}\right\|_V \le \delta_0$ for all $s > s_{k_0}$. Then by the triangle inequality we have

$$h^s \in \overline{O}_{\delta_0}\left(h^{s_{k_0}}\right) \Rightarrow h^{s_k} \in \overline{O}_{\delta_0}\left(h^{s_{k_0}}\right) \Rightarrow$$
$$h' \in \overline{O}_{\delta_0}\left(h^{s_{k_0}}\right) \Rightarrow h^s \in \overline{O}_{2\delta_0}(h')$$

for all $s > s_{k_0}$. Here we use the notation $\overline{O}_\delta\left(h^*\right) = \left\{h : \left\|h - h^*\right\|_V \le \delta\right\}$.

Put $W(\cdot) = J(\cdot)$. The assumption of the theorem implies that $W(\cdot)$ is continuous in U_{ad} functional, the set $W^* = \left\{W(h) : h \in U^*\right\}$ is at the most countable, and also $J(\cdot)$ is differentiable by Fréchet functional and the derivative $J_h(\cdot)$ obeys the Lipschitz condition with index $\gamma = \min\{\alpha, \beta\}$ in the set U_{ad} . Consider the expression

$$W\left(h^{s+1}\right) - W\left(h^s\right) = \left(J'\left(h^s + \theta_s\left(h^{s+1} - h^s\right)\right), h^{s+1} - h^s\right)_V =$$
$$= \left(J'\left(h^s\right), h^{s+1} - h^s\right)_V + \left(J'\left(h^s + \theta_s\left(h^{s+1} - h^s\right)\right) - J'\left(h^s\right), h^{s+1} - h^s\right)_V \le$$
$$\le \rho_s\left(J'\left(h^s\right), \overline{h}^s - h^s\right)_V + \rho_s^{1+\gamma} C d^{1+\gamma}, \tag{1}$$

where $s > s_{k_0}$, $\theta_s \in [0,1]$, $d = \operatorname{diam}\left(U_{ad}\right)$ is the diameter of U_{ad}.

By the inequality (1), we shall estimate the value $\left(J'\left(h^s\right), \overline{h}^s - h^s\right)_V$. To do this put $\overline{h}' = \ddot{I}\ (h' - J'(h'))$ and write

$$\left(J'\left(h^s\right), \overline{h}^s - h^s\right)_V = \left(J'(h'), \overline{h}^s - h^s\right)_V + \left(J'\left(h^s\right) - J'(h'), \overline{h}^s - h^s\right)_V =$$
$$= \left(J'(h'), \overline{h}' - h'\right)_V + \left(J'(h'), h' - h^s\right)_V + \left(J'(h'), \overline{h}^s - \overline{h}'\right)_V +$$
$$+ \left(J'\left(h^s\right) - J'(h'), \overline{h}^s - h^s\right)_V.$$

Since $h' \notin U^*$, there exists such $\lambda > 0$ that $\left\|h' - \overline{h}'\right\|_V^2 \ge \lambda > 0$. The following inequality holds true:

$$\left(J'(h'), \overline{h}' - h'\right)_V \le -\left\|h' - \overline{h}'\right\|_V^2,$$

whence, we have

$$\left(J'(h^s), \bar{h}^s - h^s\right)_V \le -\lambda + \|J'(h')\|_V \left(\|h' - h^s\|_V + \|\bar{h}^s - \bar{h}'\|_V\right) +$$
$$+ \|J'(h^s) - J'(h')\|_V \|\bar{h}^s - h^s\|_V \le$$
$$\le -\lambda + \|J'(h')\|_V \left(\|h' - h^s\|_V + \|\bar{h}^s - \bar{h}'\|_V\right) + Cd\|h' - h^s\|_V^\gamma .$$

The value $\|\bar{h}^s - \bar{h}'\|_V$ we estimate in the following way

$$\|\bar{h}^s - \bar{h}'\|_V = \left\|\ddot{I}\left(h^{s'} - J'_{\varepsilon_s}(h^s)\right) - \ddot{I}\left(h' - J'(h')\right)\right\|_V \le$$
$$\le \|h^s - h'\|_V + \|J'_{\varepsilon_s}(h^s) - J'(h')\|_V .$$

Estimate the second summand in the right-hand side of the last inequality

$$\|J'_{\varepsilon_s}(h^s) - J'(h')\|_V \le \|J'_{\varepsilon_s}(h^s) - J'(h^s)\|_V + \|J'(h^s) - J'(h')\|_V \le$$
$$\le \|J'_{\varepsilon_s}(h^s) - J'(h^s)\|_V + C\|h^s - h'\|_V^\gamma .$$

Next,

$$\|J'_{\varepsilon_s}(h^s) - J'(h^s)\|_V =$$
$$= \sup_{\|\Delta h\|_V = 1} \left| \left\langle f_{h^s}^{\varepsilon_s}(\Delta h), v(\varepsilon_s, h^s)\right\rangle_{W_+} - \left\langle f_{h^s}(\Delta h), v(h^s)\right\rangle_{W_+} \right| \le$$
$$\le \|f_{h^s}^{\varepsilon_s} - f_{h^s}\| \|v(\varepsilon_s, h^s)\|_{W_+^+} + \|f_{h^s}\| \|v(\varepsilon_s, h^s) - v(h^s)\|_{W_+^+} \le$$
$$\le \varepsilon_s'' \|v(\varepsilon_s, h^s)\|_{W_+^+} + \left(\|f_{h^s} - f_{h'}\| + \|f_{h'}\|\right) \|v(\varepsilon_s, h^s) - v(h^s)\|_{W_+^+} \le$$
$$\le \varepsilon_s'' \|v(\varepsilon_s, h^s)\|_{W_+^+} + \left(C_1\|h^s - h'\|_V^\alpha + \|f_{h'}\|\right) \|v(\varepsilon_s, h^s) - v(h^s)\|_{W_+^+} ,$$

where $v(\varepsilon_s, h^s) \in W_+^+$, $v(h^s) \in W_+^+$ are solutions of the adjoint problems

$$L^+ v(\varepsilon_s, h^s) = \Phi_u\left(u(\varepsilon_s, h^s)\right),\ L^+ v(h^s) = \Phi_u\left(u(h^s)\right).$$

The following estimations are valid (we use the a priori estimations of the solutions of the original and adjoint problems, the formula of finite increments and the fact that the derivatives $f_h(\cdot)$, $\Phi_u(\cdot)$ obey the Lipschitz condition):

$$\|v(\varepsilon_s,h^s)\|_{W_+^+} \le C_2\|\Phi_u(u(\varepsilon_s,h^s))\|_{H^-} \le$$

$$\le C_3\left(\|u(\varepsilon_s,h^s)-u(h^s)\|_{H^+}^{\beta} + \|u(h^s)-u(h')\|_{H^+}^{\beta}\right) + C_2\|\Phi_u(u(h'))\|_{H^-} \le$$

$$\le C_4\left(\|f^{\varepsilon_s}(h^s)-f(h^s)\|_{W_+^-}^{\beta} + \|f(h^s)-f(h')\|_{W_+^-}^{\beta}\right) + C_2\|\Phi_u(u(h'))\|_{H^-} <$$

$$< C_4\left(\varepsilon_s'^{\beta} + C_5 d^{\beta}\right) + C_2\|\Phi_u(u(h'))\|_{H^-} \le C_6 < +\infty,$$

$$\|v(\varepsilon_s,h^s)-v(h^s)\|_{W_+^+} \le C_2\|\Phi_u(u(\varepsilon_s,h^s))-\Phi_u(u(h^s))\|_{H^-} \le$$

$$\le C_3\|u(\varepsilon_s,h^s)-u(h^s)\|_{H^+}^{\beta} \le C_4\|f^{\varepsilon_s}(h^s)-f(h^s)\|_{W_+^-}^{\beta} < C_4\varepsilon_s'^{\beta}.$$

We have

$$\left(J'(h^s),\bar{h}^s-h^s\right)_V < -\lambda +$$
$$+\|J'(h')\|_V\left(2\|h'-h^s\|_V + C\|h^s-h'\|_V^{\gamma} +\right.$$
$$+\varepsilon_s'' C_6 + \left(C_1\|h^s-h'\|_V^{\alpha} + \|f_{h'}\|\right)C_4\varepsilon_s'^{\beta}\Big)$$
$$+Cd\|h'-h^s\|_V^{\gamma} \le -\lambda +$$
$$+\|J'(h')\|_V\left(4\delta_0 + C(2\delta_0)^{\gamma} + \varepsilon_s'' C_6 + \left(C_1(2\delta_0)^{\alpha} + \|f_{h'}\|\right)C_4\varepsilon_s'^{\beta}\right) +$$
$$+Cd(2\delta_0)^{\gamma}.$$

Choosing sufficiently small $\delta_0 > 0$ and large k_0, we obtain

$$\left(J'(h^s),\bar{h}^s-h^s\right)_V \le -\frac{\lambda}{2},\ s > s_{k_0}.$$

Thus, ultimately we obtain

$$W(h^{s+1}) - W(h^s) \le -\frac{\lambda}{2}\rho_s + \rho_s^{1+\gamma} C d^{1+\gamma},\ s > s_{k_0}. \tag{2}$$

Adding the inequalities (2) for $s = \overline{s_k, p}$, $k > k_0$, we obtain

$$W(h^p)-W(h^{s_k})\leq -\frac{\lambda}{2}\sum_{s=s_k}^{p-1}\rho_s + Cd^{1+\gamma}\sum_{s=s_k}^{p-1}\rho_s^{1+\gamma}.$$

Since $\rho_s^{1+\gamma}=o(\rho_s)$, there exist $s'\geq s_k$: $\forall s\geq s'$ $\rho_s^{1+\gamma}\leq \frac{\lambda}{8Cd^{1+\gamma}}\rho_s$. For all $p>s'$ the estimation $\sum_{s=s'}^{p-1}\rho_s^{1+\gamma}\leq \frac{\lambda}{8Cd^{1+\gamma}}\sum_{s=s'}^{p-1}\rho_s\leq \frac{\lambda}{8Cd^{1+\gamma}}\sum_{s=s_k}^{p-1}\rho_s$ is valid. The relation $\sum_{s=s_k}^{\infty}\rho_s=+\infty$ implies the existence of such number $s''\geq s_k$ that $\forall p>s''$ $\sum_{s=s_k}^{s''}\rho_s^{1+\gamma}\leq \frac{\lambda}{8Cd^{1+\gamma}}\sum_{s=s_k}^{p-1}\rho_s$.

If $p>\max\{s',s''\}$, then $\sum_{s=s_k}^{p-1}\rho_s^{1+\gamma}\leq \frac{\lambda}{4Cd^{1+\gamma}}\sum_{s=s_k}^{p-1}\rho_s$, i.e.

$$W(h^p)-W(h^{s_k})\leq -\frac{\lambda}{4}\sum_{s=s_k}^{p-1}\rho_s. \tag{3}$$

Passing to the limit in (3) as $p\to\infty$ and taking into account the condition $\sum_{s=s_k}^{\infty}\rho_s=+\infty$, we arrive to contradiction with the fact that the continuous functional $W(\cdot)$ is bounded below on the compact U_{ad}.

Thus, there exists $\delta_0>0$ such that for all k and $\delta\in(0,\delta_0]$

$$\tau_k=\min_{s>s_k}\left\{s:\left\|h^s-h^{s_k}\right\|_V>\delta\right\}<+\infty.$$

However, taking sufficiently small $\delta_0>0$ and large k_0, it is possible to repeat the proof of the estimation (3) for $s_k\leq s\leq\tau_k$. On the other hand

$$\delta_0<\left\|h^{\tau_k}-h^{s_k}\right\|_V\leq\sum_{s=s_k}^{\tau_k-1}\left\|h^{s+1}-h^s\right\|_V\leq d\sum_{s=s_k}^{\tau_k-1}\rho_s.$$

Therefore,

$$\sum_{s=s_k}^{\tau_k-1}\rho_s > \frac{\delta_0}{d}.$$

Granting the last inequality in (3), we have

$$W\left(h^{\tau_k}\right)-W\left(h^{s_k}\right)< -\frac{\lambda\delta_0}{4d}.$$

Whence,

$$\overline{\lim_{k\to\infty}} W\left(h^{\tau_k'}\right)< \lim_{k\to\infty} W\left(h^{s_k}\right).$$

Thus, the sufficient conditions of the convergence hold true, and hence, all limit points of the sequence $\left(h^s\right)$ belong to U^* and the numerical sequence $\left(J\left(h^s\right)\right)$ has a limit. The connectedness and compactness of the set of limit points $\left(h^s\right)$ follows from the fact that $\left\|h^{s+1}-h^s\right\|_V=\rho_s\left\|\overline{h}^s-h^s\right\|_V\le\rho_s d\xrightarrow[s\to\infty]{}0$ and from the following assertion: if the sequence of the points $\left(x_k\right)$ of the metric space X is imbedded into some compact K, $\rho_X\left(x_{k+1},x_k\right)\xrightarrow[k\to\infty]{}0$, then the set of limit points $\left(x_k\right)$ is a connected compact.

If the assumption that the set of values of the functional $J(\cdot)$ is at the most countable in the set U^* does not hold true, it is possible to prove weaker assertion:

Theorem 2. *If the assumptions 1) and 2) of Theorem 1 hold true, then the sequence $\left(h^s\right)$ has at least one limit point belonging to the set U^*.*

4.2 Analogy of the conditional gradient method with averaging

Consider the sequence of controls $\left(h^s\right)$ generated by the following procedure:

(i) Start from $h^0 \in U_{ad}$, $q^0 = J'_{\varepsilon_0}(h^0)$, $\bar{h}^0 = \arg\inf_{h \in U_{ad}} (q^0, h)_V$.

Put $s = 0$.

(ii) For all integer positive s compute

$$h^{s+1} = h^s + \rho_s(\bar{h}^s - h^s),$$
$$q^{s+1} = q^s + \alpha_s\left(J'_{\varepsilon_{s+1}}(h^{s+1}) - q^s\right),$$
$$\bar{h}^{s+1} = \arg\inf_{h \in U_{ad}} (q^{s+1}, h)_V,$$

where s is the number of iteration, $h^0 \in U_{ad}$ is the initial approximation, ρ_s, α_s is the multipliers selected by the conditions $\sum_{s=0}^{\infty} \rho_s = +\infty$, $\rho_s \xrightarrow[s\to\infty]{} 0$, $\cdot\, \rho_s \in (0,1)$, $\sum_{s=0}^{\infty} \alpha_s = +\infty$, $\alpha_s \in (0,1)$, $\rho_s^\gamma / \alpha_s \xrightarrow[s\to\infty]{} 0$.

Theorem 3. *Let the assumptions 1) and 2) of Theorem 1 holds true. If the functional $J(\cdot)$ takes on at the most countable number of its values in the set $U^* = \{h^* \in U_{ad} : (J'(h^*), h - h^*)_V \geq 0, \forall\, h \in U_{ad}\}$, then all limit points (which exist necessarily) of the sequence (h^s) generated by the method (i), (ii) belong to the compact connected subset U^* and the numerical sequence $(J(h^s))$ has a limit.*

Proof is similar to the proof of Theorem 1. Let us prove only that for the sequence (h^{s_k}) which converges to the control $h' \notin U^*$ there exists such $\delta_0 > 0$ that for all k and $\delta \in (0, \delta_0]$: $\tau_k = \min_{s > s_k} \{s : \|h^s - h^{s_k}\|_V > \delta\} < +\infty$.

Suppose the contrary. Let for all $\delta_0 > 0$ there exist such $k_0 = k_0(\delta_0)$ that $\|h^s - h^{s_{k_0}}\|_V \leq \delta_0$ for all $s > s_{k_0}$. Then we have

$h^s \in \overline{O}_{\delta_0}\left(h^{s_{k_0}}\right) \Rightarrow \qquad h^{s_k} \in \overline{O}_{\delta_0}\left(h^{s_{k_0}}\right) \Rightarrow \qquad h' \in \overline{O}_{\delta_0}\left(h^{s_{k_0}}\right) \Rightarrow$
$h^s \in \overline{O}_{2\delta_0}(h')$ for all $s > s_{k_0}$.

Put $W(\cdot) = J(\cdot)$. Consider

$$
\begin{aligned}
& W(h^s) - W(h^{s_k}) = J(h^s) - J(h') - J(h^{s_k}) + J(h') = \\
&= \left(J'(h' + \theta'(h^s - h')), h^s - h'\right)_V - \left(J'(h' + \theta''(h^{s_k} - h')), h^{s_k} - h'\right)_V = \\
&= \left(J'(h'), h^s - h'\right)_V + \left(J'(h' + \theta'(h^s - h')) - J'(h'), h^s - h'\right)_V - \\
&\quad - \left(J'(h'), h^{s_k} - h'\right)_V + \left(J'(h') - J'(h' + \theta''(h^{s_k} - h')), h^{s_k} - h'\right)_V \le \\
&\le \left(J'(h'), h^s - h^{s_k}\right)_V + 2^{2+\gamma} C \delta_0^{\,1+\gamma} = \\
&= \sum_{p=s_k}^{s-1} \rho_p \left(J'(h'), \overline{h}^p - h^p\right)_V + 2^{2+\gamma} C \delta_0^{\,1+\gamma}, \qquad (4)
\end{aligned}
$$

where $s > s_k$, $k > k_0$, $\{\theta', \theta''\} \subset [0,1]$.

In the inequality (4) we estimate the value $\left(J'(h'), \overline{h}^p - h^p\right)_V$. For this purpose put $\overline{h}' = \arg\inf\limits_{h \in U_{ad}} (J'(h'), h)_V$ and write

$$
\begin{aligned}
\left(J'(h'), \overline{h}^p - h^p\right)_V &= \left(J'(h') - q^p, \overline{h}^p - h^p\right)_V + \left(q^p, \overline{h}^p - h^p\right)_V \le \\
&\le \left(J'(h') - q^p, \overline{h}^p - h^p\right)_V + \left(q^p, \overline{h}' - h^p\right)_V .
\end{aligned}
$$

But

$$
\begin{aligned}
\left(q^p, \overline{h}' - h^p\right)_V &= \left(q^p - J'(h'), \overline{h}' - h^p\right)_V + \left(J'(h'), \overline{h}' - h'\right)_V + \\
&+ \left(J'(h'), h' - h^p\right)_V .
\end{aligned}
$$

Thus,

$$
\begin{aligned}
&\left(J'(h'), \overline{h}^p - h^p\right)_V \le \left(q^p - J'(h'), \overline{h}' - \overline{h}^p\right)_V + \left(J'(h'), \overline{h}' - h'\right)_V + \\
&+ \left(J'(h'), h' - h^p\right)_V .
\end{aligned}
$$

Since $h' \notin U^*$, there exists such $\lambda > 0$ that $\left(J'(h'), \overline{h}' - h'\right)_V \le -\lambda$, whence

$$\left(J'(h'),\bar{h}^p - h^p\right)_V \le -\lambda + d\left\|q^p - J'(h')\right\|_V + 2\|J'(h')\|_V \delta_0 .$$

The value $\left\|q^p - J'(h')\right\|_V$ we shall estimate in the following way

$$\left\|q^p - J'(h')\right\|_V \le \left\|q^p - J'\left(h^p\right)\right\|_V + \left\|J'\left(h^p\right) - J'(h')\right\|_V \le$$

$$\le \left\|q^p - J'\left(h^p\right)\right\|_V + C2^\gamma \delta_o^\gamma .$$

Further (in Lemma 1) it will be proved that

$$\pi_p = \left\|q^{p\,'} - J'\left(h^p\right)\right\|_V \xrightarrow[p\to\infty]{} 0 .$$

We have

$$\left(J'\left(h^p\right),\bar{h}^p - h^p\right)_V \le -\lambda + d\pi_p + dC2^\gamma \delta_o^\gamma + 2\|J'(h')\|_V \delta_0 .$$

Choosing sufficiently small $\delta_0 > 0$ and large k_0, we obtain

$$\left(J'(h'),\bar{h}^p - h^p\right)_V \le -\frac{\lambda}{2},\ p > s_{k_0} .$$

Thus, ultimately we have

$$W\left(h^s\right) - W\left(h^{s_k}\right) \le -\frac{\lambda}{2}\sum_{p=s_k}^{s-1}\rho_p + 2^{2+\gamma}C\delta_0^{\ 1+\gamma},\ s > s_k \ge s_{k_0} . \quad (5)$$

Passing to the limit in (5) as $s \to \infty$ and taking into account the condition $\sum_{p=s_k}^{\infty}\rho_p = +\infty$, we arrive to contradiction with the fact that the continuous functional $W(\cdot)$ is bounded below in the compact U_{ad}. Thus, there exists $\delta_0 > 0$ such that for all k and $\delta \in \left(0,\delta_0\right]$

$$\tau_k = \min_{s > s_k}\left\{s : \left\|h^s - h^{s_k}\right\|_V > \delta\right\} < +\infty .$$

Further, proof is similar to the proof of Theorem 1.

Lemma 1. *Let the assumptions of Theorem* 3 *hold true. Then*

$$\left\|q^s - J'\left(h^s\right)\right\|_V \xrightarrow[s\to\infty]{} 0 .$$

Proof. Let us introduce the following notation

$$\pi_s = \left\|q^s - J'\left(h^s\right)\right\|_V .$$

Consider

$$\pi_{s+1} = \left\| q^{s+1} - J'(h^{s+1}) \right\|_V =$$

$$= \left\| q^s + \alpha_s \left(J'_{\varepsilon_{s+1}}(h^{s+1}) - q^s \right) - J'(h^{s+1}) \right\|_V \le$$

$$\le (1-\alpha_s) \left\| q^s - J'(h^s) \right\|_V +$$

$$+ \left\| (1-\alpha_s)\left(J'(h^s) - J'(h^{s+1})\right) + \alpha_s \left(J'_{\varepsilon_{s+1}}(h^{s+1}) - J'(h^{s+1}) \right) \right\|_V \le$$

$$\le (1-\alpha_s)\pi_s + (1-\alpha_s)\left\| J'(h^s) - J'(h^{s+1}) \right\|_V +$$

$$+ \alpha_s \left\| J'_{\varepsilon_{s+1}}(h^{s+1}) - J'(h^{s+1}) \right\|_V .$$

Since $J'(\cdot)$ obeys the Lipschitz condition with index $\gamma = \min\{\alpha, \beta\}$ in the set U_{ad},

$$\left\| J'(h^s) - J'(h^{s+1}) \right\|_V \le C_0 \left\| h^s - h^{s+1} \right\|_V^\gamma \le C_0 \rho_s^\gamma d^\gamma .$$

Let us estimate the value $\left\| J'_{\varepsilon_{s+1}}(h^{s+1}) - J'(h^{s+1}) \right\|_V$. Reasoning as in the proof of Theorem 1, we have

$$\left\| J'_{\varepsilon_{s+1}}(h^{s+1}) - J'(h^{s+1}) \right\|_V \le C_1 \left(\varepsilon''_{s+1} + \varepsilon'^{\beta}_{s+1} \right).$$

Thus,

$$0 \le \pi_{s+1} \le (1-\alpha_s)\pi_s + \psi_s , \tag{6}$$

where $\psi_s = (1-\alpha_s) C_0 \rho_s^\gamma d^\gamma + \alpha_s C_1 \left(\varepsilon''_{s+1} + \varepsilon'^{\beta}_{s+1} \right)$.

It is clear that $\psi_s / \alpha_s \xrightarrow[s\to\infty]{} 0$. Indeed,

$$\frac{\psi_s}{\alpha_s} = \left(\frac{\rho_s^\gamma}{\alpha_s} - \rho_s^\gamma \right) C_0 d^\gamma + C_1 \left(\varepsilon''_{s+1} + \varepsilon'^{\beta}_{s+1} \right) \xrightarrow[s\to\infty]{} 0 .$$

This relation and (6) imply that $\pi_s \xrightarrow[s\to\infty]{} 0$.

If the assumption that the set of the values of the functional $J(\cdot)$ in the set U^* is at most countable does not hold true, then it is possible to prove the following weaker assertion.

Theorem 4. *If the assumptions 1) and 2) of Theorem 1 holds true, then the sequence* (h^s) *generated by the method (i), (ii) has at least one limit point which belongs to the set* U^*

5. OPTIMIZATION OF PARABOLIC SYSTEMS WITH GENERALIZED COEFFICIENTS

The solvability of parabolic systems with discontinuous coefficients were investigated in [173,174,178] and in many others papers. In partucular, these problem arise as a result of investigation of heat and mass transport in heterogeneous media with non-ideal contact between subdomains, with external condensed source and so on. In this subsection the results obtained in [173,174] are extended and the issue of optimization of parabolic systems with discontinuous solutions are considered.

5.1 Main notations

Let the system state described by the function $u(t, \xi_1, \xi_2, \dots, \xi_n)$ The heat and mass transport take place in a tube $Q = \{(0, T) \times \Omega\}$, where $\Omega \subset R^n$ is bounded regular domain of variation of space variables $\xi = (\xi_1, \dots, \xi_n)$ with boundary $\partial\Omega$.

Introduce the following notations. Let $C^\infty_{bd}(\overline{Q})$ be a set of infinitely differentiable in the domain $\overline{Q}$ functions satisfying the boundary conditions

$$u|_{t=0} = 0, u|_{\xi \in \partial\Omega} = 0,$$

$W_2^{1,0}$ be a completion of the set $C^\infty_{bd}(\overline{Q})$ in the norm

$$\|u\|^2_{W_2^{1,0}} = \int_Q u^2 + u_t^2 dQ, \qquad (1)$$

$W_2^{0,1}$ be a completion (the set of smooth in $C^\infty(\overline{Q})$ functions) in the norm

$$\|u\|^2_{W_2^{0,1}} = \int_Q u^2 + \sum_{i=1}^n \left(\frac{\partial u}{\partial \xi_i}\right)^2 dQ.$$

Denote also by $W_{2,T}^{1,0}$ a completion of the set $C^\infty_{bd^*}(\overline{Q})$ consisting of functions, which are smooth in the domain and satisfy the adjoint conditions

$$u\big|_{t=T} = 0, u\big|_{\xi\in\partial\Omega} = 0,$$

in the same norm (1).

$W_2^{-1,0}, W_2^{0,-1}, W_{2,T}^{-1,0}$ are negative spaces constructed on the basis of pairs $W_2(Q) \subset L_2(Q)$ with corresponding indices.

In addition, introduce the following notations:

$$X = W_2^{1,0} \times \left(W_2^{0,1}\right)^n, Y = W_{2,T}^{-1,0} \times \left(W_2^{0,-1}\right)^n,$$

$$X_1 = W_2^{1,0} \times \left(L_2(Q)\right)^n, Y_1 = W_{2,T}^{-1,0} \times \left(L_2(Q)\right)^n,$$

$$X_2 = L_2(Q) \times \left(W_{2,T}^{-1,0}\right)^n, Y_2 = L_2(Q) \times \left(W_2^{1,0}\right)^n.$$

We shall also consider the spaces adjoint to X, Y, X_1, Y_1, X_2, Y_2. For example, $X_1^* = W_2^{-1,0} \times \left(L_2(Q)\right)^n$. For every Cartesian product of the original space and its adjoint space (for example, X and X^*) the bilinear form $\langle \cdot,\cdot \rangle_{X\times X^*}$, which is obtained by extension the inner product in the space $\left(L_2(Q)\right)^{n+1}$ by continuity, is defined

Consider the problem of optimization of the system governed by the equations of heat and mass transport in several heterogeneous media with the conditions of non-ideal contact, external condensed source and so on., which generalized the similar problem considered in [174] (the Dirichlet problem)

$$Lx = F(\varphi). \tag{2}$$

The operator L is defined by symbolic matrix

$$L = \left(\begin{array}{c|ccc} \frac{\partial}{\partial t} & b\frac{\partial}{\partial \xi_1} & \dots & b\frac{\partial}{\partial \xi_n} \\ \hline \frac{\partial}{\partial \xi_1} & \sigma_{11} & \dots & \sigma_{1n} \\ \dots & \dots & \dots & \dots \\ \frac{\partial}{\partial \xi_n} & \sigma_{n1} & \dots & \sigma_{nn} \end{array} \right), x = \begin{pmatrix} u \\ w_1 \\ \dots \\ w_n \end{pmatrix}, F = \begin{pmatrix} f_0 \\ f_1 \\ \dots \\ f_n \end{pmatrix},$$

where u is a function describing the heat and mass transport, $\vec{w} = (w_1, \dots, w_n)$ is a vector of the substance flux.

The operator L maps X into Y with the domain of definition $D(L) = C_{bd}^{\infty}(\overline{Q}) \times (C^{\infty}(\overline{Q}))^n$.

The mapping F (possibly, non-linear) maps the Banach space of controls V_φ (with the domain of definition U_φ) into the space Y

The system coefficients obey the conditions: $b \in C^1([0, T]), b > 0, \sigma_{ij} = \sigma_{ji}$, in the equations (2) $\sigma_{ij} w_j$ is understood to be a linear functional defined in $W_2^{0,1}$ by the following rule

$$l_{ij}(\eta) = \rho_{ij}\left(\int_0^T a_{ij}(\tau) w_j(\tau, \xi) \eta(\tau, \xi) d\tau \right), \eta \in W_2^{0,1}, w_j \in W_2^{0,1}, \tag{3}$$

where $\sigma_{ij} = \{(\rho_{ij}, a_{ij})\}, \rho_{ij} \in W_1^{-1}(\Omega), a_{ij} \in C^1([0, T]), W_1^{-1}(\Omega)$ is a space adjoint to the Sobolev space $W_1^1(\Omega)$.

The functional (3) is defined correctly, since

$$\int_0^T a_{ij}(\tau) w_j(\tau, \xi) \eta(\tau, \xi) d\tau \in W_1^1(\Omega).$$

We shall suppose that the coefficient matrix $M = \left\{\sigma_{ij}\right\}_{ij=1}^{n}$ for all vector-functions $\eta = \left(\eta_1, \ldots, \eta_n\right)$ obey the conditions

$$-\sum_{i,j=1}^{n} \rho_{ij}\left(\int_0^T \frac{d}{dt}\left(\frac{a_{ij}}{b}\right)(\tau)\cdot\eta_i(\tau,\xi)\eta_j(\tau,\xi)\,d\tau\right) \geq \alpha_M \sum_{i=1}^{n}\int_Q \eta_i^2\,dQ,$$

$$\sum_{i,j=1}^{n} \rho_{ij}\left(\int_0^T \frac{da_{ij}}{dt}(\tau)\cdot\eta_i(\tau',\xi)\eta_j(\tau,\xi)\,d\tau\right) \geq \alpha_M \sum_{i=1}^{n}\int_Q \eta_i^2\,dQ,$$

$$\sum_{i,j=1}^{n} \rho_{ij}\left(a_{ij}(T)\cdot\eta_i(T,\xi)\eta_j(T,\xi)\right) \geq$$

$$\geq 0, \sum_{i,j=1}^{n} \rho_{ij}\left(a_{ij}(0)\cdot\eta_i(0,\xi)\eta_j(0,\xi)\right) \geq 0,$$

where the constant $\alpha_M > 0$ does nor depend on the functions $\eta_i \in W_2^{0,1}$.

Note that all results obtained in this section can be generalized for the functionals $\sigma_{ij}w_j$ of the form

$$l_{ij}(\eta) = \sum_{k=1}^{p} \rho_{ij}^{k}\left(\int_0^T a_{ij}^{k}(\tau)w_j(\tau,\xi)\eta(\tau,\xi)d\tau\right).$$

The oprimization problem consists in the finding out such control $\varphi \in U_\varphi \subset V_\varphi$ which provide the minimum of the performance functional

$$J(\varphi) = \Phi_1\left(x(\varphi)\right) + \Phi_2\left(\varphi\right) \to \min$$

where Φ_1, Φ_2 are some functional, $x(\varphi)$ is a solution of (2) under the control $\varphi \in U_\varphi \subset V_\varphi$.

Denote by L^* the adjoint by Lagrange operator

$$L^* y = G, \tag{4}$$

where

$L^*: Y^* \to X^*, D(L^*) = C^{\infty}_{bd^*}(\overline{Q}) \times (C^{\infty}(\overline{Q}))^n, y = (\upsilon, \eta_1, \ldots, \eta_n).$

5.2 Optimization of parabolic system with generalized coefficients

Using the technique of the a priori inequalities in negative norms [167], it is possible to prove that the following lemmas holds true.

Lemma 1. *The following inequalities are valid*

$$\|Lx\|_Y \le c\|x\|_X, \forall x \in D(L), \tag{5}$$

$$\|L^* y\|_{X^*} \le c\|y\|_Y, \forall y \in D(L^*), \tag{6}$$

where the constant $c > 0$ *does not depend on* x, y.

The inequalities in Lemma 1 do not allow to extend by continuity the operator $L(L^*)$ to an operator continuously mapping the whole space $X(Y^*)$ into the space Y (X^*, respectively). In what follows, we shall consider only extended operators denoting it by L è L^*, again.

Lemma 2. *The following inequalities hold true*

$$c^{-1}\|x\|_{X_2} \le \|Lx\|_{Y_1}, \forall x \in X, Lx \in Y_1, \tag{7}$$

$$c^{-1}\|y\|_{Y_2^*} \le \|L^* y\|_{X_1^*}, \forall y \in Y^*, L^* y \in X_1^*. \tag{8}$$

For a parabolic equation with usual coefficients the a priori inequalities in negative norms were proved in [175].

The inequalities (5)-(8) allow us to prove the unique solvability of the equations (2), (4).

In what follows, we shall suppose that the set $R(L) \cap Y_2$ is dense in the space Y_2, and also $R(L^*) \cap X_2^*$ is dense in X_2^*, where $R(L), R(L^*)$ are the ranges of values of the extended operatores (but some assertions can be proved under weaker assumptions). In the case when the coefficient matrix $M=\{\sigma_{ij}\}_{ij=1}^{n}$ is generated by a classic functions the density follows from the classic theorems of solvability of parabolic equations.

Definition 1. *A solution of the equation (2) is an element* $x\in X_1$ *such that for any* $y\in Y^*, L^*y\in X_1^*$ *the following equality holds true:*

$$\langle x,L^*y\rangle_{X_1\times X_1^*}=\langle F,y\rangle_{Y\times Y^*}$$

Theorem 1. *For all* $F\in Y_2$ *there exists a unique solution of the equation (2) in the sense of Theorem 1.*

Definition 2. *A solution of the equation (2) is an element* $x\in X_2$ *such that for any* $y\in Y^*, L^*y\in X_2^*$ *the following equality holds true:*

$$\langle x,L^*y\rangle_{X_2\times X_2^*}=\langle F,y\rangle_{Y\times Y^*}.$$

Theorem 2. *For all* $F\in Y_1$ *there exists a unique solution if the equation (2) in the sense of the Definition 2.*

The proof of Theorems 1 and 2 are similar to [167].

Remark 1. *Similar theorems hold true for the adjoint operator. Namely, for any* $G\in X_2^*\left(G\in X_1^*\right)$ *there exists a unique solution* $y\in Y_1^*\left(y\in Y_2^*\right)$ *of the equation (4) in the sense of the analogy of Definition 1 (2, respectively) for the adjoint operator.*

Remark 2. *If* $x\in X_2$ *is a solution of the equation (2) with the right-hand side* $F\in Y_1$ *in the sense of Definition 2 and* $y\in Y_1^*$ *is a solution of the equation (4) with the right-hand side* $G\in X_2^*$ *in the sense of the analogy of Definition 1 then*

$$\langle x,G\rangle_{X_2\times X_2^*}=\langle F,y\rangle_{Y_1\times Y_1^*}.$$

Similarly, for all $F\in Y_2, G\in X_1^*$ the following equality holds true

$$\langle x,G\rangle_{X_1\times X_1^*}=\langle F,y\rangle_{Y_2\times Y_2^*}$$

where x, y are solutions of the equations (2), (4) with the right-hand sides F, G in the sense of Definitions 1 and 2, respectively.

Remark 3. *Note that since Theorems 1 and 2 hold true for arbitrary right-hand sides* F *(not only* $F = \left(f_0, 0, \ldots, 0\right)$ *considered in [173, 174]), the results proved above are applicable not only to systems with singular coefficients, but also to systems with singular right-hand sides. For example, the right-hand side* $c\psi(t)\delta\left(\xi_1 - \zeta\right)(c = const)$ *corresponds to*

$$F = \left(0, \psi(t)\,\mathrm{sgn}\left(\xi_1 - \zeta\right), 0, \ldots, 0\right) \in Y_1$$

in the standard equation of heat and mass transport with constant coefficients.

Theorem 3. *Let the system state* $x(\varphi)$ *is defined from the equations (2). If*

1) *the functionals* $\Phi_1, \Phi_2\left(D\left(\Phi_1\right) = X_1, D\left(\Phi_2\right) = V_\varphi\right)$ *are weakly lower semicontinuous in the spaces* X_1, V_φ, *respectively;*
2) *the set of the admissible controls* U_φ *is weakly compact in the Banach space* V_φ;
3) $R(F) \subset Y_2$ *and the mapping* $F: V_\varphi \to Y_2$ *is weakly continuous,*

then there exists a control $\varphi^* \in U_\varphi \subset V_\varphi$ *providing a minimum of the performance functional* J.

Proof. Let us select a sequence $\varphi_k \in U_\varphi$ minimizing the functional J,

$$\inf_{\varphi \in U_\varphi} J(\varphi) = \lim_{k \to \infty} J\left(\varphi_k\right).$$

Granting the weak compactness of U_φ, we can consider that the sequence φ_k weakly converges to $\varphi^* \in U_\varphi$ in the space $y \in Y^{*}$. Taking into account that F is weakly continuous, we have that $F(\varphi_k)$ weakly converges to $F(\varphi^*)$ in Y_2. Similarly to [167] we prove that there exists such constant $C > 0$ (common for all $F \in Y_2$) that

$$\|x\|_{X_1} \leq C\|F\|_{Y_2},$$

where x is a solution of the equation (2) with the right-hand side F, whence we conclude that the sequence $x(\varphi_k)$ of solutions of (2) under the control φ_k is bounded in X_1, and hence, we can extract from it a subsequence, which weakly converges to $x^* \in X_1$ (we denote it by $x(\varphi_k)$ again). Passing in the equality

$$\langle x(\varphi_k), L^* y\rangle_{X_1 \times X_1^*} = \langle F(\varphi_k), y\rangle_{Y_2 \times Y_2^*}, \forall y \in Y^*, L^* y \in X_1^*$$

to the limit as $k \to \infty$, we obtain

$$\langle x^*, L^* y\rangle_{X_1 \times X_1^*} = \langle F(\varphi^*), y\rangle_{Y_2 \times Y_2^*}, \forall y \in Y^*, L^* y \in X_1^*,$$

whence, we conclude that x^* is a solution of (2) under the control φ^*. Taking into account that functionals Φ_1, Φ_2 are weakly lower semicontinuous we conclude that φ^* is an optimal control.

Remark 4. *À similar theorem holds true in the case when functional Φ_1 is weakly lower semicontinuous in the space X_2 and the mapping $F: V_\varphi \to Y_1$ is weakly continuous.*

Remark 5. *Assuming that U_φ is a compact set, we can prove that any minimizing sequence φ_k converges to the set V^* of optimal controls $\left(\rho(h_k, V^*) \to 0\right)$.*

Remark 6. *As far as the mapping F may be non-linear and the optimization problem may be non-convex, an optimal control may be not unique.*

If we suppose that there exist the Fréchet derivatives $\Phi_1'(x(\varphi)), \Phi_2'(\varphi)$ of the functionals $\Phi_1 : X_1 \to R, \Phi_2 : V_\varphi \to R$ at the points $x(\varphi) \in X_1, \varphi \in U_\varphi$, and also there exists the derivative $F'(\varphi)$ of the mapping $F : V_\varphi \to Y_2$ at the point $\varphi \in U_\varphi$, then we can investigate the differential properties of the functional J.

Theorem 4. *Let the system state be defined by the equations (2) and there exist the Fréchet derivatives* $\Phi_1'(x(\varphi)), \Phi_2'(\varphi), F'(\varphi)$. *Then there exists the Fréchet derivative* $J'(\varphi)$ *if the functional* J *defined by the following expressions*

$$\frac{\partial J}{\partial \varphi}(\Delta\varphi) = \langle F'(\varphi)(\Delta\varphi), y(\varphi) \rangle_{Y_2 \times Y_2^*} + \Phi_2'(\varphi)(\Delta\varphi), \quad (9)$$

where $x(\varphi) = (u, w_1, \ldots, w_n) \in X_1, y(\varphi) = (\upsilon, \eta_1, \ldots, \eta_n) \in Y_2^*$ *are the solutions of the operator equations*

$$Lx = F(\varphi), L^* y = G(x(\varphi)), \quad (10)$$

the function $G(x(\varphi)) \in X_1^*$ *is defined by the Fréchet derivative* $\Phi_1'(x(\varphi))$ *by the formula (the Riesz theorem)*

$$\Phi_1'(x(\varphi))(\Delta x) = \langle G(x(\varphi)), \Delta x \rangle_{X_1^* \times X_1}.$$

Remark 7. *A similar theorem holds true in he case of the functional* $\Phi_1 : X_2 \to R$ *and the mapping* $F : V_\varphi \to Y_1$.

Remark 8. *Developing the technique proposed in [167] it is possible to investigate the smoothness of the functional gradient* J *on the basis of the relations (9), (10) and inequalities (5)-(8) and to construct gradient type numerical methods of optimization.*

REFERENCES

1. Pontryagin L.S., Boltyansky V.G., Gamkrelidze R.V., Mischenko Ye.A. *Mathematical Theory of Optimal Processes.* Moscow, Nauka, 1969.
2. Gamcrelidze R.V. On some extremal problems in the theory of differential equations with applications to the theory of optimal control. SIAM J. Control 1965; 3: 106—128.
3. Feldbaum A.A. *Basis of Optimal Autimatic Systems Theory.* Moscow: Nauka, 1966.
4. Bellman R.E., Dreyfus S.E. *Applied Dynamic Programming.* Princeton, New Jersey: Princeton University press. 1962.
5. Tikhomirov V.M. *Some issues of Approximation Theory.* Moscow: Moscow State University, 1976.
6. Krasovski N.N. *The Theory of Motion Control.* Moscow: Nauka, 1968 (in Russian).
7. Pschenichnyi B.N. *Convex Analysis and Extremum Problems.* Moscow: Nauka, 1980. (In Russian).
8. Pschenichnyi B.N. *Necessary Conditions of Extremum.* New York: Marcel Decker, 1972.
9. Mordukhovich B.S. Approximation Mathods in Problems of Optimization and Control. Moscow: Nauka, 1988.
10. Pschenichnyi B.N., Danilin Yu.M. *Numerical Methods in Extremum Problems.* Moscow: Nauka, 1975. (In Russian.)
11. Warga J. *Optimal Control of Differential and Functional Equations.* New York and London: Academic press, 1974.
12. Bensoussan A., Lions J.L. *Controle Impulsionnel et Inequations Quasivariationnelles.* Paris: Dunod. Bordas, 1982.
13. Bublik B.N., Kirichenko N.F. *Foundations of Control Theory.* Kiev: Vyscha shkola, 1975.
14. Butkovski A.G. *Theory of Optimal Control of Systems with Distributed Parameters.* Moscow: Nauka, 1965.
15. Butkovski A.G. *Methods of Control of Systems with Distributed Parameters.* Ì oskow: Nauka, 1975.
16. Butkovski A.G., Samoylenko Yu.I. *Control of Quantum-Mechanics Processes.* Moscow: Nauka, 1984.
17. Vasilyev A.G. *Methods to Solve Extremal Problems.* Moscow: Nauka, 1981.
18. Egorov A.I. *Optimal Control of Linear Systems.* Kiev: Vyscha shkola, 1988.
19. Egorov A.I. *Optimal Control of Thermal and Diffusion Processes.* Moscow: Nauka, 1978.
20. Egorov A.I., Kapustyan V.E. "Synthesis of pulse optimal controls in distributed systems." In *Control and optimization of thermal and diffusion processes*, Preprint 84-34, Kiev, Inst. of Cybernetics, 1984.

21. Ermoliev Yu.M. *Methods of Stochastic Programming*. Moscow: Nauka, 1979.
22. Ermoliev Yu.M., GulenkoV.P., Tzarenko T.I. *Finite-Difference Method in Optimal Control Problems*. Kiev: Naukova Dumka, 1978.
23. Ermoliev Yu.M., Lyashko I.I. Mikhalevich V.S., Tyuptya V.I. *Mathematical Methods of Operation Research*. Kiev: Vyscha shkola, 1979.
24. Ermoliev Yu.M., Verchenko P.I. On method of linearization in limit extremal problems. Cybernetics 1976; 2:73-79.
25. Ermoliev Yu.M., Nurminsky Ye.A. Limit extremal problems. Cybernetics 1973; 4: 130—132.
26. Ivanenko V.I., Mel'nik V.S. *Variational Methods in Control Problems of Systems with Distributed Parameters*. Kiev: Naukova dumka, 1988.
27. Lions J.-L. Controlabilite exacte des system distibutes. Comptes Rendue der Scances de L'Academic des Sciences. Ser. s. 302; 13:471—475.
28. Lions J.-L. Controle de systemes gouvernes par des equations aux derivees partielles. Paris: Dunod Gauthier-Villars, 1968.
29. Lions J.-L. Exact controllability, stabilization and perturbation for distributed systems. SIAM Review 1988; 1:1-68.
30. Lions J.-L. Quelques methodes de resolution de problemes aux limities non lineaires. Dunod Gauthier-Villars Paris. 1969.
31. Lions J.-L. Remarks on some new non-linear boundary value problem. Lecture Notes in Nath. 1975; 446:301-328.
32. Lions J.-L. Some remarks on free boundary problems and optimal control. In *Free boundary problems*. Proc. Semin. Vol. 2: Pavia, 1979.
33. Lions J.-L. Some issues of optimal control of distributed systems. Uspekhi matem. nauk 1985; 4:55-68.
34. Lions J.-L., Magenes E. Problemes aux limities non homogenes et applications. Paris: Dunod, 1968.
35. Lurje K.A. *Optimal Control in Problems of Mathematical Physics*. Moscow: Nauka, 1975.
36. Nakonechny A.G. Minimax estimation of functionals of solutions of variations equations in Hilbert space. Kiev: KGU, 1985.
37. Osipov Yu.S., Suyetov A.P. On one Lions's problem. Dokl. AN SSSR 1984; v.276, 2:288-291.
38. Samoilenko A.M. Realization of fast-acting negative feedback on the basis of dynamic system with accumulated energy. Avtomatika i telemekhanika 1978; 12:12-23.
39. Sirazetdinov T.K. *Optimization of Systems with Distributed Parameters*. Moscow: Nauka, 1977.

40. Fedorenko R.P. *Approximate Solution of Optimal Control Problem*. Moscow, Nauka, 1978.
41. Dykhta V.A., Samsonyuk O.N. Optimal impulse control with applications. Moscow: Fizmatlit, 2000.
42. Novosel'sky S.N., Shul'gin D.F. Computation of non-stationary water transport under trickle and underground irrigation. Izvestiya of AN SSSR. Meckhanika zhidkosti i gaza 1984; 4:74-81.
43. Obraztsov N.N. *Mathematical Simulation and Optimization of Environment.* Moscow: Nauka, 1985.
44. Nents M.F. Controlling a Lethål Growth Process. Math. Biosciences 1968; 1/2: 41-55.
45. Pavlidis T. Optimal control of pulse frequency modulated systems. JEEE Trans. 1966; 4:678-684.
46. Chernous'ko F.L., Melikyan A.A. *Game Problems of Searching and Control.* Mockow: Nauka, 1978.
47. Aslanian A.A. Necessary conditions of optimality in problems of control with pulse impact at non-fixed time moments. Dokl. AN USSR 1982; 9:58-61.
48. Miller B.M. Optimality condition in control problem for system described by differential equations with measure. Avtomatika i telemekhanika 1982; 6:60-73.
49. Menaldi J.L. On the optimal stopping time problem for degenerate diffusions. SIAM J. Control 1980; 6:697-721.
50. Orlov Yu.V. *Theory of Optimal System with Generalized Control.* Moscow: Nauka, 1988.
51. Balass M.I. Sub-optimality and stability of linear distributed-parameter systems with finite-dimensional controllers. J. of Optimization Theory and Appl. 1985; 1:1-19.
52. White L.W. Optimal bang-bang controls arising in a Sobolev impulse control problem. J. Math. Anal. and Appl. 1984; 1:237-247.
53. White L.W. Point control of pseudo-parabolic problems. J. Differ. Equat. 1981; 3:366-374.
54. Anisovich V.V., Kryukov B.M. Optimal control of periodic movement of linear systems with pulse impact, Prikladnaya matematika i mekhanika 1982; 2:346-348.
55. Gaskevich I.V. Investigation of some continuous-descrete problems of oscillation theory. *In Controllable dynamical systems with contiunous-descrete parameters*. Kiev: Naukova dumka, 1984.
56. Kurzhanski A.B. *Control and Observation under Conditions of Uncertainty.* Moscow: Nauka, 1977.

57. Maksudov M.V., Aliyev M.V. Optimization of pulse systems with non-distributed boundary conditions. Dokl. AN SSSR 1985; 4:796-798.
58. Miller B.M. Non-linear pulse control of objects described by ordinary differential equations. Avtomatika i telemekhanika 1978; 1:75-86.
59. Samoilenko A.M., Perestyuk N.A. *Differential Equations with Impulse Impact.* Kiev: Vischa shkola, 1980.
60. Tzypkin Ya.Z., Popkov Ya.C. *Theory of Non-Linear Impulse Systems.* Moscow: Nauka, 1973.
61. Blaquire A. Impulsive optimal control with finite or infinite time horizons. J. Optimiz. Theory and Appl. 1985; 4:431—439.
62. Didenko V.P. Variational method of solving boundary value problems with non-simmetric operator. Dokl. AN SSSR 1978; 6:1277-1280.
63. Didenko V.P. On boundary problems for multidimensional hyperbolic boundare value problems with singularity. Dokl. AN SSSR 1972; 4:763-766.
64. Didenko V.P., Popova A.A. "Inequalities in. negative norms and variational problems for non-simmetric differential iperators." Preprint 81-29, Kiev, Inst. of Cybernetics, 1981.
65. Komkov V. *Optimal Control Theory for the Damping of Vibrations of Simple Elastic Systems.* Berlin, Heidelberg, New York:Springer-Verlag. 1972.
66. Krotov V.F., Gurman V.I. *Methods and Problems of Optimal Control Problems.* Moscow: Nauka, 1973.
67. Samborsky S.N. On controllability of systems described by partial differential. Sibirsky matem. zhurnal 1983; 2:150-159.
68. Silin D.B. Discontinuous optimal controls in linear problems. Dokl. AN SSSR 1985; 2:314-316.
69. Goodson R.E., Klein R.E. A definition and some results for distributed system observability. JEEE Trans. on aut. control 1970; 2:165-174.
70. Meirovich L., Silverberg L.M. Control of non-selfadjoint distributed-parameter systems, J. of Optimization Theory and Appl. 1985; 1:77-90.
71. Sokolowski J. On optimal parametric control of parabolic control of parabolic systems. Lect. Notes Comp. Sci. 1976; 41:623-633.
72. Teo K.L. Convergence of a strong variational algorithm for relaxed controls involving a distributed optimal control problem of parabolic type. Numer. Funct. Anal. and Optim. 1984-1985; 2-3:125-144.
73. Teo K.L., Reid D.W. First-order variation algorithm for optimal problems involving parabolic systems. Numer. Func. Anal. and Ottim. 1982; 2:141-171.
74. Berezanski Yu.M. *Expansion of Self-Adjoint Operators by Eigenfunctions.* Kiev: Naukova Dumka, 1965.

75. Krein S.G., Petunin Yu.I. Scales of Banach spaces. Uspekhi Matem. Nauk 1966; 2:89-169.
76. Banach S.S. *Course of Functional Analysis*. Kiev: Radyanska shkola, 1948.
77. Kantorovitch L.V., Akilov G.P. *Functional Analisys*. Moscow: Nauka, 1977.
78. Yosida K. *Functional Analysis*. Berlin, Gottingen, Heidelberg: Springer-Verlag, 1965.
79. Belov Yu.A., Didenko V.P., Kozlov N.N. et al. *Mathematical Software of Complex Experiment*. Vol.1. Kiev: Naukova Dumka, 1982.
80. Lyashko S.I. Pulse control of linear systems with distributed parameters. Dokl. An SSSR 1984; v.276, 2:285-287.
81. Lyashko S.I. Approximate solving pulse optimal control problems. Dokl. AN USSR 1986; 2:60-62.
82. Lyashko S.I. Numerical methods of solving pulse optimal control problems J. Vychislitel'noy Matematiki i Matematicheskoy Physiki 1987; 3:470-475.
83. Lyashko S.I., Man'kovsky A.A. Simultaneous optimization of impulse moments and intensities in control problems for parabolic equations. Cybernetics 1983; 5:687-690.
84. Lyashko S.I., Man'kovsky A.A. Controllability of parabolic systems with impulse action. Soviet Math. Dokl. 1989; v.39, 3:476-479.
85. Lyashko S.I., Red'ko S.E. Optimal pulse-point control of the dynamics of a viscous stratified fluid. Differentsial'nye Uravneniya 1987; 11:1890-1897.
86. Lyashko S.I., Red'ko S.E. On the controllability of distributed systems in a class of generalized actions. Kibernetika 1988; 6:120-121.
87. Chuprun B.Ye. Problems with constraints on changing of vector control. Avtomatika i telemekhanika 1976; 12:20-31.
88. Lavrentyev M.M., Romanov V.G., Shishatsky S.P. *Non-correct Problems of Mathematical Physics and Analysis*. Moscow: Nauka, 1980.
89. Lyashko I.I., Didenko V.P. On Tikhonov type regularization of problems with symmetric operator in a Hilbert space. Dokl. AN SSSR 1976; v.226, 3: 510—512.
90. Morozov V.A. Regular methods of solving non-correctly posed problems. Preprint of Computational Center of Moscow State University, 1974.
91. Prilepko A.I. Inverse problems of potential theory. Elliptic, parabolic, hyperbolic equations and transport equation. Mat. zametki 1973; 5:757-767.
92. Tikhonov A.N., Arsenin V.Ya. *Methods of Solving Non-Correct Problems*. Moscow: Nauka, 1979.
93. Sobolev S.L. *Some Applications of Functional Analysis in Mathematical Physics*. Leningrad: Leningrad State University, 1950.

94. Lyashko S.I. On regularization of pulse-point control in pseudo-parabolic systems. Kibernetika i vychiclitel'naya tekhnika 1986; 71:25-27.
95. Gupal A.M. *Stochactic Methods of Solving Nonsmooth Extremal Problems.* Kiev: Naukova dumka, 1979.
96. Karmanov V.G. *Mathematical Programming.* Moscow: Nauka, 1986.
97. Katkovnik V.Ya. *Linear Estimations and Stochastic Optimization Problems.* Moscow: Nauka, 1976.
98. Levitin Ye.S., Polyak B.T. Methods of Minimization with Constraints. J. Vychislitel'noy Matematiki i Matematicheskoy Physiki, 1966; 5:787-823.
99. Moiseev N.N. *Mathematical Problems of System Analysis.* Moscow: Nauka, 1981.
100. Shor N.Z. *Methods of Minimization of Non-Differentiable Functions and Their Applications.* Kiev: Naukova dumka, 1979.
101. Gaivoronski A.A., Ermoliev Yu.M. Stochatic programming problems with sequential estimating. Izv. AN SSSR, Techninal cybernetics 1979; 4:81-83.
102. Lyashko S.I., Golodnikov A.N. Method of minimizing functionals with respect to distrbution functions and its application in probability theory Cybernetics 1979; 4:532-536.
103. Mikhalevich V.S., Gupal A.M., Bolgarin I.V. Numerical methods of solving non-stationary and limit extremal problems without computation of gradient. Kibernetika, 1984; 5:56-57.
104. Zoutendijk G. *Methods of Feasible Directions. A Study in Linear and Non-Linear Programming.* Amsterdam, London, New York, Princeton: Elsevier publishing company, 1960.
105. Moiseev N.N. *Mathematical Methods in Optimal Control Systems.* Moscow: Nauka, 1971.
106. Ekeland I., Temam R. *Convex Analysis and Variational Problems.* North-Holland publishing company. Amsterdam, Oxford. American elsevier publishing company, inc. New York. 1976.
107. Nurminsky Ye.A. Numerical methods of solving determinate and stochastic minimax problems. Kiev: Naukova dumka, 1979.
108. Lyashko S.I. The method of feasible directions in the solution of exrtemal problems with unknown parameters. In *Mathematical methods in operations research and reability theory.* Akad. Nauk Ukrain. SSR, Inst. Kibernet., Kiev, 1978. (In Russian)

109. Lyashko S.I. A method of linearization in the solution of exrtemal problems for functions with unknown parameters. In *Mathematical methods in operations research and reability theory*. Akad. Nauk Ukrain. SSR, Inst. Kibernet., Kiev, 1978. (In Russian.)

110 Lyashko S.I., Mirzoahmedov F. Some methods of simultaneous identification and optimization. Dokl. Tadjikskoy AN SSR. 1984; 12:721-724.

111. Rubinstein L.M. On issue about heat transport in heterogeneous media. Izvestiyta AN SSSR. Ser. georgaph. i geophys. 1948; 1: 27-46.

112. Barenblatt G.I. On some boundary values problems for equations of liquid filtration in fissured rock. Prikladnaya matematika i mekhanika 1963; 2:348-350

113. Barenblatt G.I., Entov V.M., Ryzhik V.M. *Theory of Non-stationary Filtration of Liquid and Gas.* Moscow: Nauka, 1978.

114. Barenblatt G.I., Zheltov Yu.P., Kochina I.N. On basic idea of theory of homogeneous liquid filtration in fissured rocks. Prikladnaya matematika i mekhanika, 1960; 5:852-864.

115. Dzektzer Ye.S. Generalization of equation of grounwater mowement with free surface. Dokl. AN SSSR 1972; 5:27-46.

116. Nao R.C., Ting T.W. Initial value problems for pseudoparabolic differential equations. Indian uniq. Math. J. 1973; 25:348-368.

117. Belova G.P, Bodrov V.A., Globus A.M. et al. Hydrophysical properties of ions and mathematical simulations of salt movement in soil. Pochvovedeniye 1984; 4:113-119.

118. Lyubonova A.Sh. *Some Issues of Boundary Problems Theory for Pseudo-Parabolic Equations.* Krasnoyarsk, 1989.

119. Atamanov E.R., Mamayusupov M.Sh. *Non-classical problems for pseudo-parabolic equations*. Frunze: Ilim, 1990.

120. Lyashko S.I. Some issues of pulse-point control in pseudo-parabolic systems. Ukr. Matem. J. 1985; 3:368-371.

121. Lyashko S.I. Pulse-point control of pseudo-parabolic systems. Cybernetics 1986; 2:122-123.

122. Lyashko S.I. On solvability of pseudo-parabolic equations. Izv. vysshykh uchebnykh zavedeniy, Matematika 1985; 9:71—72.

123. Malovichko V.O. Non-local boundary value problem for singular pseudo-parabolic system. Doklady AN USSR, Ser. À. 1990; 3:25-27.

124. Matveev N.M., Svirdyuk G.A. On one linear pseudo-parabolic equation of non-stationary filtration. Mathematical physics 1985: 276.

125. Pavlov D.A., Potapov M.M. On iniform boundenness of solutions of distributed model of intra-cell mobility. Vestnik Moskovckogo Universiteta. Ser. 15, Vichisl. matem. i kibern. 1995; 2:62-64.
126. Suveyka I.V. On Cauchy problem for non-stationary equations with small viscosity. Mat. invest. 1988; 79:89-100.
127. Suveyka I.V. Mixed problems for non-stationary equations. Mat. invest. 1980; 59:27-36.
128. Showalter R.E., Ting T.W. Pseudo-parabolic partial differential equations. SIAM T. Math. Anal. 1970; 1:1-26.
129. Garding L. *Cauchy's Problem for Hyperbolic Equations.* University of Chicago. 1957.
130. Il'in V.A. On solvability of mixed problems for hyperbolic and parabolic equations. Uspekhi Matem. Nauk 1960; 2:97-154.
131. Ladyzhenskaya O.A. *Boundary Value Problems of Mathematical Physics.* Moscow: Nauka, 1973.
132. Lyashko S.I., Man'kovsky A.A. Optimal impulse control of a system with distributed parameters of hyperbolic type..Dokl. AN USSR Ser. A 1983; 4: 69-71.
133. Tikhonov A.N., Samarsky A.A. *Equations of Mathematical Physics.* Moscow: Nauka, 1977.
134. Plumb O.A., Whitaker S. Dispersion in heterogeneous porous media. Water Resour. Res. 1988; 7:913-938.
135. Shemiakin E.I. Propagation of non-stationary perturbations in viscoelastic medium. Dokl. AN SSSR 1955; 1:34-37.
136. Kachanov L.M. *Foundations of Plasticity Theory.* Moscow: Nauka, 1981.
137. Ismatov M. About one mixed problem for equation describing the sound propagation in viscous gas. Differentzialniye uravneniya 1984; 6:.1023-1024.
138. Malovischko V.A. On boundary value problems for singular pseudo-parabolic and pseudo-hyperbolic systems. Differentzialniye uravneniya 1991; 12:2120-2124.
139. Suveyka I.V. Mixed problems for equations of perturbations propagation in viscous nedia. Differentsizlniye Uravneniya 1983; 2:337-347.
140. Kozhanov A.I. Problem with directional derivative for some pseudo-parabolic and similar equations. Sibirian mathematical journal 1996; 6: 1335-1346.
141. Krein S.G. *Linear Differential equations in a Banach space.* Moscow: Nauka, 1967.
142. Lyashko I.I., Didenko V.P., Tzitritzky O.Ye. *Filtration of Noises.* Kiev: Naukova dumka, 1979.

143. Lyashko S.I. Method to solve approximately the Cauchy problem with operator coefficient of hyperbolic type. Dokl. AN SSSR 1979; v.247, 3:546-549.
144. Lyashko S.I. Approximate solving control problems with operator coefficient of hyperbolic type. Preprint 82-34, Kiev, Inst. of Cybernetics, 1982.
145. Kato T. *Theory of Linear Operators Perturbations.* Moscow: Mir, 1972.
146. Vishik M.I. Cauchy problem for equations with operator coefficients, mixed boundary value problem for systém of differential equations and approximate method of its solving. Matematichesky sbornik 1956; 1:51-148.
147. Sobolev S.L. About one new problem of mathematical physics. Izvestiya An SSSR, Saer. Mat. 1954; 1:3-50.
148. Zeleniak T.I. On invariant subspaces of one operator. Sibir. Matem. J. 1962; 3:471-475.
149. Demidenko G.V. Estimations as $t \to \infty$ of solutions of one S.L.Sobolev's problem. Sibir. Matem. J. 1984; 2:112-120.
150. Skazka V.V. Asimptotics as $t \to \infty$ of solutions of one problem of mathematical physics. Matem. sbornik 1985; 1:3-40.
151. Gabov S.A., Malysheva G.O. On Cauchy problem for one class of movements of viscous stratified liquid. J. Vychislitel'noy Matematiki i Matematicheskoy Physiki 1984; 3:467-471.
152. Gabov S.A., Malysheva G.O., Sveshnikov A.G. About one equation of dynamics of viscous stratified liquid. Differentsial'nye Uravneniya 1984; 7:1156-1164.
153. Louis J.-Cl., Wexler D. On exact controllability in Hilbert spaces. Lect. Notes in Cotr. and Inf. Sei. 1984; 58:676-685.
154. Niane M.T. Controlability exacte de l'equation des plaques vibrantes dans un polygone. Comptes Rendus de L'Academie des sciences 1988; tome 307, Serie 1, 10:517-521.
155. Astrovski A.I., Gayshun I.V. Controllability of linear non-stationary systems in class of finite order distributions. Izvestiya AN. Teoriya i sistemy upravleniya 1998; 2:24-30.
156. Semenov V.V. Impulse controllability of some linear distributed systems of Sobolev type. Dopovidi NAN Ukrayiny (to appear).
157. Lyashko S.I., Semenov V.V. On controllability of linear distributed systems in classes of generalized impacts. Kibernetika i sistemny analiz 2001; 1:18-41.
158. Berezansky Yu.M. *Self-Adjoint Operators in Spaces of Functions of Infinite Variable Number*. Kiev: Naukova Dumka, 1987.

159. Ladyzhenskaya O.A., Ural'tzeva N.N. *Linear and Quasi-linear Elliptic Equations*. Moscow: Nauka, 1964.
160. Garkusha V.I. On one approach to solving equation of first kind with simmmetric operator of Fredhol type. Ukr. Matem. J. 1979; 6:666-674.
161. Lyashko S.I., Red'ko S.E. An approximate solution for a problem of the dynamics of a viscous stratified fluid. J. Vychislitel'noy Matematiki i Matematicheskoy Physiki 1987; 5:720-729.
162. Lyashko S.I., Red'ko S.E. Approximate solution of equation of viscous stratified liquid dynamics. Dokl. AN SSSR 1986; v.290, 1:15-17.
163. Sakawa Y. Controllability for partial differential equations of parabolic type. SIAM Journal on Control 1977; 3:383-400.
164. Toshihiro K. Discrete-time controllability for distributed-parameter systems. J. L. System Sciences 1980; 9:1063-1074.
165. Klyushin D.A., Kuschan A.A., Lyashko S.I., Nomirovsky D.A., Petunin Yu.IG eneralized sopution of some operator equations in Banach spaces. J. Obchyslyuvalnoyi i Pryladnoyi Matematiki 2001 (to appear).
166. Hille E., Phillips R. *Functional Analysis and Semigroups*. Revised edition. American mathematical society colloquivem publications, v. XXXI, Providence. 1957.
167. Lyashko S.I. *Generalized Control of Linear Systems*. Kiev: Naukova dumka, 1998.
168. Petunin Yu.I., Plichko A.N. *Theory of Characteristics of Subspaces and its application*. Kiev: Vischa shkola, 1980.
169. Semenov V.V. Differentiability of performance criterion in problems of generalized optimal control of distributed systems. Dopovidi NAN Ukrayiny 2000; 10:111-114.
170. Semenov V.V., Bubnov V.V. Derivative of performance criterion in problems of generalized optimal control of linear distributed systems J. Obchyslyuval'noyi i prykladnoyi matematyky 1999; 1(84):129-136.
171. Nomirovsky D.A On convergence of gradient type methods in solving singular optimal control problems. J. Obchyslyaval'noyi ta prykladnoyi matematyky 1998; 1(83):93-98.
172. Nomirovsky D.A. Control in pseudo-hyperbolic systems. Obchyslyuval'na ta prykladna matematyka 1996; 80:68-77.
173. Deyneka V.S., Sergienko I.V., Skopetzky V.V. *Mathematical Models and Methods to Solve Problems with Discontinuous Solutions*. Kiev: Naukova Dumka, 1995.

174. Lyashko I.I., Demchenko L.I., Mistetzky G.Ye. *Numerical Solution of Heat and Mass Transport Problems in Porous Media.* Kiev: Naukova dumka, 1991.

175. Berdimuratov M.K., Lyashko S.I., Nomirovsky D.A. *Simulation and optimization of the pickling processes of the irrigated area.* Obchyslyuval'na ta prykladna matematyka 1996; 80:3-11.

176. Nomirovsky D.A. *Approximate methods of the control of the pseudo-hyperbolic systems.* Vesnik KSU. 1997, 3: Ñ. 187-198.

177. Lyashko I.I., Lyashko S.I., Nomirovsky D.A. *Controllabilaty of hyperbolic and pseudo-hyperbolic systems by the singular impacts.* Dopovidi NAN Ukrayiny 2000; 11: 131-134.

178. Nomirovsky D.A. *Optimization of the parabolic systems with generalized coefficients.* Dopovidi NAN Ukrayiny 2000; 12: 77-82.

Applied Optimization

36. G. Di Pillo and F. Giannessi (eds.): *Nonlinear Optimization and Related Topics*. 2000 ISBN 0-7923-6109-1
37. V. Tsurkov: *Hierarchical Optimization and Mathematical Physics*. 2000 ISBN 0-7923-6175-X
38. C. Zopounidis and M. Doumpos: *Intelligent Decision Aiding Systems Based on Multiple Criteria for Financial Engineering*. 2000 ISBN 0-7923-6273-X
39. X. Yang, A.I. Mees, M. Fisher and L.Jennings (eds.): *Progress in Optimization*. Contributions from Australasia. 2000 ISBN 0-7923-6286-1
40. D. Butnariu and A.N. Iusem: *Totally Convex Functions for Fixed Points Computation and Infinite Dimensional Optimization*. 2000 ISBN 0-7923-6287-X
41. J. Mockus: *A Set of Examples of Global and Discrete Optimization*. Applications of Bayesian Heuristic Approach. 2000 ISBN 0-7923-6359-0
42. H. Neunzert and A.H. Siddiqi: *Topics in Industrial Mathematics*. Case Studies and Related Mathematical Methods. 2000 ISBN 0-7923-6417-1
43. K. Kogan and E. Khmelnitsky: *Scheduling: Control-Based Theory and Polynomial-Time Algorithms*. 2000 ISBN 0-7923-6486-4
44. E. Triantaphyllou: *Multi-Criteria Decision Making Methods*. A Comparative Study. 2000 ISBN 0-7923-6607-7
45. S.H. Zanakis, G. Doukidis and C. Zopounidis (eds.): *Decision Making: Recent Developments and Worldwide Applications*. 2000 ISBN 0-7923-6621-2
46. G.E. Stavroulakis: *Inverse and Crack Identification Problems in Engineering Mechanics*. 2000 ISBN 0-7923-6690-5
47. A. Rubinov and B. Glover (eds.): *Optimization and Related Topics*. 2001 ISBN 0-7923-6732-4
48. M. Pursula and J. Niittymäki (eds.): *Mathematical Methods on Optimization in Transportation Systems*. 2000 ISBN 0-7923-6774-X
49. E. Cascetta: *Transportation Systems Engineering: Theory and Methods*. 2001 ISBN 0-7923-6792-8
50. M.C. Ferris, O.L. Mangasarian and J.-S. Pang (eds.): *Complementarity: Applications, Algorithms and Extensions*. 2001 ISBN 0-7923-6816-9
51. V. Tsurkov: *Large-scale Optimization – Problems and Methods*. 2001 ISBN 0-7923-6817-7
52. X. Yang, K.L. Teo and L. Caccetta (eds.): *Optimization Methods and Applications*. 2001 ISBN 0-7923-6866-5
53. S.M. Stefanov: *Separable Programming Theory and Methods*. 2001 ISBN 0-7923-6882-7

Applied Optimization

54. S.P. Uryasev and P.M. Pardalos (eds.): *Stochastic Optimization: Algorithms and Applications.* 2001 ISBN 0-7923-6951-3
55. J. Gil-Aluja (ed.): *Handbook of Management under Uncertainty.* 2001 ISBN 0-7923-7025-2
56. B.-N. Vo, A. Cantoni and K.L. Teo: *Filter Design with Time Domain Mask Constraints: Theory and Applications.* 2001 ISBN 0-7923-7138-0
57. S. Zlobec: *Stable Parametric Programming.* 2001 ISBN 0-7923-7139-9
58. M.G. Nicholls, S. Clarke and B. Lehaney (eds.): *Mixed-Mode Modelling: Mixing Methodologies for Organisational Intervention.* 2001 ISBN 0-7923-7151-8
59. F. Giannessi, P.M. Pardalos and T. Rapcsák (eds.): *Optimization Theory. Recent Developments from Mátraháza.* 2001 ISBN 1-4020-0009-X
60. K.M. Hangos, R. Lakner and M. Gerzson: *Intelligent Control Systems.* An Introduction with Examples. 2001 ISBN 1-4020-0134-7
61. D. Gstach: *Estimating Output-Specific Efficiencies.* 2002 ISBN 1-4020-0483-4
62. J. Geunes, P.M. Pardalos and H.E. Romeijn (eds.): *Supply Chain Management: Models, Applications, and Research Directions.* 2002 ISBN 1-4020-0487-7
63. M. Gendreau and P. Marcotte (eds.): *Transportation and Network Analysis: Current Trends.* Miscellanea in Honor of Michael Florian. 2002 ISBN 1-4020-0488-5
64. M. Patriksson and M. Labbé (eds.): *Transportation Planning.* State of the Art. 2002 ISBN 1-4020-0546-6
65. E. de Klerk: *Aspects of Semidefinite Programming.* Interior Point Algorithms and Selected Applications. 2002 ISBN 1-4020-0547-4
66. R. Murphey and P.M. Pardalos (eds.): *Cooperative Control and Optimization.* 2002 ISBN 1-4020-0549-0
67. R. Corrêa, I. Dutra, M. Fiallos and F. Gomes (eds.): *Models for Parallel and Distributed Computation.* Theory, Algorithmic Techniques and Applications. 2002 ISBN 1-4020-0623-3
68. G. Cristescu and L. Lupşa: *Non-Connected Convexities and Applications.* 2002 ISBN 1-4020-0624-1
69. S.I. Lyashko: *Generalized Optimal Control of Linear Systems with Distributed Parameters.* 2002 ISBN 1-4020-0625-X

KLUWER ACADEMIC PUBLISHERS – DORDRECHT / BOSTON / LONDON

MIX
Papier aus verantwortungsvollen Quellen
Paper from responsible sources
FSC® C105338

If you have any concerns about our products,
you can contact us on
ProductSafety@springernature.com

In case Publisher is established outside the EU,
the EU authorized representative is:
Springer Nature Customer Service Center GmbH
Europaplatz 3, 69115 Heidelberg, Germany

Printed by Libri Plureos GmbH
in Hamburg, Germany